Modular Representation Theory of Finite Groups

Modular Representation Theory of Finite Groups

Proceedings of a Symposium held at the
University of Virginia, Charlottesville, Virginia
May 8–15, 1998

Editors

Michael J. Collins
Brian J. Parshall
Leonard L. Scott

Walter de Gruyter · Berlin · New York 2001

Editors

Michael J. Collins
University of Oxford
Mathematical Institute
24–29 St. Giles'
Oxford OX1 3LB
UK

Brian J. Parshall
University of Virginia
Department of Mathematics
Kerchoff Hall
Charlottesville, VA 22903
USA

Leonard L. Scott
University of Virginia
Department of Mathematics
Kerchoff Hall
Charlottesville, VA 22903
USA

Mathematics Subject Classification 2000:
20-02; 20-06; 18E30, 20C05, 20C08, 20C20, 20C33, 20G05, 20G10, 20G40, 20J05

♾ Printed on acid-free paper which falls within the guidelines of the ANSI to ensure permanence and durability.

Library of Congress — Cataloging-in-Publication Data

Modular representation theory of finite groups : proceedings of a symposium held at the University of Virginia, Charlottesville, May 8–15, 1998 / editors, Michael J. Collins, Brian J. Parshall, Leonard L. Scott
p. cm.
ISBN 3110163675 (acid-free paper)
1. Finite groups — Congresses. 2. Representations of groups — Congresses. I. Collins, Michael J. (Michael John), 1945– II. Parshall, Brian, 1945– III. Scott, L. L. (Leonard L.)
QA177 .M66 2001
512′.2–dc21
2001047190

Die Deutsche Bibliothek — Cataloging-in-Publication Data

Modular representation theory of finite groups : proceedings of a symposium, held at the University of Virginia Charlottesville, May 8–15, 1998 / ed. Michael J. Collins … . – Berlin ; New York : de Gruyter, 2001
ISBN 3-11-016367-5

Printed in Germany.
Cover design: Thomas Bonnie, Hamburg.
Typeset using the authors' $\TeX$ files: I. Zimmermann, Freiburg.
Printing and binding: Hubert & Co. GmbH & Co. KG, Göttingen.

Preface

One of the most important problems in mathematics, stated in its simplest form, is the determination of group actions. In the case of finite groups, a series of reductions starting from theorems of O'Nan–Scott and Aschbacher indicates that a crucial step will always be to determine pairs (G, M) where M is a maximal subgroup of G in the cases where G is "almost simple" (i.e., a subgroup of the automorphism group of a nonabelian simple group which contains the inner automorphism group) or else the semidirect product of a quasisimple group H with one of its irreducible modules over a prime field. In the former case, in the situation where G is a classical group and M is not in a "known" list, M will be the normaliser of a quasisimple subgroup acting irreducibly on the underlying vector space. Thus in general, even assuming the classification of finite simple groups, one needs a precise knowledge of their representation theory, and that of the groups of Lie type in particular.

Here, there is a natural division between the representation theory for fields of characteristic that of the group (the "defining characteristic") and where the characteristic is different ("nondefining characteristic"). That this should be so is not surprising since the Sylow ℓ-subgroups of a group of Lie type when ℓ is not the characteristic p of the group have a relatively simple structure (indeed, they are sometimes abelian) compared with Sylow p-subgroups.

This book represents an outgrowth of a Symposium on the Modular Representation Theory of Finite Groups held at the University of Virginia, 8–15 May, 1998. In our proposals to funding bodies for support, we had written

> At the present time, relatively more progress has been made in the defining characteristic case. The dominant problem here has been to prove the Lusztig Conjecture concerning character values, and this has been achieved for sufficiently large primes ℓ by the combined work of Kazhdan and Lusztig, Kashiwara and Tanisaki, and Andersen, Jantzen and Soergel, though how large ℓ must be remains unknown. In particular, their work covers all families of groups of Lie type. By contrast, progress in studying the nondefining characteristic is, paradoxically in view of the remarks above, less advanced in that fully comparable results exist only for groups of type A; it is our view as organisers, in part influenced by personal interests, that it is this area to which we want to devote our attention.
>
> We have identified the representation theory in nondefining characteristic as one ripe for detailed examination because of its influence not only for the problem [described above], but also because of its interaction on the one hand with questions which are more in the sphere of the "pure group theorist", and on the other with more general questions arising from algebras introduced by those completely outside the field.

Specifically, we set as our goals to explore

(i) the modular representations of groups of Lie type in nondefining characteristic, in particular the relationship with q-Schur algebras, James' conjecture for groups of type A (an analogue of the Lusztig conjecture), and generalisations to the representations of the groups of other types,

(ii) the relationship of q-Schur algebras to the quantisation of other algebras,

(iii) connections with modular representations of symmetric groups, and

(iv) Broué's conjectures on equivalences of derived categories of modules in blocks with abelian defect group in groups of Lie type, derived and Rickard equivalences, and the corresponding problems for abstract finite groups.

We decided at an early stage to invite Meinolf Geck, Alexander Kleshchev and Raphaël Rouquier to give series of lectures in the first half of the Symposium at a level which would both instruct advanced graduate students and take the audience to the frontiers of their fields, and then to move to a "workshop format". In the event, almost all participants attended for the entire period; consequently, the distinction between the two parts of the Symposium was allowed to blur. We had also written

> By combining the lecture mini-series with a workshop, it is our intention to draw students to this important area, in an atmosphere where the Symposium is not merely instructional but is attended by many of those most intimately involved at the forefront of this work. In this way, we hope to inspire the next generation as well as allow those with a wide range of perspectives on the problems that arise to set out a programme for the next few years. The problems that we have highlighted will not have immediate solutions, nor will it be immediately evident what all the connections are; consequently, this is an area of mathematics ripe for the involvement of the next generation of mathematicians. This is one reason behind our decision to ask younger mathematicians to give the lecture series, and to give some of the workshop talks.

This, then, also is the ethos of this book. We are especially grateful to the three main lecturers for the way in which they have expanded upon their lectures. Alexander Kleshchev, since his first lectures covered background material on which he has reported elsewhere, has concentrated on a full account of his work with Jonathan Brundan, which formed the goal of his lectures, while Raphaël Rouquier has written on work that has been very much ongoing since the Symposium; indeed, in our Report, we wrote

> The range of mathematics covered is well described by the lecture titles. If any particular feature deserves mention, it would be the series of lectures

given by Rouquier in which he described a method he has developed, using an idea of Puig, for gluing together Rickard equivalences constructed locally; this may prove the major breakthrough in any proof of the Broué Conjecture. Some of this work was carried out in the week preceding the Symposium, after he had arrived in Charlottesville early.

We invited all participants, in particular those who gave survey lectures, to contribute to this volume, without the restriction that such articles need reflect lectures given, but bearing in mind the goal that we had set for this book. For the record, since only a few took up that invitation, a list of all lectures given at the Symposium is included on page ix.

We should like to thank the University of Virginia for its support both for the main lecturers from the Pratt Fund and towards that of other participants, in particular graduate students, by way of a direct grant from the Dean of the College of Arts and Sciences, and to acknowledge the support given by the National Science Foundation (grant DMS 9804881) and by the National Security Agency (grant MDA904-98-1-0045). The first named of the organisers would also like personally to thank the University for its support with a Visiting Professorship during the academic year 1997–98.

Finally, we should like to thank the staff of the Mathematics Department of the University of Virginia, without whose help this Symposium could never have taken place, and Dr. Manfred Karbe of de Gruyter for his patience and assistance in producing this volume.

Oxford and Charlottesville

Michael Collins
Brian Parshall
Leonard Scott

Lectures given during the Symposium

The Lecture Series

Meinolf Geck
The theory of modular Harish-Chandra series, modular Hecke algebras and (generalized) q-Schur algebras, I–IV

Alexander Kleshchev
Representations of symmetric groups via algebraic groups and Schur algebras, I–IV

Raphaël Rouquier
Block theory via stable and Rickard equivalences, I–V

Main Workshop Lectures

Michel Broué
Spetsoids

Richard Dipper
Morita theorems for Ariki–Koike algebras

Jie Du
Generalized q-Schur algebras and the ways to approach them

Meinolf Geck
Kazhdan–Lusztig cells and decomposition numbers

Gordon James
The cyclotomic Jantzen–Schaper theorem

Alexander Kleshchev
On translation functors for general linear groups

Jeremy Rickard
Infinite dimensional representations

Raphaël Rouquier
The Broué Conjecture – a possible approach

Bhama Srinivasan
Some arithmetic properties of Green functions

Peter Webb
Stratifications in the context of Mackey functors

Short Talks

Frauke Bleher
Ext groups for cyclic blocks – a quiver approach

Robert Boltje
Linear source modules II

Anton Cox
Infinitesimal q-Schur algebras and their blocks

Charles Curtis
Exponential sums and representations of finite reductive groups

Charles Eaton
On finite groups of p-local rank one and a variant of Alperin's conjecture

Gilles Gnacadja
A longer Milnor sequence: applications to phantom maps

Corneliu Hoffman
On small cross characteristic representations of finite groups of Lie type

Burkhard Külshammer
Linear source modules I

John Murray
On the blocks of defect zero and products of elements of order p

Daniel Nakano
Extensions of modules for Schur algebras, symmetric groups and Hecke algebras

Pham Huu Tiep
Tensor products of representations of finite groups of Lie type

Table of Contents

Preface v

Lectures given during the Symposium ix

Meinolf Geck

Modular Harish-Chandra series, Hecke algebras
and (generalized) q-Schur algebras 1

Jonathan Brundan and Alexander S. Kleshchev

Tensor products and restrictions in type A 67

Raphaël Rouquier

Block theory via stable and Rickard equivalences 101

Robert Boltje

Alperin's weight conjecture in terms of linear source modules
and trivial source modules 147

Marc Cabanes and Jeremy Rickard

Alvis–Curtis duality as an equivalence of derived categories 157

Stephen R. Doty and Daniel K. Nakano

Relating the cohomology of general linear groups and symmetric groups 175

Jie Du

Generalized q-Schur algebras and the ways to approach them 189

Meinolf Geck and Raphaël Rouquier

Filtrations on projective modules for Iwahori–Hecke algebras 211

Corneliu Hoffman

Projective representations for some exceptional finite groups of Lie type 223

Nicholas J. Kuhn

Splitting fields and twisted group rings for the finite general linear groups 231

Kay Maagard and Pham Huu Tiep

Irreducible tensor products of representations of finite quasi-simple groups of Lie type 239

Modular Harish-Chandra series, Hecke algebras and (generalized) q-Schur algebras

Meinolf Geck

Contents

1 Introduction 2
1.A Some general conventions . . . 3

2 Hom functors 3
2.A The Hom functor F_M . . . 3
2.B Dipper's hypothesis (D) . . . 6
2.C Green's hypotheses (A) and (B) . . . 10

3 Harish-Chandra series 14
3.A Split BN-pairs and the functors R_J^S and ${}^*R_J^S$ . . . 14
3.B Cuspidal modules . . . 17
3.C Hecke algebras . . . 22

4 Decomposition numbers and Brauer reciprocity 25
4.A The decomposition matrix of a module . . . 25
4.B Hypothesis (D) and Frobenius algebras over $\mathcal{O}$ . . . 28
4.C A strategy for finding the decomposition matrix of A . . . 33

5 Lusztig series and blocks 36
5.A Blocks and basic sets . . . 36
5.B ℓ-regular quotients . . . 40
5.C Iwahori–Hecke algebras . . . 45

6 (Generalized) q-Schur algebras 49
6.A Linear primes . . . 49
6.B Unipotent blocks . . . 54
6.C Concluding remarks . . . 60

References 63

1. Introduction

A fundamental step in the modular representation theory of finite groups of Lie type in non-defining characteristic was achieved in the 1982 paper of Fong and Srinivasan [24] in which the blocks of finite general linear and unitary groups are determined. The subsequent work of Dipper and James on the problem of finding the decomposition matrices of $\mathrm{GL}_n(q)$ culminates in their 1989 paper [16] which leads to a reduction of that problem to a similar problem about so-called q-Schur algebras (see also [17]). The point about q-Schur algebras is that they can be defined "generically" (i.e., in such a way that q appears as a formal parameter) in terms of Hecke algebras of type A, and their representation theory is linked to various other "generic" objects like affine Hecke algebras or quantum groups.

By these links and Lusztig's conjectures (see [63]), the representations of Hecke algebras and q-Schur algebras of type A at roots of unity over a field of characteristic 0 are now relatively well-understood. Lascoux, Leclerc and Thibon [54, 56] have spelled this out explicitly by formulating conjectures which yield efficient and purely combinatorial algorithms for computing decomposition numbers (see also Goodman–Wenzl [40]). These conjectures are now proved by Ariki [1] (in the Hecke algebra case) and Varagnolo and Vasserot [65] (in the q-Schur algebra case); moreover, Leclerc [55] proves the remarkable fact that the decomposition matrix of the q-Schur algebra (of degree n) occurs as a submatrix of the decomposition matrix of a suitable Hecke algebra of degree $\geq n$.

A simple "asymptotic" argument (see [33, (2.7)]) shows that representations at roots of unity in characteristic 0 behave in the same way as representations in prime characteristic as long as the prime is not too small. Thus we can now say that the ℓ-modular decomposition matrices of $\mathrm{GL}_n(q)$ are explicitly known for all primes ℓ which do not divide q and which are "large enough", where the precise bound is not known. (By James' conjecture [53], $e\ell > n$ would be sufficient, where e is the smallest $i \geq 2$ such that ℓ divides $1 + q + q^2 + \cdots + q^{i-1}$.)

The aim of this text is to describe the conditions and the concepts underlying the strategy of reducing the problem of finding decomposition numbers of finite groups of Lie type to a similar problem about (generalized) q-Schur algebras. These concepts include:

- Hom functors; see Section 2.
- Harish-Chandra philosophy; see Section 3.
- Decomposition numbers and Brauer reciprocity; see Section 4.

In the above sections a certain attempt at self-containedness is made. Finally, in Section 6, we explain how the above concepts are applied to $\mathrm{GL}_n(q)$ (following Dipper and James [16]) and to finite classical groups in the so-called linear prime case (following Gruber and Hiss [43]). We shall work with a definition of "linear" primes which is valid for all finite groups of Lie type and introduce (generalized) q-Schur algebras in such a general framework.

In Section 6, we will have to assume a certain familiarity with the Deligne–Lusztig theory of representations of finite groups of Lie type (see [7], [10], [59]). Especially important are the results about the compatibility of Lusztig series with blocks (see [3]). This will be explained in Section 5.

We emphasize the fact that the above methods lead to complete results in the case of linear primes. Otherwise, we still have some basic unsolved problems even for groups of BN-rank 1 (see Example 6.24)! It seems likely that, in order to solve these problems, it will be necessary to develop some completely new methods.

1.A. Some general conventions

All rings will have identity elements and ring homomorphisms respect them. All modules will be left modules and maps are written on the left unless explicitly stated otherwise.

If A is any ring we denote by A-Mod the category of left A-modules. Mostly we will assume that A is finitely generated over some commutative ground ring k; in that case we denote by A-mod the full subcategory of A-Mod whose objects are the A-modules which are finitely generated over k.

If G is any finite group and k is a commutative ring, all representations of G over k will be tacitly assumed to be of finite rank over k. In this case we denote by k_G the trivial module of G over k.

2. Hom functors

Let A be a ring and M be an A-module. In order to investigate the structure of M, we consider its endomorphism algebra H. First, on a completely general level, we have the "Fitting correspondence" between direct summands of M and projective modules for H. Dipper and Green have formulated, from different points of view and with different applications in mind, some hypotheses under which one can say more about the structure of the direct summands of M. For this purpose, one works with a functor F_M which sends an A-module V to the H-module $\mathrm{Hom}_A(M, V)$. The aim of this section is to present some basic results about that functor in a systematic way, following and combining the work of Green [41], Dipper [11] and Cabanes [5] (see also Vigneras [66, §1]).

2.A. The Hom functor F_M

Let A be a k-algebra (over some commutative ring k) and M be an A-module. We set $H := \mathrm{End}_A(M)^\circ$ (the opposite algebra of the endomorphism algebra of M). Consider any decomposition $M = M_1 \oplus \cdots \oplus M_n$ as a direct sum of A-modules. For $1 \leq i \leq n$, let $e_i \in H$ be the idempotent which projects M onto M_i. Then the following hold (see [9, §6A and Exercise 6.14]).

(F1) We have $\mathrm{id}_M = e_1 + \cdots + e_n$ and this is a decomposition of id_M into orthogonal idempotents.

(F2) The A-module M_i is indecomposable if and only if He_i is an indecomposable H-module, and this holds if and only if e_i is a primitive idempotent.

(F3) We have $M_i \cong_A M_j$ if and only if $He_i \cong_H He_j$, and this holds if and only if e_i, e_j are conjugate by a unit in H.

This is called the *Fitting correspondence* between M and its endomorphism ring H.

2.1. In order to be able to study the above correspondence in a more conceptual framework, we work with the functor

$$F_M\colon A\text{-Mod} \to H\text{-Mod}, \qquad V \mapsto \mathrm{Hom}_A(M, V), \tag{a}$$

where the action of H on $F_M(V)$ is given by $\varphi f = f \circ \varphi$ (for $\varphi \in H$, $f \in F_M(V)$) and F_M sends a map $\rho\colon V \to V'$ to the map $F_M(\rho) = \rho_*\colon F_M(V) \to F_M(V')$, $f \mapsto \rho \circ f$. In this set-up, we have

$$He_i = \{e_i \circ \varphi \mid \varphi \in \mathrm{End}_A(M)\} = \mathrm{Hom}_A(M, M_i) = F_M(M_i). \tag{b}$$

Thus, if the Krull–Schmidt–Azumaya Theorem holds, then (F1)–(F3) can be rephrased by saying that the map $V \mapsto F_M(V)$ defines a bijection between the isomorphism classes of indecomposable direct summands of M and the isomorphism classes of projective indecomposable H-modules.

Example 2.2. Assume that M is of the form $M = Ae$ where $0 \neq e \in A$ is an idempotent. Then, for any $V \in A$-mod, we have

$$F_M(V) = \mathrm{Hom}_A(M, V) = \mathrm{Hom}_A(Ae, V) \cong eV,$$

where the isomorphism is given by sending $\varphi \in \mathrm{Hom}_A(M, V)$ to $\varphi(e) = e\varphi(e) \in eV$. Applying this to $V = M$ also yields that we have a ring isomorphism $H \cong eAe$ (see [23, (I.8.2)]). Thus, identifying $F_M(V)$ with eV and H with eAe, we see that the action of $H \subseteq A$ on $F_M(V) \subseteq V$ is simply given by restriction. This situation is studied in detail in Green [42, §6.2]. For example, we have that F_M defines a bijection between the isomorphism classes of simple A-modules V with $eV \neq \{0\}$ and the isomorphism classes of simple H-modules.

The following special case is discussed in [9, §11D]. Let G be a finite group, $H \subseteq G$ be a subgroup and $e \in \mathbb{C}H$ be an idempotent. We consider the $\mathbb{C}H$-module $V := \mathbb{C}He$. Then, by [9, (11.21)], the induced module $\mathrm{Ind}_H^G(V)$ can be identified with $M := \mathbb{C}Ge$. The endomorphism algebra $H \cong e\mathbb{C}Ge$ is called the corresponding *Hecke algebra*; see also [9, p. 281] for some historical remarks.

The following result will prove useful in several places below.

Lemma 2.3 (See [41, (2.1a)]). *Let $M = M_1 \oplus \cdots \oplus M_n$ be a decomposition of M as a direct sum of A-modules. Let V be any A-module and $i \in \{1, \ldots, n\}$. Then we have an isomorphism of k-modules*

$$\mathrm{Hom}_A(M_i, V) \to \mathrm{Hom}_H(F_M(M_i), F_M(V)), \quad \varphi \mapsto \varphi_* = (f \mapsto \varphi \circ f).$$

Proof. Set $F = F_M$. If $\varphi \in \mathrm{Hom}_A(M_i, V)$ is such that $\varphi_* = 0$, then $\varphi \circ f = 0$ for all $f \in F(M_i)$. Taking $f = e_i$, this implies $\varphi(e_i(m)) = 0$ for all $m \in M$ and so $\varphi = 0$. Thus, for each i, the map $\varphi \mapsto \varphi_*$ is injective.

Now, since Hom_A and F preserve direct sums, we obtain an injection

$$F(V) = \mathrm{Hom}_A(\bigoplus_i M_i, V) \hookrightarrow \mathrm{Hom}_H(\bigoplus_i F(M_i), F(V)) = \mathrm{Hom}_H(H, F(V))$$

which is given by $f \mapsto f_*$. But that map has an inverse, namely the map $\mathrm{Hom}_H(H, F(V)) \to F(V)$, $\psi \mapsto \psi(1_H)$. Hence all the injective maps above must be isomorphisms. □

2.4. Suppose the following finiteness assumptions are satisfied:

The ring k is local and noetherian (for example, a field or a discrete valuation ring), and the algebra A is finitely generated over k.

Under these assumptions, the following hold:

(a) We have $\mathrm{rad}(k)A \subseteq \mathrm{rad}(A)$ and so $A/\mathrm{rad}(A)$ is a finite dimensional semisimple algebra over the field $k/\mathrm{rad}(k)$.

(b) If V, V' are A-modules which are finitely generated over k, then $\mathrm{Hom}_A(V, V')$ is also finitely generated over k. In particular this applies to H (if M is finitely generated over k), and F_M takes A-mod into H-mod.

(c) Let V be a finitely generated A-module. Then $\mathrm{rad}(V) = \mathrm{rad}(A)V$ and $\mathrm{hd}(V) := V/\mathrm{rad}(V)$ is a semisimple A-module.

(d) Every finitely generated A-module can be written as a finite direct sum of indecomposable A-modules.

For (a), see [9, (5.22)]. (b) is shown in [9, Exercise 2.6]. The first statement in (c) is proved in [9, (5.29)]; then the second statement follows from (a). Finally, see [23, (I.3.8)]
for (d).

2.B. Dipper's hypothesis (D)

Let A be a k-algebra as above, $M \in A$-mod and $H = \mathrm{End}_A(M)^\circ$. The finiteness conditions in (2.4) are assumed to be satisfied. We consider the following hypothesis, formulated and studied in detail by Dipper [11]:

(D) There exists an epimorphism $\beta\colon P \to M$, where $P \in A$-mod is projective and $\ker(\beta)$ is invariant under $\mathrm{End}_A(P)$.

In order to avoid any technical problems about projective modules, we will also assume that idempotents can be lifted from $A/\operatorname{rad}(A)$ to A and from $H/\operatorname{rad}(H)$ to H so that A and H are *semiperfect rings* (see [9, p. 132]). Then we can use the results in [9, §6C] about projective covers, besides the various standard characterizations of projective modules in [9, (2.22)].

Example 2.5. (D) is satisfied if one of the following holds.

(a) M is simple and $\beta\colon P \to M$ is a projective cover.

(b) We have $\mathrm{Hom}_A(\ker(\beta), M) = \{0\}$.

This is clear in case (a), since then M is the quotient of P by its unique maximal submodule. Now assume that (b) holds. Let $\psi \in \mathrm{End}_A(P)$ and set $N := \ker(\beta)$. Assume, if possible, that $\psi(N) \not\subseteq N$. Then we have a non-trivial map $N \to \psi(N) \to (\psi(N)+N)/N \subseteq P/N \cong_A M$, contradicting (b).

The following result gives a necessary and sufficient condition for (D) to hold. The argument for proving sufficiency is due to Geck–Hiss–Malle [38, (2.3)]. The necessity is already proved by Dipper [11, (2.8)].

Lemma 2.6. *Assume that $\beta \in \mathrm{Hom}_A(P, M)$ is surjective where $P \in A$-mod is projective. Consider the map $\beta^*\colon \mathrm{End}_A(M) \to \mathrm{Hom}_A(P, M)$, $\varphi \mapsto \varphi \circ \beta$, which is obviously injective (since β is surjective). Then* (D) *holds if and only if β^* is surjective.*

In particular, in the case where k is a field, we see that (D) holds if and only if $\dim_k \mathrm{Hom}_A(P, M) = \dim_k \mathrm{End}_A(M)$.

Proof. Assume first that β^* is surjective. Let $\psi \in \mathrm{End}_A(P)$ and $p \in \ker(\beta)$. By assumption, we can find some $\varphi \in \mathrm{End}_A(M)$ such that $\beta \circ \psi = \varphi \circ \beta$. Then we have $\beta(\psi(p)) = \varphi(\beta(p)) = 0$, and so also $\psi(p) \in \ker(\beta)$. Thus (D) is satisfied.

Conversely, assume that (D) holds and let $\varphi \in \mathrm{Hom}_A(P, M)$. Since P is projective, there exists some $\psi \in \mathrm{End}_A(P)$ such that $\beta \circ \psi = \varphi$. Now, since $\ker(\beta)$ is invariant under $\mathrm{End}_A(P)$, we obtain an induced map $\psi^\bullet \in \mathrm{End}_A(M)$ such that $\varphi = \beta \circ \psi = \psi^\bullet \circ \beta = \beta^*(\psi^\bullet)$. Hence β^* is surjective. □

Remark 2.7. Assume that $\beta\colon P \to M$ satisfies (D). Then we can write $P = P_M \oplus P'$ where the restriction of β to P_M is a projective cover of M; denote that restriction by β_M. Since β_M is surjective, we have an injection $\mathrm{End}_A(M) \hookrightarrow \mathrm{Hom}_A(P_M, M)$, $\varphi \mapsto \varphi \circ \beta_M$. Since Hom_A preserves direct sums, we obtain a sequence of maps

$$\begin{aligned}\mathrm{Hom}_A(P, M) &\cong_k \mathrm{End}_A(M) \hookrightarrow \mathrm{Hom}_A(P_M, M)\\ &\subseteq \mathrm{Hom}_A(P_M, M) \oplus \mathrm{Hom}_A(P', M) \cong_k \mathrm{Hom}_A(P, M)\end{aligned}$$

where the left-most isomorphism is given by Lemma 2.6 and the composition of all maps is the identity. Hence, using again Lemma 2.6, we conclude that

(D) also holds for $\beta_M\colon P_M \to M$ and we have $\mathrm{Hom}_A(P', M) = \{0\}$.

Lemma 2.8 (Dipper [11]). *Assume that* (D) *holds and set* $T := \mathrm{End}_A(P)^\circ$. *Then* β *induces a surjective* k*-algebra homomorphism* $\beta^\bullet\colon T \to H$ *such that* $\beta \circ \psi = \beta^\bullet(\psi) \circ \beta$ *for* $\psi \in T$. *If* β *is a projective cover, then* $\ker \beta^\bullet \subseteq \mathrm{rad}(T)$.

Proof. Thanks to hypothesis (D), for any $\psi \in T$ we have a well-defined map $\psi^\bullet \in H$ such that $\beta \circ \psi = \psi^\bullet \circ \beta$. Thus the assignment $\psi \mapsto \psi^\bullet$ gives the required map $\beta^\bullet$. That map is surjective since P is projective. Indeed, given $\varphi \in \mathrm{End}_A(M)$, we consider the map $\varphi \circ \beta \in \mathrm{Hom}_A(P, M)$ which is equal to $\beta \circ \psi$ for some $\psi \in \mathrm{End}_A(P)$.

Now assume that $\beta\colon P \to M$ is a projective cover. Let $\psi \in \ker(\beta^\bullet)$. Then $\beta \circ \psi = \beta^\bullet(\psi) \circ \beta = 0$ and so $\psi(P) \subseteq \ker(\beta)$. By [9, (6.25)(i)], we have $\ker(\beta) \subseteq \mathrm{rad}(A)P$ and so $\psi(P) \subseteq \mathrm{rad}(A)P$. But this implies $\psi \in \mathrm{rad}(T)$. Indeed, by the characterization of radicals in [9, (5.10)], we must show that $\psi' := \mathrm{id}_P - \alpha \circ \psi \circ \beta$ is an isomorphism for all $\alpha, \beta \in T$. Now we also have $(\alpha \circ \psi \circ \beta)(P) \subseteq \mathrm{rad}(A)P$ and so $\psi'(P) + \mathrm{rad}(A)P = P$. Hence, by the Nakayama Lemma, ψ' is surjective and, by [9, (5.8)], ψ' is also injective. □

In general we cannot expect that the functor F_M is exact. However, assuming that (D) holds, we can relate F_M with the analogous functor defined with respect to P. This is made more precise as follows.

Remark 2.9. Assume that $\beta\colon P \to M$ satisfies (D). We set $T := \mathrm{End}_A(P)^\circ$ and consider the functor $F_P\colon A\text{-mod} \to T\text{-mod}$ defined with respect to P.

(a) For any $V \in A$-mod, we have a natural embedding $F_M(V) \subseteq F_P(V)$ given by $f \mapsto f \circ \beta$. This is an embedding of T-modules where $F_M(V)$ is regarded as a T-module via the map $\beta^\bullet\colon T \to H$ of Lemma 2.8.

(b) Let $M = M_1 \oplus \cdots \oplus M_n$ be a decomposition as a direct sum of A-modules. Then, with the identification of (a), we have $F_P(M_i) = F_M(M_i)$ for all i.

There is nothing to prove in (a). As far as (b) is concerned, note that we have embeddings $F_M(M_i) \subseteq F_P(M_i)$ for all i. Since Hom_A preserves direct sums, we also obtain an embedding $F_M(M) = F_M(M_1 \oplus \cdots \oplus M_n) \subseteq F_P(M_1 \oplus \cdots \oplus M_n) = F_P(M)$ which is, in fact, an equality by Lemma 2.6. Hence all the inclusions $F_M(M_i) \subseteq F_P(M_i)$ must also be equalities.

Parts (a) and (b) of the following result are due to Dipper [11]; note that (b) also appears (in a more general setting where only weaker finiteness assumptions on A are required) in Vigneras [66].

Theorem 2.10. *Assume that* (D) *holds. Let* $M = M_1 \oplus \cdots \oplus M_n$ *be a decomposition into indecomposable* A*-modules* (*which exists by Remark* 2.4(d)).

(a) *Each* M_i *has a simple head and we have* $F_M(\mathrm{hd}(M_i)) \cong_H \mathrm{hd}(F_M(M_i))$. *Moreover, we have* $M_i \cong_A M_j$ *if and only if* $\mathrm{hd}(M_i) \cong_H \mathrm{hd}(M_j)$.

(b) *The map* $V \mapsto F_M(V)$ *defines a bijection between the isomorphism classes of simple* A*-modules* V *with* $\mathrm{Hom}_A(M, V) \neq \{0\}$ *and the isomorphism classes of simple* H*-modules.*

(c) *We have* $\mathrm{End}_A(V) \cong_k \mathrm{End}_H(F_M(V))$ *for every simple* A*-module* V *with* $\mathrm{Hom}_A(M, V) \neq \{0\}$.

Proof. We use the notation and the conventions of Remark 2.9, and proceed in several steps.

Step 1. First we show that if $V \in A$-mod is simple with $\mathrm{Hom}_A(M, V) \neq \{0\}$, then $F_M(V)$ is a simple H-module and we have $F_M(V) = F_P(V)$ (under the identification in Remark 2.9(a)).

This is seen as follows. Since M is a factor module of P, we have $F_P(V) \neq \{0\}$. We claim that $F_P(V)$ is a simple T-module. Indeed, let $0 \neq f \in F_P(V)$. Then f is surjective and, since P is projective, the induced map $f_*\colon F_P(P) \to F_P(V)$ is surjective. This means that, as a T-module, we have $F_P(V) = Tf$ and so $F_P(V)$ is simple, as claimed.

Now we use the natural embedding of T-modules $F_M(V) \subseteq F_P(V)$ as in Remark 2.9(a). Since $F_P(V)$ is simple, we must have $F_M(V) = F_P(V)$. Since the action of T on $F_M(V)$ is defined via the map $\beta^\bullet$, we conclude that $F_M(V)$ is also simple as an H-module.

Step 2. We prove that $\mathrm{hd}(M_i)$ is simple and that $F_M(\mathrm{hd}(M_i)) \cong_H \mathrm{hd}(F_M(M_i))$ for all i. For this purpose, we consider the natural map $\pi : M_i \to \mathrm{hd}(M_i)$. Since P is projective, the functor F_P is exact and so we obtain a surjective map $F_P(\pi)\colon F_P(M_i) \to F_P(\mathrm{hd}(M_i))$. By Remark 2.4(c), we can write $\mathrm{hd}(M_i) \cong_A \bigoplus_V n_V V$ where the sum is over all simple A-modules V (up to isomorphism) and n_V are non-negative integers. Since F_P preserves direct sums, we thus have a surjective map of T-modules

$$F_P(M_i) \to F_P(\mathrm{hd}(M_i)) \cong_T \bigoplus_V n_V F_P(V)$$

where each $F_P(V)$ with $n_V > 0$ is non-zero and simple by Step 1. Now, using Remark 2.9(b) and Step 1, we conclude that we also have a surjective map of H-modules

$$F_M(M_i) \to \bigoplus_V n_V F_M(V)$$

where each $F_M(V)$ with $n_V > 0$ is a simple H-module. But $F_M(M_i)$ is a projective indecomposable H-module and hence has a simple head. It follows that in the direct sum on the right hand side only one non-trivial term can occur. This means that $\operatorname{hd}(M_i)$ is simple and that $F_M(\operatorname{hd}(M_i)) \cong_H \operatorname{hd}(F_M(M_i))$, as desired.

Step 3. In order to complete the proof of (a), it remains to show that each M_i is determined by $\operatorname{hd}(M_i)$ up to isomorphism. Now, if $M_i \cong_A M_j$, then we certainly have $\operatorname{hd}(M_i) \cong_A \operatorname{hd}(M_j)$. Conversely, assume that the latter condition holds. Then, using the isomorphism of Step 2, we conclude that $\operatorname{hd}(F_M(M_i)) \cong_H \operatorname{hd}(F_M(M_j))$. But this implies that $F_M(M_i) \cong_H F_M(M_j)$ (since M_i, M_j are projective indecomposable H-modules) and hence that $M_i \cong_A M_j$ (using (F3)).

Step 4. We now complete the proof of (b). Note that the simple A-modules V with $\operatorname{Hom}_A(M, V) \neq \{0\}$ are precisely those of the form $\operatorname{hd}(M_i)$ for some i and the simple H-modules are the modules $\operatorname{hd}(F_M(M_i))$ for some i. Therefore the required bijection between isomorphism classes follows from (a).

Step 5. We now prove (c) by an argument similar to that in Step 4 of the proof of Theorem 2.13 below (which I learned from [57]). Let $V \in A$-mod be simple with $\operatorname{Hom}_A(M, V) \neq \{0\}$. By Step 1, we may assume that $V = \operatorname{hd}(M_i)$ for some i. We show that the map

$$\operatorname{End}_A(\operatorname{hd}(M_i)) \to \operatorname{End}_H(F_M(\operatorname{hd}(M_i))), \quad \varphi \mapsto \varphi_*,$$

is an isomorphism. Since $\operatorname{hd}(M_i)$ and $F_M(\operatorname{hd}(M_i))$ are simple by (a), we already know that $\operatorname{End}_A(\operatorname{hd}(M_i))$ and $\operatorname{End}_H(F_M(\operatorname{hd}(M_i)))$ are division rings. Moreover, the identity on $\operatorname{hd}(M_i)$ is sent to the identity on $F_M(\operatorname{hd}(M_i))$. Hence the map $\varphi \mapsto \varphi_*$ is injective. Now let $\psi \in \operatorname{End}_H(F_M(\operatorname{hd}(M_i)))$. Using the isomorphism $F_M(\operatorname{hd}(M_i)) \cong_H \operatorname{hd}(F_M(M_i))$ of (a), we regard ψ as an endomorphism of $\operatorname{hd}(F_M(M_i))$. Let $\pi\colon F_M(M_i) \to \operatorname{hd}(F_M(M_i))$ be the canonical map. Since $F_M(M_i)$ is projective, there exists some $\hat{\psi} \in \operatorname{End}_H(F_M(M_i))$ such that $\pi \circ \hat{\psi} = \psi \circ \pi$. Using Lemma 2.3 (with $V = M_i$), there exists a unique $\varphi \in \operatorname{End}_A(M_i)$ such that $\varphi_* = \hat{\psi}$. Now φ induces a map $\bar{\varphi} \in \operatorname{End}_A(\operatorname{hd}(M_i))$ and we have $\bar{\varphi}_* = \psi$. □

Theorem 2.11 (Cf. Dipper [11, (2.32)]). *Assume that k is a field and that* (D) *holds. Let $M = M_1 \oplus \cdots \oplus M_n$ be a decomposition into indecomposable A-modules and set $V_j = \operatorname{hd}(M_j)$ for all j. Then V_j is a simple A-module and $F_M(V_j)$ is a simple H-module* (*see Theorem* 2.10), *and we have*

$$[M_i : V_j] = [F_M(M_i) : F_M(V_j)] \quad \textit{for } 1 \leq i, j \leq n.$$

Here, $[M_i : V_j]$ stands for the multiplicity of V_j as a composition factor of M_i. The expression $[F_M(M_i) : F_M(V_j)]$ has a similar meaning; note that the latter multiplicity is an entry in the Cartan matrix of H (see [23, p. 55]).

Proof. Write $F = F_M$. Recall that $F(M_j)$ is a projective indecomposable H-module with head isomorphic to $F(V_j)$, by Theorem 2.10(a). Hence, by [23, (I.16.4)], we have

$$[F(M_i) : F(V_j)] \dim_k \operatorname{End}_H(F(V_j)) = \dim_k \operatorname{Hom}_H(F(M_j), F(M_i)).$$

On the other hand, we let $\beta_j : P_j \to M_j$ be a projective cover of M_j. Since $V_j = \operatorname{hd}(M_j)$ is simple, P_j is indecomposable with head isomorphic to V_j (see [9, (6.25)(ii)]). Hence, again using [23, (I.16.4)], we also have

$$[M_i : V_j] \dim_k \operatorname{End}_A(V_j) = \dim_k \operatorname{Hom}_A(P_j, M_i).$$

Thus, taking into account Theorem 2.10(c) and using Lemma 2.3 (with $V = M_i$), we see that it enough to prove that

$$\dim_k \operatorname{Hom}_A(P_j, M_i) = \dim_k \operatorname{Hom}_A(M_j, M_i) \quad \text{for } 1 \leq i, j \leq n. \qquad (*)$$

For this purpose, we set $P := P_1 \oplus \cdots \oplus P_n$ and let $\beta = \beta_1 \oplus \cdots \oplus \beta_n : P \to M$. Then β is a projective cover of M, using the "additivity of projective covers" in [9, (6.25)(iii)]. By Remark 2.7, we conclude that (D) holds for $\beta : P \to M$.

Now β_j defines an injective map $\operatorname{Hom}_A(M_j, M_i) \to \operatorname{Hom}_A(P_j, M_i), \varphi \mapsto \varphi \circ \beta_j$, which means that we have $\dim_k \operatorname{Hom}_A(M_j, M_i) \leq \dim_k \operatorname{Hom}_A(P_j, M_i)$ for all i, j. Since Hom_A preserves direct sums, we conclude that

$$\dim_k \operatorname{Hom}_A(M, M_i) \leq \dim_k \operatorname{Hom}_A(P, M_i) \quad \text{for } 1 \leq i \leq n.$$

But we have in fact equality by Remark 2.9(b). Hence the previous inequalities must also be equalities; i.e., $(*)$ is proved. □

2.C. Green's hypotheses (A) and (B)

Let A be a k-algebra, $M \in A$-mod and $H = \operatorname{End}_A(M)^\circ$. We now assume that k is field and that A and M have finite dimension over k. We consider the following two hypotheses, formulated by Green [41]:

(A) Every composition factor of $\operatorname{soc}(M)$ is also a composition factor of $\operatorname{hd}(M)$, and vice versa. (Recall that $\operatorname{soc}(M)$ is the sum of all simple submodules of M and that $\operatorname{hd}(M) = M/\operatorname{rad}(M)$.)

(B) The algebra H is quasi-Frobenius (or, self-injective); i.e., H regarded as a left module over itself is an injective module (see [8, §58 and §61]).

Note that (B) holds, in particular, if H is symmetric (see [8, §66]).

Example 2.12. Let $A = kG$ where G is a finite group and k is a field of characteristic $p > 0$. Let U be a p-subgroup of G and M be the permutation module on the cosets of U. Then (A) is certainly satisfied. Furthermore, if $U = \{1\}$, then kG is known to be symmetric and so (B) also holds.

If G is a finite group with a split BN-pair of characteristic p and U is a Sylow p-subgroup of G then, by a result of Tinberg (see Green [41]), H is quasi-Frobenius and so (B) is satisfied.

Theorem 2.13 (Green [41], Cabanes [5], Linckelmann [57]). *Let us assume that both* (A) *and* (B) *hold. Let $M = M_1 \oplus \cdots \oplus M_n$ be a decomposition into indecomposable A-modules. Then the following hold.*

(a) *Each M_i has a simple socle and we have $F_M(\mathrm{soc}(M_i)) \cong_H \mathrm{soc}(F_M(M_i))$. Moreover, we have $M_i \cong_A M_j$ if and only if $\mathrm{soc}(M_i) \cong_A \mathrm{soc}(M_j)$.*

(b) *Each M_i has a simple head and we have $F_M(\mathrm{hd}(M_i)) \cong_H \mathrm{hd}(F_M(M_i))$. Moreover, we have $M_i \cong_A M_j$ if and only if $\mathrm{hd}(M_i) \cong_A \mathrm{hd}(M_j)$.*

(c) *We have $\mathrm{hd}(M_i) \cong_H \mathrm{soc}(M_i)$ for all i if H is a symmetric algebra.*

(d) *The map $V \mapsto F_M(V)$ is a bijection between the isomorphism classes of simple A-modules V with $\mathrm{Hom}_A(V, M) \neq \{0\}$ and isomorphism classes of simple H-modules. We have $\mathrm{End}_A(V) \cong_k \mathrm{End}_H(F_M(V))$ for every simple A-module V with $\mathrm{Hom}_A(V, M) \neq \{0\}$.*

Proof. We follow the proof given by Linckelmann [57]. To abbreviate notation, we set $F = F_M$. We will frequently use the following convention. Whenever V is an A-module and V' is a submodule of V, we can naturally identify $F(V')$ with a submodule of $F(V)$ by composing a map in $F(V')$ with the embedding $V' \subseteq V$.

Let $e_i \in H$ be the idempotent which projects M onto M_i. Note that, for any i, we have $F(M_i) = He_i$ and this is a projective indecomposable H-module (see (F2)). Now hypothesis (B) means, in particular, that projective H-modules are injective. Thus, the following two properties hold:

(B1) Let Q be a projective indecomposable H-module. Then $\mathrm{soc}(Q)$ is simple and Q is uniquely determined by $\mathrm{soc}(Q)$. Moreover, all simple H-modules arise in this way (see [8, (58.12), (58.13)]).

(B2) Let Q be a projective H-module and E', E be H-modules with $E' \subseteq E$. Then every $\psi \in \mathrm{Hom}_H(E', Q)$ can be extended to a map $\hat{\psi} \in \mathrm{Hom}_A(E, Q)$ (see [9, (2.25)(iii)]).

We now proceed in a number of steps.

Step 1. First we show that $\mathrm{soc}(M_i)$ is simple. To see this, assume if possible that there are simple A-submodules $S, S' \subseteq M_i$ such that $S \cap S' = \{0\}$. Then $F(S)$, $F(S')$ are H-submodules of $F(M_i)$ which are non-zero by (A); note also that $F(S) \cap F(S') = \{0\}$. But this would mean that $\mathrm{soc}(F(M_i))$ is not simple, contradicting (B1).

Step 2. Next we claim that $F(\mathrm{soc}(M_i)) \cong_H \mathrm{soc}(F(M_i))$. This is proved as follows. We have an embedding $F(\mathrm{soc}(M_i)) \subseteq F(M_i)$ and, by (B1), we already know that

$\mathrm{soc}(F(M_i))$ is simple. Hence it is enough to show that $F(\mathrm{soc}(M_i))$ is non-zero and simple. This is done as follows.

By hypothesis (A), we have $F(\mathrm{soc}(M_i)) \neq \{0\}$. So $F(\mathrm{soc}(M_i))$ has a simple quotient which then must be isomorphic to $\mathrm{soc}(F(M_j))$ for some j (see (B1)). Thus, for some j, there exists a surjective homomorphism of H-modules

$$\alpha\colon F(\mathrm{soc}(M_i)) \to \mathrm{soc}(F(M_j)).$$

Applying (B2) yields that there exists some $\hat{\alpha} \in \mathrm{Hom}_H(F(M_i), F(M_j))$ such that the restriction of $\hat{\alpha}$ to $F(\mathrm{soc}(M_i))$ is equal to the composition of α with the embedding $\mathrm{soc}(F(M_j)) \subseteq F(M_j)$. Using Lemma 2.3 (with $V = M_j$), we see that there exists a unique $\varphi \in \mathrm{Hom}_A(M_i, M_j)$ such that $\hat{\alpha}(f) = \varphi \circ f$ for $f \in F(M_i)$. If $\hat{\alpha}$ were not injective, then φ could not be injective either and so $\mathrm{soc}(M_i) \subseteq \ker(\varphi)$ by Step 1. But then $F(\mathrm{soc}(M_i))$ would also lie in the kernel of α, a contradiction. So $\hat{\alpha}$ and, hence also α, are injective. It follows that α is an isomorphism. This means in particular that $F(\mathrm{soc}(M_i))$ is simple, as desired.

Step 3. In order to complete the proof of (a), it remains to show that each M_i is determined by $\mathrm{soc}(M_i)$ up to isomorphism. Now, if $M_i \cong_A M_j$, then we certainly also have $\mathrm{soc}(M_i) \cong_A \mathrm{soc}(M_j)$. Conversely, assume that the latter condition holds. Then, by Step 2, we also have $\mathrm{soc}(F(M_i)) \cong_H \mathrm{soc}(F(M_j))$. Using (B1), this implies $F(M_i) \cong_H F(M_j)$ and so $M_i \cong_A M_j$ (see (F3)).

Step 4. Let V be a simple A-module with $\mathrm{Hom}_A(V, M) \neq \{0\}$. By (a), we may assume that $V = \mathrm{soc}(M_i)$ for some i. We now show that the map

$$\mathrm{End}_A(\mathrm{soc}(M_i)) \to \mathrm{End}_H(F(\mathrm{soc}(M_i))), \quad \varphi \mapsto F(\varphi) = \varphi_*,$$

is an isomorphism. Using (a), we already know that both $\mathrm{soc}(M_i)$ and $F(\mathrm{soc}(M_i))$ are simple. Hence both $\mathrm{End}_A(\mathrm{soc}(M_i))$ and $\mathrm{End}_H(F(\mathrm{soc}(M_i)))$ are division algebras over k. Since, clearly, the identity on $\mathrm{soc}(M_i)$ is sent to the identity on $F(\mathrm{soc}(M_i))$ by F, the map $\varphi \to \varphi_*$ is injective. It remains to prove that it is surjective. Let $\psi \in \mathrm{End}_H(F(\mathrm{soc}(M_i)))$. Using (B2), we see that there exists some $\hat{\psi} \in \mathrm{End}_H(F(M_i))$ such that the restriction of $\hat{\psi}$ to $F(\mathrm{soc}(M_i))$ is equal to the composition of ψ with the embedding $F(\mathrm{soc}(M_i)) \subseteq F(M_i)$. Applying Lemma 2.3 (with $V = M_i$), we find a unique $\varphi \in \mathrm{End}_A(M_i)$ such that $\varphi_* = \hat{\psi}$. Now φ restricts to a map $\varphi_1 \in \mathrm{End}_A(\mathrm{soc}(M_i))$ and we have $(\varphi_1)_* = \psi$.

Step 5. We now complete the proof of (d). The endomorphism algebras of simple modules have already been considered in Step 4. Now note that the simple A-modules V with $\mathrm{Hom}_A(V, M) \neq \{0\}$ are precisely those of the form $\mathrm{soc}(M_i)$ for some i. On the other hand, by (B1), the simple H-modules are the modules $\mathrm{soc}(F(M_i))$ for some i. Therefore the required bijection between isomorphism classes follows from (a).

Step 6. We claim that $\mathrm{hd}(M_i)$ is simple for all i. To prove this, consider the natural map $\pi: M_i \to \mathrm{hd}(M_i)$. Then the map

$$\mathrm{End}_A(\mathrm{hd}(M_i)) \to \mathrm{Hom}_A(M_i, \mathrm{hd}(M_i)), \quad \varphi \mapsto \varphi \circ \pi,$$

is an isomorphism. (The injectivity is clear. To see the surjectivity, just note that since $\mathrm{hd}(M_i)$ is semisimple, any map $M_i \to \mathrm{hd}(M_i)$ has $\mathrm{rad}(M_i)$ in its kernel.) Using Lemma 2.3 (with $V = \mathrm{hd}(M_i)$) we see that

$$\mathrm{End}_A(\mathrm{hd}(M_i)) \cong_k \mathrm{Hom}_H(F(M_i), F(\mathrm{hd}(M_i))).$$

We now count dimensions on both sides of the above isomorphism. Let us write $\mathrm{hd}(M_i) \cong_A \bigoplus_V n_V V$ where the sum is over all simple A-modules V (up to isomorphism) and n_V are non-negative integers. By hypothesis (A), we have $\mathrm{Hom}_A(V, M) \neq \{0\}$ for each V with $n_V > 0$. Using (d) and the fact that F preserves direct sums, we obtain $F(\mathrm{hd}(M_i)) \cong_H \bigoplus_V n_V F(V)$, where each $F(V)$ with $n_V > 0$ is non-zero and simple. Since $F(M_i)$ has a simple head, we have $\mathrm{Hom}_H(F(M_i), \mathrm{hd}(F(M_i))) \cong_k \mathrm{End}_H(\mathrm{hd}(M_i))$, and so

$$\begin{aligned}\dim_k \mathrm{Hom}_H(F(M_i), F(\mathrm{hd}(M_i))) &= \sum_V n_V \dim_k \mathrm{Hom}_H(F(M_i), F(V))\\ &= n_{V_0} \dim_k \mathrm{End}_H(F(V_0)),\end{aligned}$$

where V_0 is such that $F(V_0) \cong_H \mathrm{hd}(F(M_i))$. Note also that $F(V) \ncong_H F(V_0)$ if $n_V > 0$ and $V \ncong_A V_0$, by (d). On the other hand, we have

$$\begin{aligned}\dim_k \mathrm{End}_A(\mathrm{hd}(M_i)) &= \sum_V n_V^2 \dim_k \mathrm{End}_A(V)\\ &= n_{V_0}^2 \dim_k \mathrm{End}_A(V_0) + \sum_{V \neq V_0} n_V^2 \dim_k \mathrm{End}_A(V).\end{aligned}$$

The left hand sides of the above two dimension formulas must be equal. Hence so are the right hand sides. On the other hand, by (d), we know that $\dim_k \mathrm{End}_A(V_0) = \dim_k \mathrm{End}_H(F(V_0))$. Inserting this into the above formulas, we conclude that $n_{V_0} = 1$ and $n_V = 0$ for all $V \neq V_0$.

Step 7. Now we can also complete the proof of (b). We begin by showing that $\mathrm{hd}(F(M_i)) \cong_H F(\mathrm{hd}(M_i))$. First note that $\mathrm{hd}(M_i)$ is simple by Step 6 and so $F(\mathrm{hd}(M_i))$ is non-zero and simple, using (A) and (d). Now consider the natural map $\pi\colon M_i \to \mathrm{hd}(M_i)$. Then the induced map $\pi_*\colon F(M_i) \to F(\mathrm{hd}(M_i))$ is non-zero (since $\pi_*(e_i) = \pi \circ e_i \neq 0$) and hence must be surjective. Consequently, the simple head of $F(M_i)$ is isomorphic to $F(\mathrm{hd}(M_i))$.

Finally, the fact that each M_i is determined by $\mathrm{hd}(M_i)$ up to isomorphism is proved as the analogous statement in Theorem 2.10(a).

Step 8. Finally we prove (c). If H is symmetric then $\mathrm{hd}(F(M_i)) \cong_H \mathrm{soc}(F(M_i))$ for all i (see [9, (9.12)]). Using (a) and (b), we deduce that $\mathrm{hd}(M_i) \cong_A \mathrm{soc}(M_i)$, as desired. □

Finally, we only cite the following result which shows exactly which part of the module category of A is controlled by H.

Theorem 2.14 (Cabanes [5]). *Let A-mod$_M$ be the full subcategory of A-mod whose objects are the modules $V \in A$-mod such that V is isomorphic to a quotient of M^t (for some $t \geq 1$) and also to a submodule of M^t (for some $t \geq 1$). Assume that* (B) *holds. Then the Hom functor F_M induces an equivalence of additive categories from A-mod$_M$ to H-mod.*

3. Harish-Chandra series

In this section, we first introduce some basic definitions and results concerning finite groups with a BN-pair and the operations of Harish-Chandra induction and restriction. We show that the set of simple modules of such a group is partitioned into so-called Harish–Chandra series which are indexed by suitable equivalence classes of pairs consisting of a Levi subgroup and a cuspidal simple module. Over fields of characteristic 0, this is a classical theory; see Carter [7, §9.2]. Roughly speaking, the theory also works over fields of positive characteristic, but some care is needed since modules need no longer be semisimple; this has first been worked out by Hiss [47]. The general results on Hom functors in Section 2 will then be used to relate the simple modules in a Harish–Chandra series with the simple modules of the corresponding Hecke algebra, following Dipper [11], Dipper–James [16], Geck–Hiss–Malle [38] and Geck–Hiss [36].

3.A. Split BN-pairs and the functors R_J^S and $^*R_J^S$

Let G be a finite group with a split BN-pair of characteristic p which satisfies Chevalley's commutator relations; see [7, §2.5–2.6]. We don't repeat all the axioms here, but just recall some basic results. We have $B = UH$ where $U = O_p(B)$ is the largest normal p-subgroup of B and H is an abelian p'-subgroup such that $H = B \cap N$. Let $W = N/H$ be the Weyl group of G. For any $w \in W$ we denote by $\dot{w}$ a representative in N. The group W is a Coxeter group with respect to the set of generators

$$S = \{w \in W \mid w \neq 1 \text{ and } B \cup B\dot{w}B \text{ is a subgroup of } G\}.$$

For any subset $J \subseteq S$, we have a corresponding standard parabolic subgroup $W_J = \langle J \rangle \subseteq W$. Moreover, J also defines a standard parabolic subgroup $P_J = BN_J B \subseteq G$, where $N_J = \{H\dot{w} \mid w \in W_J\}$. We have a Levi decomposition $P_J = U_J L_J$ where $U_J = O_p(P_J)$ is the largest normal p-subgroup of P_J and L_J is a complementary subgroup which is uniquely determined by the condition that $H \subseteq L_J$. The group L_J is called a standard Levi subgroup of G; it is itself a finite group with a split BN-pair of characteristic p corresponding to the subgroups $B_J = B \cap L_J$ and $N_J = \{H\dot{w} \mid w \in W_J\}$.

Example 3.1. Let $G = \mathrm{GL}_n(q)$, where q is a power of a prime p. We can take B to be the subgroup of all upper triangular matrices and N to be the subgroup of all monomial matrices in G. Then $B \cap N = H$ is the subgroup of all diagonal matrices and $W = N/H$ can be naturally identified with the symmetric group $\mathfrak{S}_n$. The generating set S consists of the basic transpositions $(1, 2), \ldots, (n-1, n)$.

3.2. Now let k be a commutative ring with 1 such that p is invertible in k. For any $J \subseteq S$, let kL_J-mod be the category of finitely generated kL_J-modules. We define functors

$$R_J^S\colon kL_J\text{-mod} \to kG\text{-mod} \quad \text{and} \quad {}^*R_J^S\colon kG\text{-mod} \to kL_J\text{-mod}$$

as follows. Let $X \in kL_J$-mod. Then we can regard X as a kP_J-module via the canonical map $P_J \to L_J$ with kernel U_J; denote that kP_J-module by $\tilde{X}$. Then we set $R_J^S(X) = \mathrm{Ind}_{P_J}^G(\tilde{X})$ (where Ind denotes the usual induction of modules from subgroups).

Conversely, let $Y \in kG$-mod. Then, since U_J is normalized by L_J, the fixed point set $\mathrm{Fix}_{U_J}(Y)$ is naturally a kL_J-module, which we denote ${}^*R_J^S(Y)$. Note that $\mathrm{Fix}_{U_J}(Y)$ is finitely generated over k. This follows from the following alternative description. Let

$$e_J = |U_J|^{-1} \sum_{u \in U_J} u \in kG;$$

i.e., e_J is the central idempotent of kU_J corresponding to the trivial kU_J-module. Then the map $\mathrm{Fix}_{U_J}(Y) \to e_J Y$, $y \mapsto e_J y$, is an isomorphism of kL_J-modules.

Remark 3.3. By definition, we have $R_J^S(X) = kG \otimes_{kP_J} \tilde{X}$. At some places, it will be convenient to use the following alternative interpretations of $R_J^S(X)$. First, consider the (kG, kL_J)-bimodule $k[G/U_J]$, where the actions of G and L_J are given by left and right multiplication, respectively. Then, using the decomposition $gP_J = \bigcup_{l \in L_J} gU_J l$ $(g \in G)$, we have a canonical isomorphism

$$R_J^S(X) \xrightarrow{\sim} k[G/U_J] \otimes_{kL_J} X, \quad g \otimes x \mapsto gU_J \otimes x. \tag{a}$$

(This description is due to Broué; see [10, Chapter 4].)

Further, $\mathrm{Hom}_{kP_J}(kG, X)$ is a kG-module, where $g \in G$ acts via right translation on kG. Then, denoting by $T_J \subseteq G$ a set of right coset representatives of P_J in G, we also have an isomorphism (see [14, (1.3)])

$$\mathrm{Hom}_{kP_J}(kG, \tilde{X}) \xrightarrow{\sim} R_J^S(X), \quad f \mapsto \sum_{g \in T_J} g^{-1} \otimes f(g). \tag{b}$$

If $\varphi \in \mathrm{Hom}_{kL_J}(X, X')$ for some $X, X' \in kL_J$-mod, we have an induced map $R_J^S(\varphi)\colon R_J^S(X) \to R_J^S(X')$. In terms of the interpretation in (b), we have

$$R_J^S(\varphi)(f) = \varphi \circ f \quad \text{for all } f \in \mathrm{Hom}_{kP_J}(kG, \tilde{X}). \tag{c}$$

Lemma 3.4. *Harish-Chandra induction and restriction are exact and preserve direct sums and k-torsion modules. Moreover, if $P \in kL_J$-mod is projective, then so is $R_J^S(P)$.*

Proof. The fact that R_J^S is exact and preserves direct sums follows from Remark 3.3(a) since the analogous statements hold for tensor products (note that $k[G/U_J]$ is free as a kL_J-module). For ${}^*R_J^S$, it is enough to observe that ${}^*R_J^S$ is given by multiplication with an idempotent; see (3.2). The fact that k-torsion modules are preserved follows from the analogous properties of ordinary induction and restriction. Finally, the assertion about projectivity is proved as follows. Suppose that P is a direct summand of a direct sum of n copies, say, of kL_J. Then, since R_J^S is additive, $R_J^S(P)$ is a direct summand of a direct sum of n copies of $R_J^S(kL_J)$. Now the definition shows that $R_J^S(kL_J) = \mathrm{Ind}_{U_J}^G(k)$, which is projective since $|U_J|$ is invertible in k. □

Lemma 3.5 (Transitivity). *Let $J \subseteq I \subseteq S$. Then we have, for any $X \in kL_J$-mod and any $Y \in kG$-mod,*

$$R_J^S(X) \cong_{kG} R_I^S \circ R_J^I(X) \quad \textit{and} \quad {}^*R_J^S(Y) \cong_{kL_J} {}^*R_J^I \circ {}^*R_I^S(Y).$$

Proof. This uses some standard results about intersections of parabolic subgroups. For details, see Dipper–Fleischmann [14, (1.12)] and Digne–Michel [10, (4.7)]. □

Lemma 3.6 (Adjointness). *Let $X \in kL_J$-mod and $Y \in kG$-mod. Then*

$$\mathrm{Hom}_{kG}\left(R_J^S(X), Y\right) \cong_k \mathrm{Hom}_{kL_J}\left(X, {}^*R_J^S(Y)\right)$$
$$\mathrm{Hom}_{kG}\left(Y, R_J^S(X)\right) \cong_k \mathrm{Hom}_{kL_J}\left({}^*R_J^S(Y), X\right).$$

Proof. The first isomorphism follows form a general adjoint property, while the proof of the second isomorphism uses the fact that p is invertible in k. For details see Dipper–Fleischmann [14, (1.8)]. □

Lemma 3.7 (Mackey formula). *Let $I, J \subseteq S$ and D_{IJ} be the set of distinguished double coset representatives of W with respect to W_I and W_J (see* [9, §64C]*). Then we have, for any $X \in kL_J$-mod,*

$${}^*R_I^S \circ R_J^S(X) \cong_{kL_I} \bigoplus_{w \in D_{IJ}} R_{I\cap {}^wJ}^I \circ \mathrm{ad}(\dot{w}) \circ {}^*R_{I^w \cap J}^J(X),$$

where $\mathrm{ad}(\dot{w})$ stands for the functor on the appropriate module categories induced by the conjugation map $x \mapsto \dot{w}x\dot{w}^{-1}$.

Proof. The proof is an adaptation of that for ordinary induction and restriction of modules for finite groups. For details see Dipper–Fleischmann [14, (1.14)] and Digne–Michel [10, (5.1)]. □

Definition 3.8. Let $I, J \subseteq S$ and $Z \in kL_I$-mod, $X \in kL_J$-mod. We say that the pairs (I, Z) and (J, X) are *associate* if there exists some $w \in W$ such that ${}^wJ = I$ and $\mathrm{ad}(\dot{w})(X) \cong_{kL_I} Z$.

By [7, (9.2.2)], we have ${}^wJ = I$ if and only if L_I, L_J are conjugate by some element of N. Note that this does not necessarily mean that P_I, P_J are also conjugate. As an example, consider $G = \mathrm{GL}_3(q)$. Then $S = \{s_1, s_2\}$ where $s_1 = (1, 2)$ and $s_2 = (2, 3)$ are the standard generators of the symmetric group $\mathfrak{S}_3$. It is readily checked that $P_{\{s_1\}}$ and $P_{\{s_2\}}$ are not conjugate in G while s_1 and s_2 certainly are conjugate in $\mathfrak{S}_3$.

The following result is clear in the case of ordinary induction, but in the case of Harish-Chandra induction it has only recently been established.

Theorem 3.9 (Dipper–Du [13], Howlett–Lehrer [51]). *Let (I, Z) and (J, X) be associate pairs, as in Definition* 3.8. *Then we have $R_J^S(X) \cong_{kG} R_I^S(Z)$.*

3.B. Cuspidal modules

For the rest of this section, we will assume that k is a field. Recall that we assume throughout that p is invertible in k. We denote by $\mathrm{Irr}_k(G)$ the set of simple kG-modules (up to isomorphism); for each $Y \in \mathrm{Irr}_k(G)$, we denote by P_Y a projective cover of Y.

Definition 3.10. Let $X \in kG$-mod. We say that X is *cuspidal* if ${}^*R_J^S(X) = \{0\}$ for every proper subset $J \subset S$. Thus, X is cuspidal if and only if $\mathrm{Fix}_{U_J}(X) = \{0\}$ for every proper $J \subset S$. We set

$$\mathrm{Irr}_k^\circ(G) = \{Y \in \mathrm{Irr}_k(G) \mid Y \text{ is cuspidal}\}.$$

Similar notations are used for Levi subgroups of G.

This definition, for any field k, first appeared in [14]. Note that it really only makes sense if we work over a field of characteristic different from p since otherwise every p-subgroup has a non-trivial fixed point on every non-zero kG-module.

Remark 3.11. Let $Y \in \mathrm{Irr}_k(G)$ and $X \in \mathrm{Irr}_k(L_J)$ for some $J \subseteq S$. Then ${}^*R_J^S(Y)$ is non-zero and X is a composition factor of ${}^*R_J^S(Y)$ if and only if

$$\mathrm{Hom}_{kG}(R_J^S(P_X), Y) \cong_k \mathrm{Hom}_{kL_J}(P_X, {}^*R_J^S(Y)) \neq \{0\},$$

and this holds if and only if P_Y is a direct summand of $R_J^S(P_X)$. Note that $R_J^S(P_X)$ is projective by Lemma 3.4.

Lemma 3.12. *Let $Y \in \mathrm{Irr}_k(G)$ and $J \subseteq S$ be a minimal subset (with respect to inclusion) such that ${}^*R_J^S(Y) \neq \{0\}$. Then all composition factors of ${}^*R_J^S(Y)$ are cuspidal.*

Proof. Let $X \in \mathrm{Irr}_k(L_J)$ be a composition factor of ${}^*R_J^S(Y)$ and assume, if possible, that X is not cuspidal. Then, by Remark 3.11, there exists a proper subset $I \subset J$ and

$Z \in \mathrm{Irr}_k(L_I)$ such that P_X is a direct summand of $R_I^J(P_Z)$. Using Lemma 3.5 and Lemma 3.6, we now obtain

$$\begin{aligned}\mathrm{Hom}_{kL_I}(P_Z, {}^*R_I^S(Y)) &\cong_k \mathrm{Hom}_{kL_I}(P_Z, {}^*R_I^J \circ {}^*R_J^S(Y))\\ &\cong_k \mathrm{Hom}_{kL_J}(R_I^J(P_Z), {}^*R_J^S(Y)),\end{aligned}$$

and the space on the right hand side is non-zero since P_X is a summand of $R_I^J(P_Z)$ and X is a composition factor of ${}^*R_J^S(Y)$. In particular, we deduce that ${}^*R_I^S(Y) \neq \{0\}$, contradicting the minimality of J. □

Definition 3.13. Let $X \in \mathrm{Irr}_k^\circ(L_J)$ for some subset $J \subseteq S$. Then (J, X) is called a *cuspidal pair*. Moreover, the corresponding *Harish-Chandra series* $\mathrm{Irr}_k(G, (J, X))$ is defined as the set of all $Y \in \mathrm{Irr}_k(G)$ such that

(a) J is a minimal subset of S such that ${}^*R_J^S(Y) \neq \{0\}$, and

(b) X is a composition factor of ${}^*R_J^S(Y)$.

By Lemma 3.12, we have that

$$\mathrm{Irr}_k(G) = \bigcup_{(J,X)\ \text{cuspidal}} \mathrm{Irr}_k(G, (J, X)).$$

Note that the union need not be disjoint; furthermore, at this stage, it is not even clear if every Harish–Chandra series $\mathrm{Irr}_k(G, (J, X))$ is non-empty.

Lemma 3.14. *Let $Y \in \mathrm{Irr}_k(G)$ and $I, J \subseteq S$ be such that ${}^*R_I^S(Y) \neq \{0\}$ and ${}^*R_J^S(Y) \neq \{0\}$. Assume that ${}^*R_J^S(Y)$ has a composition factor $X \in \mathrm{Irr}_k(L_J)$ and Y is a composition factor of $R_I^S(Z)$ where $Z \in \mathrm{Irr}_k^\circ(L_I)$.*

(a) *We have* $\mathrm{Hom}_{kG}(R_J^S(P_X), R_I^S(Z)) \neq \{0\}$.

(b) *There exists some $w \in W$ with $I \subseteq {}^wJ$.*

Proof. (a) We have $\mathrm{Hom}_{kG}(P_Y, R_I^S(Z)) \neq \{0\}$, since Y is a composition factor of $R_I^S(Z)$. On the other hand, we also have that P_Y is a direct summand of $R_J^S(P_X)$ by Remark 3.11.

(b) We have $\mathrm{Hom}_{kG}(R_J^S(P_X), R_I^S(Z)) \cong_k \mathrm{Hom}_{kL_I}({}^*R_I^S \circ R_J^S(P_X), Z)$ by using Lemma 3.6. We apply the Mackey formula in Lemma 3.7 and find that the latter space is isomorphic to

$$\bigoplus_{w \in D_{IJ}} \mathrm{Hom}_{kL_I}(R_{I\cap {}^wJ}^I \circ \mathrm{ad}(\dot{w}) \circ {}^*R_{I^w\cap J}^J(P_X), Z).$$

Applying Lemma 3.6 once more, we rewrite this as

$$\bigoplus_{w \in D_{IJ}} \mathrm{Hom}_{kL_{I\cap {}^wJ}}(\mathrm{ad}(\dot{w}) \circ {}^*R_{I^w\cap J}^J(P_X), {}^*R_{I\cap {}^wJ}^I(Z)).$$

By (a), at least one term in the above sum must be non-zero. Now, since Z is cuspidal, this must be a term for which $I \subseteq {}^wJ$. □

The following result combines ideas from [38, §2] and [47, (5.3)].

Proposition 3.15. *Let $Y \in \mathrm{Irr}_k(G, (J, X))$ for some cuspidal pair (J, X). Assume that $I \subseteq S$ is such that ${}^*R_I^S(Y)$ is non-zero and has a simple quotient $Z \in \mathrm{Irr}_k^\circ(L_I)$. Then we have the following.*

(a) *There exists some $w \in W$ such that $I = {}^wJ$ and $\mathrm{ad}(\dot{w})(X) \cong_{kL_I} Z$.*

(b) $\mathrm{Hom}_{kG}(R_J^S(X), R_I^S(Z)) \cong_k \mathrm{Hom}_{kG}(R_J^S(P_X), R_I^S(Z)) \neq \{0\}$, *where P_X is a projective cover of X.*

(c) *If X and Z are absolutely simple, we have the following dimension equality:* $\dim_k \mathrm{Hom}_{kG}(R_J^S(X), R_I^S(Z)) = |\{w \in W \mid I = {}^wJ,\ \mathrm{ad}(\dot{w})(X) \cong_{kL_I} Z\}|$.

Proof. First we check that $I = {}^wJ$ for some $w \in W$. Indeed, the assumptions of Lemma 3.14 are satisfied since, using Lemma 3.6, we have

$$\mathrm{Hom}_{kG}(Y, R_I^S(Z)) \cong_k \mathrm{Hom}_{kL_I}({}^*R_I^S(Y), Z) \neq \{0\}.$$

Hence we have $I \subseteq {}^wJ$ for some $w \in W$. In particular, this means that $|I| \leq |J|$. To see that we have in fact equality, we argue as follows. Since J is minimal such that ${}^*R_J^S(Y) \neq \{0\}$, every composition factor of ${}^*R_J^S(Y)$ is cuspidal by Lemma 3.12. Now, just for the moment, let us choose X to be a simple quotient of ${}^*R_J^S(Y)$. Then we can repeat the whole argument with the roles of I, J reversed, and so also $|J| \leq |I|$. Thus we must have $I = {}^wJ$ for some $w \in W$, as desired.

Performing a similar computation as in the proof of Lemma 3.14 and using the fact that we only need to consider terms with $I = {}^wJ$, we obtain

$$\mathrm{Hom}_{kG}(R_J^S(P_X), R_I^S(Z)) \cong_k \bigoplus_{w \in D_{IJ}, I = {}^wJ} \mathrm{Hom}_{kL_I}(\mathrm{ad}(\dot{w})(P_X), Z).$$

Since P_X has a unique simple quotient and is uniquely determined by that quotient, we have $\mathrm{Hom}_{kL_I}(\mathrm{ad}(\dot{w})(P_X), Z) \cong_k \mathrm{Hom}_{kL_I}(\mathrm{ad}(\dot{w})(X), Z)$. We now see that if we repeat the above computation of $\mathrm{Hom}_{kG}(R_J^S(P_X), R_I^S(Z))$ with X instead of P_X, we arrive at the same result. This proves the desired isomorphism in (b) as well as the dimension formula in (c), using that $\mathrm{Hom}_{kL_I}(\mathrm{ad}(\dot{w})(X), Z)$ is isomorphic to k or $\{0\}$ according to whether $\mathrm{ad}(\dot{w})(X)$ is isomorphic to Z or not. Finally, (b) and (c) also imply (a). □

Theorem 3.16 (Hiss [47]). *Let $Y \in \mathrm{Irr}_k(G)$ and (J, X) be a cuspidal pair. Then the following are equivalent:*

(a) $Y \in \mathrm{Irr}_k(G, (J, X))$.

(b) *Y is isomorphic to a submodule of $R_J^S(X)$.*

(c) *Y is isomorphic to a quotient of* $R_J^S(X)$.

In particular, $\operatorname{Irr}_k(G, (J, X))$ *is non-empty. If* (I, Z) *is another cuspidal pair, then we have* $\operatorname{Irr}_k(G, (J, X)) \cap \operatorname{Irr}_k(G, (I, Z)) = \emptyset$ *unless* (I, Z) *and* (J, X) *are associate (in the sense of Definition* 3.8*).*

Proof. We proceed in several steps.

Step 1. First we prove that if (I, Z) is another cuspidal pair such that there exists some $Y \in \operatorname{Irr}_k(G, (J, X)) \cap \operatorname{Irr}_k(G, (I, Z))$, then (I, Z) and (J, X) are associate. This follows from a double application of Proposition 3.15(a). Indeed, let Z' be a simple quotient of ${}^*R_I^S(Y)$. Then, by Lemma 3.12, Z' is cuspidal and a first application of Proposition 3.15(a) shows that the pairs (I, Z) and (I, Z') are associate. Then a second application shows that the pairs (J, X) and (I, Z') are also associate.

Step 2. Next we prove that (a) implies (b). Let $Z \in \operatorname{Irr}_k(L_J)$ be a simple quotient of ${}^*R_J^S(Y)$. By Lemma 3.12, Z is cuspidal. Now all the assumptions of Proposition 3.15 are satisfied (with $I = J$) and so there exists some $w \in W$ such that $J = {}^wJ$ and $\operatorname{ad}(\dot{w})(X) \cong_{kL_J} Z$. It follows that

$$\begin{aligned}\operatorname{Hom}_{kG}(Y, R_J^S(\operatorname{ad}(\dot{w})(X))) &\cong_k \operatorname{Hom}_{kL_J}({}^*R_J^S(Y), \operatorname{ad}(\dot{w})(X)) \\ &\cong_k \operatorname{Hom}_{kL_J}({}^*R_J^S(Y), Z) \neq \{0\}.\end{aligned}$$

Thus Y is isomorphic to a submodule of $R_J^S(\operatorname{ad}(\dot{w})(X))$. It remains to use Theorem 3.9.

Step 3. Now assume that Y is as in (b) or in (c). We show that then (a) holds. Using Lemma 3.6, we see that ${}^*R_J^S(Y) \neq \{0\}$. Let $I \subseteq J$ be minimal such that ${}^*R_I^S(Y) \neq \{0\}$ and let $Z \in \operatorname{Irr}_k(L_I)$ be any composition factor of ${}^*R_I^S(Y)$. Now all the assumptions of Lemma 3.14 are satisfied (with the roles of I, J reversed) and so we have $J \subseteq {}^wI$ for some $w \in W$. Since $I \subseteq J$, we conclude that $I = J$ and ${}^wJ = J$. Thus $Y \in \operatorname{Irr}_k(G, (J, X))$, as desired.

Step 4. Finally we show that, if Y satisfies (b), then (c) also holds. Using Lemma 3.6, we see that ${}^*R_J^S(Y) \neq \{0\}$. Let $Z \in \operatorname{Irr}_k(L_J)$ be a simple submodule of ${}^*R_J^S(Y)$. Then we have, again using Lemma 3.6,

$$\operatorname{Hom}_{kG}(R_J^S(Z), Y) \cong_k \operatorname{Hom}_{kL_J}(Z, {}^*R_J^S(Y)) \neq \{0\},$$

which means that Y is a quotient of $R_J^S(Z)$. So, by Theorem 3.9, it is enough to show that (J, X) and (J, Z) are associate. Now, using the equivalence of (a) and (b), we have $Y \in \operatorname{Irr}_k(G, (J, X))$ which means, in particular, that J is a minimal subset of S such that ${}^*R_J^S(Y) \neq \{0\}$. Hence, by Lemma 3.12, $Z \in \operatorname{Irr}_k^\circ(L_J)$ and so $Y \in \operatorname{Irr}_k(G, (J, Z))$. By Step 1, (J, X) and (J, Z) must be associate, as desired. □

Remark 3.17. In Hiss' paper [47], the statement of the above result looks slightly more complicated. But this is just due to the fact that Theorem 3.9 was not yet available. For example, looking through the above proof, we see that if $Y \in \operatorname{Irr}_k(G, (J, X))$

then, without assuming Theorem 3.9, one can only deduce that Y is a submodule of $R_J^S(\mathrm{ad}(\dot{w}(X))$, where ${}^wJ = J$.

Remark 3.18. The idea of partitioning the simple modules of G into series can also be generalized in another direction, namely using the Lusztig functor [58] instead of the functors in (3.2). This yields the *generalized Harish-Chandra theory* of Broué, Malle and Michel; see [4] for a survey and further references.

Example 3.19. Let $G = \mathrm{GL}_2(q)$ and consider the module $M := R_{\emptyset}^S(k_H) = \mathrm{Ind}_B^G(k_B)$. We have $[G : B] = q + 1$. There are two essentially different cases:

(a) The characteristic of k does not divide $[G : B]$. Then $M \cong_{kG} k_G \oplus Y$ where Y is a simple module of dimension q. In fact, Y is the Steinberg module; see [64]. Thus we have $\mathrm{Irr}_k(G, (\emptyset, k_H)) = \{k_G, Y\}$.

(b) The characteristic of k divides $[G : B]$. Then M is indecomposable with $\mathrm{soc}(M) \cong_{kG} \mathrm{hd}(M) \cong_{kG} k_G$. Thus we have $\mathrm{Irr}_k(G, (\emptyset, k_H)) = \{k_G\}$.

Theorem 3.20 (Geck–Hiss–Malle [38, (2.3)]). *Let* (J, X) *be a cuspidal pair* (*see Definition* 3.13) *and set*

$$H = \mathrm{End}_{kG}(M)^{\circ} \quad \text{where } M := R_J^S(X) \in kG\text{-mod}\,.$$

Let $\beta_X \colon P_X \to X$ *be a projective cover and* $\beta \colon P \to M$ *be the induced map, where* $P := R_J^S(P_X)$. (*Note that* β *is surjective and* P *is projective by Lemma* 3.4.) *Then Dipper's hypothesis* (D) *holds for* β. *Consequently, we have a bijection*

$$\mathrm{Irr}_k(G, (J, X)) \xrightarrow{\sim} \mathrm{Irr}(H), \quad Y \mapsto F_M(Y),$$

where $\mathrm{Irr}(H)$ *denotes the set of simple* H*-modules* (*up to isomorphism*).

Proof. In order to show that (D) holds, we only need to check, by Lemma 2.6, that $\dim_k \mathrm{Hom}_A(P, M) = \dim_k \mathrm{End}_A(M)$. But that equality of dimensions has been already established in Proposition 3.15(b). The fact that F_M induces a bijection as above is contained in Theorem 2.10(b). □

Remark 3.21. Assume that k has characteristic 0 and is "large enough" so that kG and kL_J are split semisimple algebras. Then there exists a primitive idempotent $e \in kP_J$ such that $\tilde{X} \cong_{kP_J} kP_J e$ and we are in the setting of Example 2.2. So we can identify $H = ekGe$, and F_M is given by sending $V \in kG$-mod to $eV \in H$-mod. This is the "classical" setting of *ordinary* Harish-Chandra series and Hecke algebras. In particular, if we are in this case, all the results in [7, Chapter 10] apply; for example, we have a degree formula relating the dimension of $V \in \mathrm{Irr}_k(G, (J, X))$ with the so-called "generic degree" of the corresponding simple H-module eV (see [7, (10.9.6)]).

We have already seen in Theorem 3.16 that Green's hypothesis (A) also holds. To check that (B) is satisfied, we need some more detailed information about the structure of H; see (3.30).

To close this paragraph, we mention a result which shows that the above bijection is compatible with the transitivity of Harish–Chandra induction.

3.22. In the set-up of Theorem 3.20, assume that $I \subseteq S$ is such that $J \subseteq I$. To abbreviate notation, set $M_I := R_J^I(X)$ and $H_I := \mathrm{End}_{kL_I}(M_I)$. Then we have a natural map (see Remark 3.3(c))

$$\iota_I \colon H_I \to H, \quad \varphi \mapsto R_I^S(\varphi).$$

Using the description of H in (3.C) below, it can be shown that ι_I is in fact injective; see [43, (2.5)]. We get an induced map $\iota_I^* \colon H\text{-mod} \to H_I\text{-mod}$ by pulling back the action from H to H_I via ι_I.

Proposition 3.23 (Geck–Hiss–Malle [38, (2.7)]). *In the above notation, assume that the map $\iota_I \colon H_I \to H$ is an isomorphism. Then the map $Z \mapsto R_I^S(Z)$ defines a bijection between* $\mathrm{Irr}_k(L_I, (J, X))$ *and* $\mathrm{Irr}_k(G, (J, X))$ *such that the following diagram is commutative:*

$$\begin{array}{ccc} \mathrm{Irr}_k(G, (J, X)) & \xrightarrow{F_M} & \mathrm{Irr}(H) \\ {\scriptstyle R_I^S}\uparrow & & \downarrow{\scriptstyle \iota_I^*} \\ \mathrm{Irr}_k(L_I, (J, X)) & \xrightarrow{F_{M_I}} & \mathrm{Irr}(H_I)\,. \end{array}$$

Proof. The fact that $Z \mapsto R_I^S(Z)$ defines a bijection between $\mathrm{Irr}_k(L_I, (J, X))$ and $\mathrm{Irr}_k(G, (J, X))$ is shown in [38, (2.7)]. The commutativity of the diagram follows readily from the definition of the functors which are involved. □

3.C. Hecke algebras

Let us fix a cuspidal simple kL_J-module X for some subset $J \subseteq S$. We now consider the structure of the endomorphism algebra

$$H = \mathrm{End}_{kG}(M)^\circ \quad \text{where} \quad M := R_J^S(X) \in kG\text{-mod}\,.$$

For this purpose, we assume from now on that

$$X \text{ is absolutely simple; i.e., we have } \mathrm{End}_{kL_J}(X) = k\,\mathrm{id}_X.$$

We now collect some basic results about H from [7, Chapter 10]. These results are due to Howlett–Lehrer [50] in the case where k has characteristic 0, but the proofs of the statements in (3.24)–(3.26) are easily seen to work, word by word, in general. (Recall, however, that we assume throughout that p is invertible in k.)

3.24. Let $C_J := \{w \in W \mid {}^wJ = J\}$. Then, by [7, (9.2.1)], we have $N_W(W_J) = W_J \rtimes C_J$, and the map $N_G(L_J) \cap N \to N_W(W_J)$, $n \mapsto nH$, induces an isomorphism

$$(N_G(L_J) \cap N)L_J/L_J \cong N_W(W_J)/W_J \cong C_J\,.$$

For $w \in C_J$ we denote by $\dot{w}$ a representative in N; we have $\dot{w} \in N_G(L_J) \cap N$. The *ramification group* of X is defined as

$$W_{(J,X)} := \{w \in C_J \mid \mathrm{ad}(\dot{w})(X) \cong_{kL_J} X\}.$$

Let $K_{(J,X)}$ be the subgroup of G generated by L_J and the elements $\dot{w}$ for $w \in W_{(J,X)}$. Then, by [7, (10.3.1)], we have $K_{(J,X)}/L_J \cong W_{(J,X)}$. Moreover, Proposition 3.15(c) shows that $\dim_k H = |W_{(J,X)}|$.

3.25. Let $\rho\colon kL_J \to \mathrm{End}_k(X)$ be the representation afforded by X and let $\tilde{\rho}\colon kK_{(J,X)} \to \mathrm{End}_k(X)$ be a projective representation extending ρ. (Recall our assumption that X is absolutely simple.) Thus we have

$$\tilde{\rho}(\dot{w}_1)\tilde{\rho}(\dot{w}_2) = \lambda(w_1, w_2)\tilde{\rho}(\dot{w}_1\dot{w}_2) \quad \text{for all } w_1, w_2 \in W_{(J,X)},$$

where $\lambda\colon W_{(J,X)} \times W_{(J,X)} \to k^\times$ is a 2-cocycle (see [7, (10.3.3)]). That cocycle can be chosen such that the following conditions hold (see [7, (10.3.4)]):

(a) $\lambda(w, 1) = \lambda(1, w) = 1$ for all w;

(b) $\lambda(w, w^{-1}) = 1$ for all w;

(c) $\lambda(w_1, w_2) = \lambda(w_2^{-1}, w_1^{-1})^{-1} = \lambda(w_2^{-1}w_1^{-1}, w_1)$ for all w_1, w_2;

(d) $\lambda(w_1, w_2)$ is a root of unity for all w_1, w_2.

(In order to make sure that these conditions hold, one has to assume that k contains certain roots of elements defined in terms of X.)

3.26. We identify $R_J^S(X) = \mathrm{Hom}_{kP_J}(kG, \tilde{X})$ as in Remark 3.3(b). For each $w \in W_{(J,X)}$, we define $B_w \in \mathrm{End}_k(\mathrm{Hom}_{kP_J}(kG, \tilde{X}))$ by

$$(B_w f)(a) = \frac{1}{|U_J|} \sum_{u \in U_J} \tilde{\rho}(\dot{w}) f(\dot{w}^{-1}ua), \qquad f \in \mathrm{Hom}_{kP_J}(kG, \tilde{X}),\ a \in kG.$$

Then B_w does not depend on the choice of the representative $\dot{w}$ (see [50, Remark 1, p.45]) and we have $B_w \in H$ for all $w \in W_{(J,X)}$ (see [7, (10.1.4)]).

The fact that the elements $\{B_w\}$ form a basis of H is seen as follows. We have already noted in (3.24) that $\dim_k H = |W_{(J,X)}|$. Hence it is enough to show that the elements $\{B_w\}$ are linearly independent. For this purpose, let $0 \neq v \in X$ and define $f_v \in \mathrm{Hom}_k(kG, X)$ by

$$f_v(g) = \rho(g)(v) \text{ if } g \in P_J \text{ and } f_v(g) = 0 \text{ if } g \notin P_J. \tag{a}$$

It is shown in the proof of [7, (10.1.5)] that we have in fact $f_v \in \mathrm{Hom}_{kP_J}(kG, \tilde{X})$ and that, for any $w, w' \in W_{(J,X)}$, we have

$$(B_{w'} f_v)(\dot{w}) = \begin{cases} |U_J \cap {}^wU_J|\tilde{\rho}(\dot{w})(v) & \text{if } w = w', \\ 0 & \text{if } w \neq w'. \end{cases} \tag{b}$$

This easily implies the desired linear independence of the elements $\{B_w\}$. (See [7, (10.1.5)] and also [14, (2.9)], where such results on endomorphism algebras are proved in a more general setting.)

In the case where k has characteristic 0, multiplication rules for the basis elements $\{B_w\}$ of H were determined by Howlett-Lehrer [50] (see [7, (10.8.5)]). In [38, §3], this has been generalized to the case where k has positive characteristic. To state the main result, we need one further definition.

3.27. For each $w \in W_{(J,X)}$ we set $q_w = |U \cap U^{w_0 w}|$ where $w_0 \in W$ is the longest element, and define a new 2-cocycle $\lambda'\colon W_{(J,X)} \times W_{(J,X)} \to k^\times$ by

$$\lambda'(w_1, w_2) = q_{w_1}^{-1/2} q_{w_2}^{-1/2} q_{w_1 w_2}^{1/2} \lambda(w_1, w_2).$$

(Here, we assume that p has a square root $p^{1/2} \in k$ and, for any integer m, we set $p^{m/2} = (p^{1/2})^m$.) Note that λ' is cohomologous to λ and still satisfies the relations (a) and (b) in (3.25). In the case where G is a "finite group of Lie type", the numbers q_w can be computed using the length function on the Weyl group of the underlying algebraic group, as explained in [7, p. 74].

The main result can now be stated as follows.

Theorem 3.28 (Howlett–Lehrer [50], Geck–Hiss–Malle [38]). *Let $\{B_w\}_{w \in W_{(J,X)}}$ be the basis of H as in (3.26). Then there exists a decomposition $W_{(J,X)} = R_{(J,X)} \rtimes C_{(J,X)}$ such that, for some $D_{(J,X)} \subseteq R_{J,X)}$, the pair $(R_{(J,X)}, D_{(J,X)})$ is a Coxeter system and the following relations hold.*

(a) *We have $B_w \circ B_x = \lambda'(w, x) B_{wx}$ and $B_x \circ B_w = \lambda'(x, w) B_{xw}$ for all $w \in W_{(J,X)}$ and $x \in C_{(J,X)}$.*

(b) *Let $w \in R_{(J,X)}$ and $w = d_1 \cdots d_t$ with $d_i \in D_{(J,X)}$ be a reduced expression. Then $B_w = \gamma B_{d_1} \circ \cdots \circ B_{d_t}$, where $\gamma = \prod_{i=1}^{t} \lambda(d_i, d_{i+1} \cdots d_t)$.*

(c) *For $d \in D_{(J,X)}$, we have $B_d^2 = \alpha_d B_1 + \beta_d B_d$ where $0 \neq \alpha_d \in k$ and $\beta_d \in k$.*

Remark 3.29. The above result shows that the structure of H is determined by the following data:

(a) The group $W_{(J,X)}$ and its decomposition $W_{(J,X)} = R_{(J,X)} \rtimes C_{(J,X)}$.

(b) The 2-cocycle $\lambda\colon W_{(J,X)} \times W_{(J,X)} \to k^\times$.

(c) The constants appearing in the quadratic relations in Theorem 3.28(c).

In order to describe $W_{(J,X)}$ as in (a), one needs information about the normalizer of W_J in W; this is provided in Howlett [49], see also Borcherds [2]. If k has characteristic 0, then the cocyle λ in (b) is always trivial; see Lusztig [59, (8.6)] and Geck [31]. Moreover, the constants in (c) are expressible in terms of certain character degrees (see [7, (10.5.5)]) and hence are explicitly computable using Lusztig's results in [59, 60]. As far as (b) and (c) are concerned, the situation in positive characteristic is open except for some special cases (see [38, §3]).

3.30. Recall that we already know that $R_J^S(X)$ satisfies Dipper's hypothesis (D) (see Theorem 3.20) and Green's hypothesis (A) (see Theorem 3.16). We now check that hypothesis (B) is also satisfied, so that all the results in Section 2 can be applied.

In order to check that H is quasi-Frobenius, we show that H is in fact a symmetric algebra. (Note that such an algebra is always quasi-Frobenius; see [8, §66].) We define a k-linear map $\tau\colon H \to k$ by $\tau(B_1) = 1$ and $\tau(B_w) = 0$ for $1 \neq w \in W_{(J,X)}$. Then τ is a central function on H and we have

$$\tau(B_w B_{w'}) \neq 0 \quad \text{if and only if} \quad w' = w^{-1};$$

see [7, (10.9.2)] in characteristic 0 and [36, (2.7)] in positive characteristic (and also Cabanes' simplified argument in [6]). Thus, H is a symmetric algebra with respect to the bilinear form $H \times H \to k$, $(h, h') \mapsto \tau(hh')$.

Corollary 3.31 (Geck–Hiss [36, (2.11)]). *The head and the socle of any indecomposable direct summand of $R_J^S(X)$ are simple and isomorphic to each other.*

Proof. By Theorem 3.16 and (3.30), Green's hypotheses (A) and (B) are satisfied. The assertion then follows from Theorem 2.13. □

4. Decomposition numbers and Brauer reciprocity

Let $\mathcal{O}$ be a discrete valuation ring with residue field k and quotient field K. We consider an $\mathcal{O}$-algebra A, finitely generated and free over $\mathcal{O}$. By extension of scalars, we obtain finite-dimensional algebras $KA = K \otimes_{\mathcal{O}} A$ and $\bar{A} = k \otimes_{\mathcal{O}} A$. If we assume that KA is split semisimple, the representations or modules over K will be called "ordinary" and those over k will be called "modular". Using Brauer's concept of the "decomposition matrix" (see [23, (I.17)]) it is possible to relate "ordinary" with "modular" representations by working with suitable A-lattices and "reducing" them modulo the maximal ideal of $\mathcal{O}$. In this section, we place ourselves in a setting similar to that in Section 2 where we fix an A-lattice M and consider its endomorphism ring H. Now the aim is to show, following Dipper [11] and Cabanes [6], that the functor F_M also provides a link between the decomposition matrix of A and that of H. This leads to a general strategy for computing the decomposition matrix of A which will be explained in (4.C).

4.A. The decomposition matrix of a module

Let $\mathcal{O}$ be a discrete valuation ring as above, with field of fractions K and residue field k. (Note that we do not exclude the case that $K = \mathcal{O}$.) Let A be an $\mathcal{O}$-algebra, finitely generated and free over $\mathcal{O}$. (In particular, the finiteness assumptions in (2.4) are satisfied.) We write $KA = K \otimes_{\mathcal{O}} A$ and $\bar{A} = k \otimes_{\mathcal{O}} A$. We shall assume throughout that

KA and $\bar{A}$ are split (i.e., the endomorphism ring of any simple module just consists of scalars multiples of the identity) and KA is semisimple.

Recall that an A-lattice is an A-module which is finitely generated and free over $\mathcal{O}$. The above assumption implies (see [9, (30.18)]) that the Krull–Schmidt–Azuyama Theorem holds for A-lattices. Moreover, idempotents can be lifted from $\bar{A}$ to A (see [9, Exercise 6.16]) and so A is semiperfect (see [9, p. 132] and also the remarks at the beginning of (2.B)). Thus we are in the usual setting for Brauer's "modular representation theory". We assume that the reader is familiar with the basic results about decomposition matrices and Brauer reciprocity; see [23, (I.17)].

4.1. We use the following conventions about scalar extension.

(a) First, we may naturally identify A with an $\mathcal{O}$-submodule of KA and, denoting by $\pi \in \mathcal{O}$ a fixed generator of the maximal ideal of $\mathcal{O}$, we may also identify $\bar{A}$ with $A/\pi A$. Similar conventions apply to any $\mathcal{O}$-module Y which is finitely generated and free over $\mathcal{O}$: we write $KY = K \otimes_{\mathcal{O}} Y$ and identify Y with an $\mathcal{O}$-submodule of KY. On the other hand, we write $\overline{Y} = k \otimes_{\mathcal{O}} Y$ and identify this with $Y/\pi Y$. Under this identification, if Y is an A-module, the induced action of $\bar{A}$ on $\overline{Y}$ corresponds to the natural action of $A/\pi A$ on $Y/\pi Y$. (See [9, p. 410]).

(b) Now let V, V' be A-lattices. Then $\mathrm{Hom}_A(V, V')$ is a finitely generated $\mathcal{O}$-submodule of the free $\mathcal{O}$-module $\mathrm{Hom}_{\mathcal{O}}(V, V')$ (see (2.4)(b)) and hence is also free over $\mathcal{O}$. If $\varphi \in \mathrm{Hom}_A(V, V')$, we have induced maps $\varphi_K : KV \to KV'$ and $\bar{\varphi} : \overline{V} \to \overline{V}'$. Using [9, (2.39)] and (a), we may identify

$$K\,\mathrm{Hom}_A(V, V') = \mathrm{Hom}_{KA}(KV, KV') \quad \text{(induced by the inclusion } \mathcal{O} \subseteq K\text{)},$$

$$\overline{\mathrm{Hom}_A(V, V')} \subseteq \mathrm{Hom}_{\bar{A}}(\overline{V}, \overline{V}') \quad \text{(given by reduction modulo } \pi\text{)},$$

but note that the second inclusion need not necessarily be an equality.

Now let us fix an A-lattice M. Then M can be written as a direct sum of indecomposable A-lattices. Each summand in that decomposition gives rise to a KA-module which itself is a direct sum of absolutely simple KA-modules (since KA is split semisimple). The multiplicities arising in these decompositions are recorded in a matrix as follows.

Definition 4.2. (Dipper [11, (4.1)]) Let $\{M_1, \ldots, M_s\}$ be the set of indecomposable direct summands of M (up to isomorphism) and $\{Y_1, \ldots, Y_t\}$ be the set of simple KA-modules (up to isomorphism) which are composition factors of KM. Then we define

$$d_{ij}^M = \dim_K \mathrm{Hom}_{KA}(KM_j, Y_i) = [KM_j : Y_i] \quad \text{for } 1 \leq i \leq t, 1 \leq j \leq s.$$

Thus d_{ij}^M is the multiplicity of Y_i in a decomposition of KM_j as a direct sum of simple modules. The matrix $D_M = (d_{ij}^M)$ is called the *decomposition matrix* of M. Note that, by the Krull–Schmidt–Azumaya Theorem, D_M is well-defined up to permutations of the rows and columns.

Example 4.3. Assume that $M = A$ where A is regarded as a module via left multiplication. Then, in the notation of Definition 4.2, $\{M_1, \dots, M_s\}$ is the set of projective indecomposable A-modules (up to isomorphism) and we have $\mathrm{Irr}(\bar{A}) = \{V_1, \dots, V_s\}$, where we set $V_j = \mathrm{hd}(M_j)$ for all j (see [9, §6C]). Now "Brauer reciprocity" (see [23, (I.17.8)]) states that

$$d_{ij}^A = \dim_k \mathrm{Hom}_{\bar{A}}(\overline{M_j}, \overline{(Y_i)_0}) = [\overline{(Y_i)_0} : V_j] \quad \text{for } 1 \leq i \leq t, 1 \leq j \leq s,$$

where $(Y_i)_0$ is an A-lattice such that $K(Y_i)_0 \cong_{KA} Y_i$. In other words, d_{ij}^A is the multiplicity of V_j as a composition factor in $\overline{(Y_i)_0}$. Thus D_A as defined in Definition 4.2 is the usual "decomposition matrix" of A (see [23, (I.17)]).

Remark 4.4. Let $H := \mathrm{End}_A(M)^\circ$. As we have seen in (4.1)(b), H is like A, i.e., an $\mathcal{O}$-algebra which is finitely generated and free over $\mathcal{O}$. We shall now also bring into play our functor F_M, defined as in Section 2,

$$F_M \colon A\text{-mod} \to H\text{-mod}, \quad V \mapsto \mathrm{Hom}_A(M, V).$$

Note that F_M in fact takes A-lattices to H-lattices. Using (4.1)(b), we can identify $KH = \mathrm{End}_{KA}(KM)^\circ$ and $K\,\mathrm{Hom}_A(M, V) = \mathrm{Hom}_{KA}(KM, KV)$. Under these identifications, we have

$$F_{KM}(KV) = KF_M(V) \quad \text{for any } V \in A\text{-mod}.$$

In this sense, F_M is compatible with scalar extension from $\mathcal{O}$ to K.

Lemma 4.5. *Let $\{M_1, \dots, M_s\}$ and $\{Y_1, \dots, Y_t\}$ be as in Definition 4.2. We set $E_i := F_{KM}(Y_i)$ and $P_j := F_M(M_j)$ for all i, j. Then the following hold.*

(a) *$\{E_1, \dots, E_t\}$ is the set of simple KH-modules (up to isomorphism).*

(b) *$\{P_1, \dots, P_s\}$ is the set of projective indecomposable H-modules (up to isomorphism).*

(c) *We have $d_{ij}^M = d_{ij}^H$ for all $1 \leq i \leq t$ and $1 \leq j \leq s$.*

Thus D_M (see Definition 4.2) is equal to D_H (see Example 4.3), where the matching of the rows and columns is given by the functors F_M and F_{KM}.

Proof. Since KA is split semisimple, the module KM itself is projective. Hence Dipper's hypothesis (D) is trivially satisfied for KM and so Theorem 2.10(b) yields (a). The assertion in (b) is just (F2) in the Fitting correspondence; see Section 2. Finally, let us prove (c). By definition, we have

$$d_{ij}^M = \dim_K \mathrm{Hom}_{KA}(KM_j, Y_i) \quad \text{and} \quad d_{ij}^H = \dim_K \mathrm{Hom}_{KH}(KP_j, E_i).$$

On the other hand, an application of Lemma 2.3 to the functor F_{KM} yields

$$\dim_K \operatorname{Hom}_{KA}(KM_j, Y_i) = \dim_K \operatorname{Hom}_{KH}(F_{KM}(KM_j), F_{KM}(Y_i)).$$

Hence (c) follows by using the compatibility of F_M with scalar extension from $\mathcal{O}$ to K; see Remark 4.4. □

4.6. We shall need the notion of pure submodules; see [23, (I.17.1) and (I.17.2)]. Let V be an A-lattice and $V' \subseteq V$ be an A-submodule. Then V' is again an A-lattice but this is not necessarily the case for V/V'. We set

$$\sqrt{V'} = \{v \in V \mid av \in V' \text{ for some } 0 \neq a \in \mathcal{O}\}.$$

It is readily checked that $\sqrt{V'}$ is an A-submodule of V. Moreover, by construction, $V/\sqrt{V'}$ is torsion-free and hence free over $\mathcal{O}$. We call $\sqrt{V'}$ the *purification* of V' in V. We say that V' is *pure* in V if $\sqrt{V'} = V'$. Note that $\sqrt{V'}$ is the intersection of all pure submodules of V which contain V'. Alternatively, denoting by $\iota\colon V' \to V$ the inclusion map, we also have that

$$V' \text{ is pure in } V \text{ if and only if } \bar{\iota}\colon \overline{V}' \to \overline{V} \text{ is injective.} \tag{a}$$

The importance of pure submodules lies in the following result (see [9, (23.7)]).

We have an inclusion-preserving bijection

$$\{\text{pure } A\text{-submodules of } V\} \to \{KA\text{-submodules of } KV\}, \qquad V' \mapsto KV', \tag{b}$$

with inverse given by intersecting a KA-submodule of KV with $V \subseteq KV$.

4.B. Hypothesis (D) and Frobenius algebras over $\mathcal{O}$

We keep the above notation, where M is an A-lattice and $H = \operatorname{End}_A(M)^\circ$. We have seen in Lemma 4.5 that the decomposition matrix of H equals the decomposition matrix of M. Using Dipper's hypothesis (D), we will now relate the latter matrix with the decomposition matrix of A itself.

Lemma 4.7 (Dipper [11, (4.7) and (4.8)]). *Let $\beta\colon P \to M$ be a surjective map of A-modules where $P \in A$-mod is projective.*

(a) *Hypothesis (D) holds for β if and only if $K\ker(\beta)$ and KM have no irreducible constituent in common.*

(b) *If* (D) *holds for β, then* (D) *also holds for the induced map $\bar{\beta}\colon \overline{P} \to \overline{M}$ and we have an equality $\overline{H} = \operatorname{End}_{\bar{A}}(\overline{M})^\circ$ (cf. (4.1)(b)).*

Proof. (a) Suppose first that $K\ker(\beta)$ and KM have no irreducible constituent in common. Then we have

$$K \operatorname{Hom}_A(\ker(\beta), M) = \operatorname{Hom}_{KA}(K\ker(\beta), KM) = \{0\}$$

and so $\mathrm{Hom}_A(\ker(\beta), M) = \{0\}$. Hence Example 2.5 shows that (D) holds. Conversely, assume that (D) holds for β. Then (D) also holds for the induced map $\beta_K \colon KP \to KM$, since $\mathrm{End}_A(P)$ generates $\mathrm{End}_{KA}(KP)$ and $\ker(\beta)$ generates $\ker(\beta_K) = K\ker(\beta)$ as K-vector spaces. But we have $KP \cong_{KA} KM \oplus K\ker(\beta_K)$ and so

$$\mathrm{Hom}_{KA}(KP, KM) = \mathrm{Hom}_{KA}(KM, KM) \oplus \mathrm{Hom}_{KA}(K\ker(\beta), KM).$$

By Lemma 2.6, we conclude that $\mathrm{Hom}_{KA}(K\ker(\beta), KM) = \{0\}$, as desired.

(b) The fact that $\overline{P}$ is projective is clear. It is readily checked that $\ker(\beta)$ is an $\mathcal{O}$-pure submodule of P and so we have $\ker(\bar{\beta}) = \overline{\ker(\beta)}$. This implies that (D) also holds for $\bar{\beta}$. Now we have a commutative diagram of maps

$$\begin{array}{ccc} \mathrm{End}_A(P) & \xrightarrow{\beta^\bullet} & \mathrm{End}_A(M) \\ \downarrow & & \downarrow \\ \mathrm{End}_{\bar{A}}(\overline{P}) & \xrightarrow{\bar{\beta}^\bullet} & \mathrm{End}_{\bar{A}}(\overline{M}) \end{array}$$

where the two horizontal arrows are the surjective maps given by Lemma 2.8 and the vertical maps are given by reduction modulo π. Now, since P is projective, the left vertical map is also surjective. Hence so is the right vertical map; i.e., we have $\overline{H} = \mathrm{End}_{\bar{A}}(\overline{M})^\circ$. □

Theorem 4.8 (Dipper [11, (4.9)]). *Assume that* (D) *holds. Let* $\{M_1, \dots, M_s\}$ *and* $\{Y_1, \dots, Y_t\}$ *be as in Definition* 4.2. *For each* j, *let* $\beta_j \colon P_j \to M_j$ *be a projective cover of* M_j. *Then the following hold.*

(a) *The module* P_j *is indecomposable for each* j. *Hence* $[KP_j : Y_i]$ *is a coefficient of the decomposition matrix* D_A (*see Example* 4.3).

(b) *We have* $[KP_j : Y_i] = [KM_j : Y_i]$ *for all* i, j. *Hence* $[KP_j : Y_i]$ *is also a coefficient of the decomposition matrix* D_M (*see Definition* 4.2).

(c) *We have* $\mathrm{Hom}_{\bar{A}}(\overline{M}_i, \overline{M}_j) = \overline{\mathrm{Hom}_A(M_i, M_j)}$ *for all* i, j (*cf.* (4.1(b)), *and each* $\overline{M}_j$ *is indecomposable.*

Thus, using Lemma 4.5, *the decomposition matrix* D_H *of* H *is a submatrix of the decomposition matrix* D_A *of* A.

Proof. By Theorem 2.10(a), each M_j has a simple head. This implies (a), using [9, (6.23)]). To prove (b), set $P = P_1 \oplus \cdots \oplus P_s$ and $\beta = \beta_1 \oplus \cdots \oplus \beta_s \colon P \to M$. Then β is a projective cover of M (see [9, (6.25)(iii)]). So, by Remark 2.7, hypothesis (D) also holds for β. By Lemma 4.7(a), this implies that KM and $K\ker(\beta)$ have no irreducible constituent in common. Consequently, the same also holds for KM_j and $K\ker(\beta_j)$, which means that (b) holds. Finally, let us prove (c). Each M_j is a factor module of a projective indecomposable module. Hence so is $\overline{M}_j$. By [9, Exercise 6.13], $\overline{M}_j$ is indecomposable. Now, by (4.1)(b), reduction modulo π yields injections $\overline{\mathrm{Hom}_A(M_i, M_j)} \subseteq \mathrm{Hom}_{\bar{A}}(\overline{M}_i, \overline{M}_j)$ for all i, j. Hence, since reduction modulo π and Hom preserve direct sums, we obtain an injection $\overline{\mathrm{Hom}_A(M, M)} \subseteq \mathrm{Hom}_{\bar{A}}(\overline{M}, \overline{M})$. But, by Lemma 4.7(b), the latter injection is also surjective. Hence all the previous injections must be also surjective. □

Remark 4.9. A slightly different proof of the above result has been given by Plesken; see [45, (5.1.2)].

The above result means that if we start with an A-lattice M and take its endomorphism ring H, then we obtain at least a part of the decomposition matrix of A from that of H. The following constructions, which appear in Dipper [11, §4] and Cabanes [5], [6], will be used to obtain larger parts of D_A.

4.10. Consider the Hom functor $F_M \colon A\text{-mod} \to H\text{-mod}$. We shall use the convention that if M' is an A-submodule of M, we can naturally identify $F_M(M') = \mathrm{Hom}_A(M, M')$ with a submodule of $H = F_M(M)$, by composing a map in $F_M(M')$ with the embedding $M' \subseteq M$. Now let $I \subseteq H$ be a left ideal and set

$$IM := \sum_{\iota \in I} \iota(M) \subseteq M.$$

Note that IM is an A-submodule of M. If $I = Hy$ for some $y \in H$, we simply write yM instead of IM. Using our above convention, we have

$$I \subseteq F_M(IM) \subseteq H. \tag{a}$$

The module IM can be defined alternatively, as follows. We may naturally regard M as a right H-module (where the action is given by applying a map in H to an element of M). Then we have

$$IM = \text{image of the natural map } M \otimes_H I \to M, m \otimes \iota \mapsto \iota(m). \tag{b}$$

Lemma 4.11. *We have $I = F_M(\sqrt{IM})$ for any $\mathcal{O}$-pure left ideal $I \subseteq H$.*

Proof. We proceed in several steps.

Step 1. We claim that $F_M(\sqrt{IM}) \subseteq H$ is pure. To prove this, let $\varphi \in H$ and $0 \neq a \in \mathcal{O}$ be such that $a\varphi \in F_M(\sqrt{IM})$. This means that $a\varphi(m) \in \sqrt{IM}$ for all $m \in M$. Since $\sqrt{IM} \subseteq M$ is pure, we conclude that $\varphi(m) \in \sqrt{IM}$ for all $m \in M$, and so $\varphi \in F_M(\sqrt{IM})$. Hence $F_M(\sqrt{IM})$ is pure.

Step 2. Now we prove that $F_M(\sqrt{IM}) = \sqrt{F_M(IM)}$. By Step 1, we already know that $\sqrt{F_M(IM)} \subseteq F_M(\sqrt{IM})$. Conversely, let $\varphi \in F_M(\sqrt{IM})$. We must show that there exists some $0 \neq a \in \mathcal{O}$ such that $a\varphi \in F_M(IM)$. Let $\{m_1, \dots, m_r\}$ be an $\mathcal{O}$-basis of M. Since $\varphi \in F_M(\sqrt{IM})$, there exist $0 \neq a_i \in \mathcal{O}$ such that $a_i\varphi(m_i) \in IM$. Now set $a = a_1 \cdots a_r \in \mathcal{O} \setminus \{0\}$. Then we have $a\varphi(m_i) \in IM$ for all i and hence also $a\varphi(m) \in IM$ for all $m \in M$, as desired.

Step 3. Now both I and $\sqrt{F_M(IM)}$ are pure submodules of H and we have $I \subseteq \sqrt{F_M(IM)}$ by (4.10)(a). Using (4.6), we conclude that it is enough to prove that $KI = K\sqrt{F_M(IM)}$. But we certainly have $K\sqrt{F_M(IM)} = KF_M(IM) = F_{KM}(K(IM))$ where the second equality holds by (4.4). Finally, since KI and KM are generated as K-vector spaces by I and M, respectively, we also have $K(IM) = (KI)(KM)$. Thus we are reduced to showing the equality $KI = F_{KM}((KI)(KM))$, which will be done in the next step.

Step 4. By the discussion in Step 3, it remains to check that the assertion of the lemma holds in the case where $K = \mathcal{O}$. In that case, by our general assumption, A is semisimple and hence so is H (being the endomorphism algebra of a semisimple module). It follows that there exists some left ideal $I' \subseteq H$ such that $H = I \oplus I'$. We claim that we also have

$$M = IM \oplus I'M.$$

Indeed, the definition immediately shows that $M = IM + I'M$. Hence it is enough to prove that $\dim_K M \geq \dim_K IM + \dim_K I'M$. To see this, we argue as follows. By (4.10)(b), we have $\dim_K(M \otimes_K I) \geq \dim_K IM$ and, similarly, $\dim_K(M \otimes_H I') \geq \dim_K I'M$. But, on the other hand, the fact that the tensor product is additive implies that $M \cong_K M \otimes_H H \cong_K (M \otimes_H I) \oplus (M \otimes_H I')$, and so $\dim_K M \geq \dim_K IM + \dim_K I'M$, as desired.

Now we can complete the proof as follows. We certainly have $\sqrt{IM} = IM$ and, by (4.10)(a), we also have $I \subseteq F_M(IM)$. Similarly, we have $I' \subseteq F_M(I'M)$. Hence, since $H = I \oplus I'$, it is enough to show that $H = F_M(M) = F_M(IM) \oplus F_M(I'M)$. But this is clear since $M = IM \oplus I'M$. □

We say that H is a *Frobenius algebra* if there exists a linear map $t \colon H \to \mathcal{O}$ such that the bilinear form $H \times H \to \mathcal{O}$, $(h, h') \mapsto t(hh')$, is non-degenerate over $\mathcal{O}$; i.e., the determinant of the Gram matrix with respect to an $\mathcal{O}$-basis of H is a unit in $\mathcal{O}$ (see [8, (61.3)]). Note that a Frobenius algebra over a field is quasi-Frobenius.

The following result holds in fact for algebras which are "self-injective relative to $\mathcal{O}$" (see [6, §2]), but the version that we present here (following [6]) is technically slightly simpler and sufficient for all our applications.

Theorem 4.12 (Dipper [11, (4.17)], Cabanes [6]). *Assume that H is a Frobenius algebra. Let $I, J \subseteq H$ be $\mathcal{O}$-pure left ideals. Then the functor F_M induces an isomorphism*

$$\operatorname{Hom}_A(\sqrt{IM}, \sqrt{JM}) \to \operatorname{Hom}_H(I, J).$$

Proof. We follow the argument given by Cabanes [6]. By definition, the functor F_M induces an $\mathcal{O}$-linear map

$$\operatorname{Hom}_A(\sqrt{IM}, \sqrt{JM}) \to \operatorname{Hom}_H(F_M(\sqrt{IM}), F_M(\sqrt{JM})), \quad \varphi \mapsto \varphi_*.$$

Now we have the equalities $I = F_M(\sqrt{IM})$ and $J = F_M(\sqrt{JM})$ from Lemma 4.11. So it remains to show that the above map is injective and surjective.

Let us first check the injectivity. Let $\varphi \in \mathrm{Hom}_A(\sqrt{IM}, \sqrt{JM})$ be such that $\varphi_* = 0$. Then we must show that $\varphi(m) = 0$ for all $m \in M$ such that $am \in IM$ for some $0 \neq a \in \mathcal{O}$. Since M is $\mathcal{O}$-free, it is sufficient to show that $\varphi(m) = 0$ for all $m \in IM$. Using the definition of IM, we see that it is enough to show that $\varphi(i(m)) = 0$ for all $m \in M$ and $i \in I$. But this is now clear since $\varphi \circ i = \varphi_*(i) = 0$ by assumption.

Now we prove surjectivity. Let $\psi \in \mathrm{Hom}_H(I, J)$ and consider the map $t \circ \psi \in \mathrm{Hom}_{\mathcal{O}}(I, \mathcal{O})$, where $t \in \mathrm{Hom}_{\mathcal{O}}(H, \mathcal{O})$ is a map giving H the structure of a Frobenius algebra. Since I is pure in H, any $\mathcal{O}$-basis of I can be extended to an $\mathcal{O}$-basis of H. Consequently $t \circ \psi$ can be extended to a map $\hat{t} \in \mathrm{Hom}_{\mathcal{O}}(H, \mathcal{O})$. The fact that the bilinear form on H associated with t is non-degenerate over $\mathcal{O}$ implies that there exists some $h_\psi \in H$ such that $\hat{t}(h) = t(hh_\psi)$ for all $h \in H$. But, since I is a left ideal, we have $hi \in I$ for all $h \in H$ and $i \in I$ and so

$$\begin{aligned} t(h\psi(i)) &= t(\psi(hi)) && \text{since } \psi \in \mathrm{Hom}_H(I, J) \\ &= (t \circ \psi)(hi) = \hat{t}(hi) && \text{since } \hat{t} \text{ equals } t \circ \psi \text{ on } I \\ &= t((hi)h_\psi) = t(h(ih_\psi)) && \text{by the definition of } h_\psi. \end{aligned}$$

Using once more the fact that the bilinear form associated with t is non-degenerate, we conclude that $\psi(i) = ih_\psi$ for all $i \in I$. Thus we see that ψ can be extended to a map $\hat{\psi} \in \mathrm{Hom}_H(H, H)$, namely, the map given by right multiplication with h_ψ. Thus we have

$$h_\psi \circ i = ih_\psi = \psi(i) \quad \text{for all } i \in I. \tag{$*$}$$

The above relation shows that $h_\psi(i(m)) = (\psi(i))(m) \in JM$ for all $i \in I$ and $m \in M$, and so $h_\psi(IM) \subseteq JM$. This clearly implies that we also have $h_\psi(\sqrt{IM}) \subseteq \sqrt{JM}$. Hence, by restriction, h_ψ gives rise to a map in $\mathrm{Hom}_A(\sqrt{IM}, \sqrt{JM})$. Finally, relation $(*)$ also shows that F_M sends that map to ψ. □

Corollary 4.13. *Assume that H is a Frobenius algebra and that $\{y_\lambda\}_{\lambda\in\Lambda}$ is a collection of elements in H such that the left ideals Hy_λ are $\mathcal{O}$-pure and the A-modules $\sqrt{y_\lambda M}$ satisfy Dipper's hypothesis* (D). *Then F_M induces an isomorphism of $\mathcal{O}$-algebras*

$$\mathrm{End}_A\Big(\dotplus_{\lambda\in\Lambda} \sqrt{y_\lambda M}\Big) \cong_{\mathcal{O}} \mathrm{End}_H\Big(\dotplus_{\lambda\in\Lambda} Hy_\lambda\Big)$$

(where $\dotplus$ denotes external direct sum), and the decomposition matrix of the algebra on the right hand side is a submatrix of D_A.

Proof. Since Hom preserves direct sums, Theorem 4.12 shows that we have an $\mathcal{O}$-linear isomorphism as desired. That isomorphism is also a ring homomorphism since F_M preserves the composition of maps. The statement about the decomposition matrices follows from Theorem 4.8: since each module $\sqrt{y_\lambda M}$ satisfies (D), the same holds for their external direct sum. □

4.14. Keeping the notation and assumptions of Corollary 4.13, we set

$$\tilde{M} = \dotplus_{\lambda \in \Lambda} \sqrt{y_\lambda M} \quad \text{and} \quad \tilde{H} = \mathrm{End}_A(\tilde{M})^\circ.$$

We assume that $y_\lambda = \mathrm{id}_M$ for some $\lambda \in \Lambda$; then M is a direct summand of $\tilde{M}$. Let $\{M_1, \dots, M_s\}$ and $\{Y_1, \dots, Y_t\}$ be as in Definition 4.2. Now note that $K\sqrt{y_\lambda M} = K(y_\lambda M) = y_\lambda KM$ is a factor module of KM for all λ. It follows that

(a) Every composition factor of $K\tilde{M}$ is isomorphic to Y_i for some i.

Furthermore, let $\{M_{s+1}, \dots, M_{s+r}\}$ (for some $r \geq 0$) be indecomposable A-lattices such that

(b) $\{M_1, \dots, M_s, M_{s+1}, \dots, M_{s+r}\}$ is the set of indecomposable direct summands of $\tilde{M}$ (up to isomorphism).

Let $D_{\tilde{M}} = (d_{ij}^{\tilde{M}})$ and $D_M = (d_{ij}^M)$ be the decomposition matrices of $\tilde{M}$ and M, respectively, defined with respect to the above labellings (see Definition 4.2). Note that both $D_{\tilde{M}}$ and D_M have t rows, but $D_{\tilde{M}}$ has $r+s$ columns and D_M has s columns. By Lemma 4.5, the above matrices are also the decomposition matrices of $\tilde{H}$ and H, respectively.

(c) We have $d_{ij}^{\tilde{M}} = d_{ij}^M$ for $1 \leq i \leq t, 1 \leq j \leq s$.

Thus we see that D_M is a submatrix of $D_{\tilde{M}}$ in such a way that each column of D_M is also a (complete) column of $D_{\tilde{M}}$ but $D_{\tilde{M}}$ may have more columns than D_M.

4.C. A strategy for finding the decomposition matrix of A

Let $\mathrm{Irr}_K(A)$ and $\mathrm{Irr}_k(A)$ denote the simple KA-modules (up to isomorphism) and the simple $\bar{A}$-modules (up to isomorphism), respectively. Let D_A be the decomposition matrix of A, as in Example 4.3. Recall that D_A has rows and columns indexed by the elements of $\mathrm{Irr}_K(A)$ and $\mathrm{Irr}_k(A)$, respectively. We have seen in Corollary 4.13 that parts of D_A can be obtained from decomposition matrices of certain endomorphism algebras. Now our aim is to describe a set of conditions which ensure that, in this way, we obtain all of D_A. Such a strategy was first used by Dipper and James in order to obtain the decomposition matrices of $\mathrm{GL}_n(q)$ in non-defining characteristic; see [11], [16]. We shall make the following assumption:

The elementary divisors of D_A are all 1.

It is well-known that this is the case, for example, if $A = \mathcal{O}G$ for some finite group G (see [9, (18.14)]). Actually, we will only be interested in computing a certain square submatrix of D_A whose definition involves the following concept.

Definition 4.15. (R. Parker) A subset of $\mathrm{Irr}_K(A)$ is called an *ordinary basic set* if the submatrix of D_A with rows labelled by the elements in that set has determinant ± 1.

We also have a relative notion of basic sets with respect to a central idempotent $b \in A$ by considering only simple modules on which b acts as the identity.

For a general discussion of basic sets and their relevance to the problem of computing decomposition matrices, we refer to the introduction of [35]. (Note that it is not known if ordinary basic sets always exist, even in the case of group algebras; but see Theorem 5.8 below.) Here, we only mention that if we have found an ordinary basic set then, in order to determine all of D_A, it is enough (in a well-defined sense) to compute the square submatrix of D_A whose rows correspond to the simple modules in that basic set. The strategy we will now present aims at computing such a square submatrix of D_A. We proceed in two steps.

Step 1. Try to find a collection of A-lattices M_0^σ (where σ runs over some index set $\mathcal{S}$) such that the following conditions hold.

(1a) For each $\sigma \in \mathcal{S}$, let $\mathfrak{B}^\sigma$ be the set of composition factors of KM_0^σ. Then $\mathfrak{B}^\sigma \cap \mathfrak{B}^{\sigma'} = \emptyset$ for $\sigma \neq \sigma'$ and $\mathfrak{B} := \bigcup_{\sigma \in \mathcal{S}} \mathfrak{B}^\sigma$ is an ordinary basic set for A.

(1b) For each $\sigma \in \mathcal{S}$, the algebra $H_0^\sigma := \mathrm{End}_A(M_0^\sigma)^\circ$ is a Frobenius algebra.

Step 2. For each $\sigma \in \mathcal{S}$, try to find elements $y_\lambda^\sigma \in H_0^\sigma$ (where λ runs over some index set Λ^σ) such that the following conditions hold.

(2a) The identity on M_0^σ is among the elements $\{y_\lambda^\sigma\}$. Moreover, for each $\lambda \in \Lambda^\sigma$, the left ideal $H_0^\sigma y_\lambda^\sigma$ is $\mathcal{O}$-pure in H_0^σ.

(2b) For each $\lambda \in \Lambda^\sigma$, there exists a projective A-module P_λ^σ and a surjection $\beta_\lambda^\sigma \colon P_\lambda^\sigma \to \sqrt{y_\lambda^\sigma M_0^\sigma}$ such that $K\ker(\beta_\lambda^\sigma)$ has no irreducible constituent in $\mathfrak{B}$.

Now let us assume that the requirements in Step 1 and Step 2 are satisfied. For each $\sigma \in \mathcal{S}$, we denote by D^σ the decomposition matrix of $\tilde{H}_0^\sigma$ where

$$\tilde{H}_0^\sigma = \mathrm{End}_{H_0^\sigma}\Big(\mathop{+}_{\lambda\in\Lambda^\sigma} H_0^\sigma y_\lambda^\sigma\Big)^\circ.$$

The following result shows that we have formulated a precise set of conditions under which we have a complete reduction of the problem of computing D_A to a similar problem about the endomorphism algebras $\tilde{H}_0^\sigma$, for $\sigma \in \mathcal{S}$.

Lemma 4.16. *In the above setting, let $\dot{D}_A$ be the submatrix of D_A with rows labelled by the simple modules in $\mathfrak{B}$. Then $\dot{D}_A$ is a square matrix with determinant ± 1. Moreover, if we enumerate the elements of $\mathcal{S}$ as $\sigma_1, \sigma_2, \ldots, \sigma_m$ (for some $m \geq 1$), then*

$\dot{D}_A$ has the following shape

$$\dot{D}_A = \left[\begin{array}{cccc|c} D^{\sigma_1} & 0 & \dots & 0 & \\ 0 & D^{\sigma_2} & & \vdots & \\ \vdots & & \ddots & 0 & * \\ 0 & \dots & 0 & D^{\sigma_m} & \end{array}\right]$$

where $*$ stands for a certain number of columns corresponding to projective indecomposable modules which do not arise as the projective cover of an indecomposable direct summand of $\sqrt{y_\lambda^\sigma M_0^\sigma}$ for any $\sigma \in \mathcal{S}$, $\lambda \in \Lambda^\sigma$. If, moreover, we have that

$$D^\sigma \text{ is a square matrix for all } \sigma \in \mathcal{S}, \tag{$\square$}$$

then $\dot{D}_A$ is a block diagonal matrix with diagonal blocks $D^{\sigma_1}, \dots, D^{\sigma_m}$.

Proof. The first statement follows from the fact that $\mathfrak{B}$ is an ordinary basic set; see condition (1a). Now the set $\mathfrak{B}$ is the disjoint union of all the sets $\mathfrak{B}^\sigma$ for $\sigma \in \mathcal{S}$. Let us fix $\sigma \in \mathcal{S}$. By condition (2b) and Lemma 4.7(a), we know that each module $\sqrt{y_\lambda^\sigma M_0^\sigma}$ satisfies Dipper's hypothesis (D). Also, taking into account conditions (1b) and (2a), we see that we are in the setting of (4.14). Hence D^σ is a submatrix of $\dot{D}_A$. More precisely, the embedding of the first matrix into the latter matrix is given as follows. By Lemma 4.5, we can identify D^σ with the decomposition matrix of the module

$$\tilde{M}_0^\sigma = \dotplus_{\lambda \in \Lambda^\sigma} \sqrt{y_\lambda^\sigma M_0^\sigma}.$$

Under this identification, the rows of D^σ have a labelling by the simple modules in $\mathfrak{B}^\sigma$ and the columns of D^σ have a labelling by the indecomposable direct summands of $\tilde{M}_0^\sigma$. Let M' be such an indecomposable direct summand and $\beta' \colon P' \to M'$ be a projective cover. Then, by Theorem 4.8, P' is indecomposable and the multiplicity of $Y \in \mathfrak{B}$ in KM' (which is a coefficient of D^σ) equals that of Y in KP' (which is a coefficient in $\dot{D}_A$). Hence, in order to see that $\dot{D}_A$ has a shape as claimed, we must check that $K \ker(\beta')$ has no irreducible constituent in $\mathfrak{B}$. This is proved as follows. First note that we may assume that P' is a direct summand of P_λ^σ and that β' is obtained by restriction from β_λ^σ. Hence we have $K \ker(\beta') \subseteq K \ker(\beta_\lambda^\sigma)$, and the assertion follows from condition (2b).

Finally, if ($\square$) holds, then all columns of $\dot{D}_A$ are obtained, since we already know that $\dot{D}_A$ is a square matrix. $\square$

Remark 4.17. In the above setting, how can we check that condition ($\square$) holds? As in the proof of Lemma 4.16, we know that $D_{\tilde{H}_0^\sigma} = D_{\tilde{M}_0^\sigma}$, where $\tilde{M}_0^\sigma$ is defined as the external direct sum of the modules $\sqrt{y_\lambda^\sigma M_0^\sigma}$ for $\lambda \in \Lambda^\sigma$. As in Definition 4.2, let $\{Y_1, \dots, Y_t\}$ be the set of simple KA-modules (up to isomorphism) which are constituents of $K\tilde{M}_0^\sigma$. Then (□) certainly holds if the matrix of all multiplicities

$$[y_\lambda^\sigma K M_0^\sigma : Y_i], \quad 1 \leq i \leq t,\ \lambda \in \Lambda^\sigma.$$

is square and has a non-zero determinant. Although this is a rather weak criterion, it will turn out to be sufficient for the case $G = \mathrm{GL}_n(q)$.

Another possibility for verifying condition (□) would be to check that each algebra $\overline{H}_0^\sigma$ is "quasi–hereditary" in the sense of Cline, Parshall and Scott (see [21] and the references there).

5. Lusztig series and blocks

Throughout this section, let $\boldsymbol{G}$ be a connected reductive algebraic group over an algebraically closed field of characteristic $p > 0$. We assume that $\boldsymbol{G}$ is defined over a finite field with q elements (where q is some power of p) and denote by $F\colon \boldsymbol{G} \to \boldsymbol{G}$ the corresponding Frobenius map. Then

$$G = \boldsymbol{G}^F = \{g \in \boldsymbol{G} \mid F(g) = g\}$$

is a "finite group of Lie type"; see [7, (1.18)]. Let $\boldsymbol{B} \subseteq \boldsymbol{G}$ be an F-stable Borel subgroup and $\boldsymbol{T}_1 \subseteq \boldsymbol{G}$ be an F-stable maximal torus which is contained in $\boldsymbol{B}$. Let $\boldsymbol{W}$ be the Weyl group of $\boldsymbol{G}$ with respect to $\boldsymbol{T}_1$. Then G is a finite group with a split BN-pair of characteristic p and Weyl group W where

$$B := \boldsymbol{B}^F, \quad N := N_{\boldsymbol{G}}(\boldsymbol{T}_1), \quad H := \boldsymbol{T}_1^F, \quad W := \boldsymbol{W}^F = N/H$$

(see [7, (2.9)]). In particular, the results in Section 3 apply to G. In this section, we describe — partly without proof — some basic results concerning modular representations of G in non-defining characteristic which make essential use of the Deligne–Lusztig theory concerning the "ordinary" representations of G.

Throughout, let K be a finite but "sufficiently large" algebraic extension of $\mathbb{Q}$; we assume that K is embedded in an algebraic closure of $\mathbb{Q}_\ell$, where ℓ is a prime which is not equal to the defining characteristic p. Let $\mathcal{O} \subseteq K$ be a discrete valuation ring whose residue field k has characteristic ℓ.

5.A. Blocks and basic sets

We begin by discussing the partition of $\mathrm{Irr}_K(G)$ into "Lusztig series". Below we will see that this partition is compatible with blocks and basic sets.

5.1. Let $\boldsymbol{T} \subseteq \boldsymbol{G}$ be an F-stable maximal torus and set $T = \boldsymbol{T}^F$. Then, for any $\theta \in \mathrm{Irr}_K(T)$, we have a corresponding virtual KG-module $R_T^G(\theta)$, whose construction is due to Deligne–Lusztig and can be found, for example, in [7, Chapter 7]. These virtual modules are used to define a partition of $\mathrm{Irr}_K(G)$, as follows.

Let G^* be a finite group of Lie type which is dual to G. Note that G^* is defined in terms of a connected reductive group $\boldsymbol{G}^*$ over $\mathbb{F}_q$ which is dual to $\boldsymbol{G}$; see [59, (8.4)] or [7, (4.3)]. Then we have a natural bijective correspondence

$$\{(\boldsymbol{T}, \theta) \mid \theta \in \mathrm{Irr}(T)\} / G\text{-conjugation} \leftrightarrow \{(\boldsymbol{T}^*, s) \mid s \in T^*\}/ G^*\text{-conjugation} \quad \text{(a)}$$

where $\boldsymbol{T}, \boldsymbol{T}^*$ are F-stable maximal tori in $\boldsymbol{G}$ and $\boldsymbol{G}^*$, respectively, and where we set $T = \boldsymbol{T}^F, T^* = (\boldsymbol{T}^*)^F$ (see [10, (13.13)]).

Now fix a semisimple element $s \in G^*$. Then we define a subset $\mathrm{Irr}_K(G, s) \subseteq \mathrm{Irr}_K(G)$ as follows. A simple KG-module Y belongs to $\mathrm{Irr}_K(G, s)$ if and only Y is a constituent of $R_T^G(\theta)$, where (T, θ) corresponds to a pair of the form (T^*, s). The set $\mathrm{Irr}_K(G, s)$ is called the *Lusztig series* corresponding to s. We have a partition

$$\mathrm{Irr}_K(G) = \coprod_s \mathrm{Irr}_K(G, s), \qquad \text{(b)}$$

where s runs over a complete set of representatives for the conjugacy classes of semisimple elements in G^* (see [59, (8.4.4)] and [60, §2]).

The simple modules in the series $\mathrm{Irr}_K(G, 1)$ play a special role in the whole theory; they are called the *unipotent* simple KG-modules.

5.2. A subgroup $L \subseteq G$ is called a *regular subgroup* if we have $L = \boldsymbol{L}^F$ where $\boldsymbol{L} \subseteq \boldsymbol{G}$ is an F-stable Levi complement in some (not necessarily F-stable) parabolic subgroup of $\boldsymbol{G}$. Generalizing the construction of $R_T^G(\theta)$, Lusztig [58] has defined a functor R_L^G which sends a virtual KL-module to a virtual KG-module. (Note that, in general, R_L^G may depend on the parabolic subgroup involved in the definition, but we will not need to consider such questions here.)

(a) We have the following transitivity formula. Consider a pair (T, θ) such that $T \subseteq L$. Then we have $R_T^G(\theta) = R_L^G(R_T^L(\theta))$; see [58, Corollary 5]).

(b) The functor R_L^G is compatible with Lusztig series in the following sense. Let $L \subseteq G$ be a regular subgroup. Then L^* can be identified with a subgroup of G^*. Now let $s \in L^*$ be a semisimple element and $X \in \mathrm{Irr}_K(L, s)$. Then, regarding s as an element of G^*, we have that all irreducible constituents of $R_L^G(X)$ belong to $\mathrm{Irr}_K(G, s)$. (For a proof see [58, Corollary 6]. Note that the above formulation uses the dual group, unlike the original version in [58]. The translation is performed in the set-up of [59, (8.4)].)

(c) The functor R_L^G is a generalization of Harish-Chandra induction (as defined in (3.2)) in the following sense. Assume that $\boldsymbol{L}$ is a Levi complement of an F-stable parabolic subgroup of $\boldsymbol{G}$. Then, replacing L by a suitable G-conjugate, we can assume that $L = L_J$ for some subset $J \subseteq S$. In this case, we have $R_L^G(X) \cong_{KG} R_J^S(X)$ for all $X \in KL$-mod (see [10, p. 81]).

As an illustration, we consider the general linear groups, following [24, §1].

5.3. Let $G = \mathrm{GL}_n(q)$. First, we remark that G^* can be identified with G. This is easily seen from the definitions in [7, Chapter 4] and the description of the root datum of G in [7, p. 21]. Now let $s \in G$ be a semisimple element and set $L := C_G(s)$. Then L is again a direct product of general linear groups over extension fields of $\mathbb{F}_q$. In fact, L is a regular subgroup of G in the sense of (5.2). Moreover, as explained in [24, p. 116], s defines a simple module $\theta_s \in \mathrm{Irr}(L)$ of dimension 1 and a sign $\varepsilon_s = \pm 1$ such that we have a bijection

$$\mathrm{Irr}(L, 1) \xrightarrow{\sim} \mathrm{Irr}(G, s), \quad X \mapsto \varepsilon_s R_L^G(\theta_s \otimes X). \tag{a}$$

Thus, in order to obtain a full description of $\mathrm{Irr}_K(G)$, we are reduced to describing the series $\mathrm{Irr}_K(G, 1)$. By [10, (15.8)], we have a bijection $\mathrm{Irr}_K(\mathfrak{S}_n) \leftrightarrow \mathrm{Irr}_K(G, 1)$, $M \leftrightarrow Y_M$, such that

$$\mathrm{Ind}_B^G(K_B) = R_H^G(K_H) = \sum_{M \in \mathrm{Irr}_K(\mathfrak{S}_n)} (\dim_K M)\, Y_M. \tag{b}$$

More precisely, by [10, (15.8)], every simple module in $\mathrm{Irr}_K(G, 1)$ is a linear combination of $R_T^G(K_T)$'s. A similar statement then also holds for $\mathrm{Irr}_K(L, 1)$. Now, by [10, (12.6)], we have $\theta_s \otimes R_T^L(K_T) = R_T^L(\mathrm{Res}_T^L(\theta_s))$. Hence (a) and the transitivity formula in (5.2)(a) imply that

$$\text{every simple } KG\text{-module is a linear combination of } R_T^G(\theta)\text{'s.} \tag{c}$$

The above results also yield a characterization of the cuspidal simple modules, as follows. Indeed, in the above setting, if $\mathrm{Irr}_K(G, s)$ contains a cuspidal module, Y say, then Y occurs in $R_T^G(\theta)$ for some pair (T, θ) with $s \in T$. Combining (5.2)(a) and (5.2)(c), we see that no G-conjugate of T can be contained in L_J for a proper subset $J \subset S$. But, in GL_n, this is only possible if T is a Coxeter torus; i.e., we have $T \cong \mathbb{F}_{q^n}^\times$. Then we also see that $C_G(s) = T$ and it follows that

$$\text{every cuspidal simple } KG\text{-module is of the form } \pm R_T^G(\theta) \text{ where } T \cong \mathbb{F}_{q^n}^\times. \tag{d}$$

In fact one can show that every Lusztig series of G is a Harish-Chandra series.

The following results concerning unipotent modules will be needed in (6.B).

5.4. Assume that $\boldsymbol{G}$ has a connected center. Let $p\colon \boldsymbol{G} \to \boldsymbol{G}_{\mathrm{ad}}$ be the adjoint quotient and set $G_{\mathrm{ad}} = \boldsymbol{G}_{\mathrm{ad}}^F$. Since $\boldsymbol{G}$ has a connected center, p restricts to a surjective map $p\colon G \to G_{\mathrm{ad}}$ (see [60, §3]). For any KG_{ad}-module Y we denote by $p_*(Y)$ its pull-back to G via p. We have the following important property, due to Deligne and Lusztig (see [7, p. 380]):

$$\mathrm{Irr}_K(G, 1) = \{p_*(Y) \mid Y \in \mathrm{Irr}_K(G_{\mathrm{ad}}, 1)\}. \tag{a}$$

Thus, whenever we consider unipotent modules, we may assume that $\boldsymbol{G}$ is semisimple of adjoint type if this is convenient. Considering the simple components involved in $\boldsymbol{G}$, we may even reduce further to the case where $\boldsymbol{G}$ is simple of adjoint type. We will only formulate a precise statement in the following special case.

By analogy with [59, (9.7)], we say that G is a group of type GL, if the following conditions are satisfied. First, we assume that $\boldsymbol{G}$ has a connected center. Furthermore, we assume that $\boldsymbol{W}$ is a direct product of Weyl groups of type A, and that the smallest power of F leaving such a factor stable acts trivially on that factor. Now note that $\boldsymbol{G}_{\text{ad}}$ is the direct product of simple algebraic groups of adjoint type A, but that F may still permute these factors. By a similar reduction argument as in the proof of [60, (8.1)], it follows that

$$G_{\text{ad}} \cong \mathrm{PGL}_{n_1}(q^{d_1}) \times \cdots \times \mathrm{PGL}_{n_r}(q^{d_r}), \quad \text{where } n_i, d_i \geq 1 \text{ for all } i. \tag{b}$$

Combining (a) and (b) with the results in (5.3), we see that if G is a group of type GL then $\mathrm{Irr}_K(G, 1)$ coincides with the Harish-Chandra series $\mathrm{Irr}_K(G, (\emptyset, K_H))$; moreover, any $Y \in \mathrm{Irr}_K(G, 1)$ is a linear combination of $R_T^G(K_T)$'s and any cuspidal simple KG-module is of the form $\pm R_T^G(\theta)$, for a suitable pair (T, θ).

5.5. Now we turn to the problem of describing blocks of $\mathcal{O}G$. First recall that, for any $V \in \mathrm{Irr}_K(G)$, we have a corresponding primitive central idempotent

$$e_V := \frac{1}{|G|} \sum_{g \in G} \chi(1)\chi(g)g^{-1} \in KG,$$

where χ denotes the character of V. A *block* is primitive central idempotent $b \in \mathcal{O}G$. In KG, we can express a block b uniquely as a sum of idempotents e_V as above. If e_V occurs in that decomposition, we say that V belongs to b.

An important property of the partition of the simple KG-modules into blocks is the fact that the decomposition matrix of G has a block diagonal shape with diagonal blocks corresponding to the blocks of simple modules (see [23, (I.17.9)]).

Theorem 5.6 (Broué–Michel [3]). *Let $s \in G^*$ be a semisimple element of order prime to ℓ. Then we have $b_s := \sum_V e_V \in \mathcal{O}G$ where V runs over all simple KG-modules in $\bigcup_t \mathrm{Irr}_K(G, st)$, the union being taken over all elements $t \in C_{G^*}(s)$ whose order is a power of ℓ.*

5.7. The next result will show that the simple modules in the Lusztig series $\mathrm{Irr}_K(G, s)$, where s has order prime to ℓ, give rise to an ordinary basic set (in the sense of Definition 4.15). But this only works under some mild conditions on $\boldsymbol{G}$ and on ℓ. Recall from [7, p. 28] that the prime ℓ is called *good* for $\boldsymbol{G}$ if it is good for each simple factor involved in $\boldsymbol{G}$; the conditions for the various simple types are as follows.

$$\begin{aligned} A_n &: \text{no condition}, \\ B_n, C_n, D_n &: \ell \neq 2, \\ G_2, F_4, E_6, E_7 &: \ell \neq 2, 3, \\ E_8 &: \ell \neq 2, 3, 5. \end{aligned}$$

Theorem 5.8 (Geck–Hiss [35]). *Assume that the center of $\boldsymbol{G}$ is connected and that ℓ is good for $\boldsymbol{G}$. Let $s \in G^*$ be a semisimple element of order prime to ℓ. Then the* *simple modules in* $\operatorname{Irr}_K(G, s)$ *form an ordinary basic set with respect to* b_s *(with* b_s *as in Theorem* 5.6).

(The above result actually holds under slightly weaker assumptions on the center of $\boldsymbol{G}$; for more details see [32].)

5.B. ℓ-regular quotients

The following concept was originally introduced by Hiss [47, §6] for the case of unipotent modules. In combination with the results on blocks and basic sets in (5.A), it will prove very useful in the subsequent discussion.

Definition 5.9. (Gruber–Hiss [43, (3.3)]) Let Y be an $\mathcal{O}G$-lattice and write $KY = Y_{\ell'} \oplus Y_\ell$, where $Y_{\ell'}$ is the sum of all simple KG-submodules of KY which lie in Lusztig series corresponding to semisimple elements in G^* of order prime to ℓ, and Y_ℓ is a complement to $Y_{\ell'}$. Then, by (4.6), there exists a unique $\mathcal{O}$-pure sublattice $V \subseteq Y$ such that $K(Y/V) \cong_{KG} KY_{\ell'}$. The quotient Y/V is called the *ℓ-regular quotient of* Y and denoted by $\pi_{\ell'}(Y)$.

5.10. The operator $\pi_{\ell'}$ can also be interpreted as multiplication with a suitable idempotent in KG. Indeed, if we set $e_{\ell'} = \sum_V e_V \in KG$ where the sum is over all simple KG-modules V which lie in a Lusztig series corresponding to a semisimple element in G^* of order prime to ℓ, then we have

$$\pi_{\ell'}(Y) \cong_{\mathcal{O}G} e_{\ell'}Y \subseteq KY. \tag{a}$$

The above interpretation immediately shows that $\pi_{\ell'}$ is additive; we have

$$\pi_{\ell'}(Y_1 \oplus Y_2) \cong_{\mathcal{O}G} \pi_{\ell'}(Y_1) \oplus \pi_{\ell'}(Y_2) \quad \text{for all } \mathcal{O}G\text{-lattices } Y_1, Y_2. \tag{b}$$

Taking ℓ-regular quotients is compatible with Harish-Chandra induction and restriction: if $J \subseteq S$, we have

$$\pi_{\ell'}(R_J^S(X)) \cong_{\mathcal{O}G} R_J^S(\pi_{\ell'}(X)) \qquad \text{for all } \mathcal{O}L_J\text{-lattices } X, \tag{c}$$

$$\pi_{\ell'}({}^*R_J^S(Y)) \cong_{\mathcal{O}L_J} {}^*R_J^S(\pi_{\ell'}(Y)) \quad \text{for all } \mathcal{O}G\text{-lattices } Y. \tag{c$'$}$$

The statement in (c) follows from (5.2)(b), (c) and the fact that Harish-Chandra induction is exact and preserves $\mathcal{O}$-torsion modules. Then (c′) follows from the analogous properties of Harish-Chandra restriction and the adjointness in Lemma 3.6.

The following result shows an important property of ℓ-regular quotients.

Lemma 5.11 (Hiss). *Let Y be an $\mathcal{O}G$-lattice which is a direct summand of $\pi_{\ell'}(P)$ for some projective $\mathcal{O}G$-module P. Then there exists a projective $\mathcal{O}G$-module Q such that Q is a direct summand of P and $Y \cong_{\mathcal{O}G} \pi_{\ell'}(Q)$. In particular, Dipper's hypothesis* (D) *is satisfied for Y, and Y is indecomposable if and only if Q is indecomposable.*

Proof. Let $M := \pi_{\ell'}(P)$ and $\beta\colon P \to M$ be the canonical map. Let $P = P_1 \oplus \cdots \oplus P_n$ be a decomposition into projective indecomposable $\mathcal{O}G$-modules and set $Y_i = \pi_{\ell'}(P_i)$ for all i. Note that $Y_i \neq \{0\}$ by [46, Theorem 3.1]. Being quotients of projective indecomposable modules, the Y_i are indecomposable (see [9, Exercise 6.13]) and, by (5.10)(b), we have $M \cong_{\mathcal{O}G} \pi_{\ell'}(P) \cong_{\mathcal{O}G} Y_1 \oplus \cdots \oplus Y_n$. So Y must be the direct sum of some of the Y_i. But then Y is the ℓ-regular quotient of the direct sum of the corresponding P_i. The fact that hypothesis (D) holds for Y follows from Lemma 4.7(a). □

In the case of unipotent modules, we also have the following compatibility result with respect to adjoint quotients.

Lemma 5.12. *In the setting of* (5.4), *let Y_0 be an $\mathcal{O}G_{\mathrm{ad}}$-lattice such that all constituents of KY_0 are unipotent. Then Y_0 is the ℓ-regular quotient of a projective module for G_{ad} if and only if $p_*(Y_0)$ is the ℓ-regular quotient of a projective module for G.*

Proof. First note that — by general theory, using [23, (IV.4.26)] — we have the following relation between projective modules for G_{ad} and projective modules for G. Let Q_0 be a projective $\mathcal{O}G_{\mathrm{ad}}$-module. Then there exists a projective $\mathcal{O}G$-module Q such that $p_*(Q_0)$ is a quotient of Q and we have $KQ \cong_{KG} Kp_*(Q_0) \oplus Q'$ where Q' is a sum of simple KG-modules which do not have $Z(G)$ in their kernel. Moreover, every projective $\mathcal{O}G$-module Q such that $Z(G)$ acts trivially on $\mathrm{hd}(Q)$ arises in this way from some projective $\mathcal{O}G_{\mathrm{ad}}$-module Q_0.

The desired statement about ℓ-regular quotients easily follows from this. □

We now consider the Gelfand–Graev representation and its ℓ-regular quotients.

5.13. Assume that the center of $\mathbf{G}$ is connected. Recall that we have $B = UH$ where U is the largest normal p-subgroup of B and H is an abelian p'-group. Let $\gamma\colon U \to K^\times$ be a *non-degenerate* linear character, as defined in [7, p. 253]. Then the values of γ certainly lie in $\mathcal{O}$, so γ defines an $\mathcal{O}U$-module $\mathcal{O}_\gamma$ which is $\mathcal{O}$-free of rank 1. By analogy with the definition in [7, p. 254], we call

$$\Gamma_G := \mathrm{Ind}_U^G(\mathcal{O}_\gamma)$$

the *Gelfand–Graev lattice* of G. By [7, (8.1.2)], any two non-degenerate characters of U are conjugate by an element of H. This implies that Γ_G is unique up to isomorphism. Now note that Γ_G is a projective module since it is obtained by induction from a subgroup of G whose order is prime to ℓ. Combining this with Theorem 5.6, we conclude that we have a direct sum decomposition

$$\Gamma_G = \bigoplus_s \Gamma_{G,s}, \quad \text{where } \Gamma_{G,s} := b_s\Gamma_G, \tag{a}$$

the sum being taken over a set of representatives of the conjugacy classes of semisimple elements of G^* of order prime to ℓ.

By [7, (8.1.3)], $K\Gamma_G$ is multiplicity-free. The simple KG-modules which are constituents of $K\Gamma_G$ are called the *regular KG-modules* (see [7, p. 281]). By [7, (8.4.7)], each Lusztig series contains precisely one regular KG-module. Now let $s \in G^*$ be a semisimple element of order prime to ℓ and consider the projective $\mathcal{O}G$-module $\Gamma_{G,s}$. Then we set

$$X_{G,s} := \pi_{\ell'}(\Gamma_{G,s}). \tag{b}$$

Since $KX_{G,s}$ is a simple KG-module, $X_{G,s}$ is certainly indecomposable. Hence Lemma 5.11 implies that:

$$\text{each summand } \Gamma_{G,s} \text{ in (a) is a projective indecomposable module.} \tag{c}$$

(See also Hiss [46, (3.2)], where (c) is shown by a different argument which works without any assumption on the center of $\boldsymbol{G}$.)

The following result shows that the above constructions are compatible with Harish-Chandra restriction. Note that everything also works for L_J since the underlying Levi complement in $\boldsymbol{G}$ also has a connected center; see [7, (8.1.4)].

Lemma 5.14. *Let $s \in G^*$ be semisimple of order prime to ℓ and $J \subseteq S$. Then*

$${}^*R_J^S(X_{G,s}) \cong_{\mathcal{O}L_J} \bigoplus_{s' \in T} X_{L_J,s'}$$

where $T \subseteq L_J^*$ is a set of representatives of the L_J^*-conjugacy classes which are contained in the conjugacy class of s in G^*.

Proof. The proof consists of reducing the desired statement to the following known fact (see [7, (8.1.6)]):

$${}^*R_J^S(K\Gamma_G) \cong_{KL_J} K\Gamma_{L_J}.$$

The reduction proceeds as follows. Since ℓ-regular quotients are compatible with direct sums and Harish-Chandra restriction (see (5.10)), it is enough to show that

$${}^*R_J^S(\Gamma_{G,s}) \cong_{\mathcal{O}L_J} \bigoplus_{s' \in T} \Gamma_{L_J,s'}.$$

Using (5.13)(a), (c), we see that it is in fact enough to show that

$${}^*R_J^S(\Gamma_G) \cong_{\mathcal{O}L_J} \Gamma_{L_J}.$$

Now note that Γ_G and Γ_{L_J} are projective modules and that ${}^*R_J^S$ maps projective modules to projective modules since ${}^*R_J^S$ is given by multiplication with an idempotent; see (3.2). So ${}^*R_J^S(\Gamma_G)$ and Γ_{L_J} are projective modules which become isomorphic after scalar extension from $\mathcal{O}$ to K. But then those two projective modules must themselves be isomorphic since the Cartan matrix of $\mathcal{O}G$ has a non-zero determinant (see [23, (I.17.10), (IV.3.9)]). □

Example 5.15. Let $s = 1$ and consider the $\mathcal{O}G$-lattice $X_{G,1}$. By [7, (8.4.9)], we have $\dim_K KX_{G,1} = |U|$ where U is the largest normal p-subgroup in B. So $KX_{G,1}$ must be the Steinberg module for G, as defined in [7, (6.2)]. (Note that the Steinberg module occurs in $\mathrm{Ind}_B^G(K_B)$ and has dimension $|U|$, so it lies in $\mathrm{Irr}_K(G, 1)$ and is regular.) Therefore we will write

$$\mathrm{St}_G := X_{G,1} = \pi_{\ell'}(\Gamma_{G,1})$$

and call this lattice the *Steinberg lattice* for G. We also write $\mathrm{St}_{G,K} := K\,\mathrm{St}_G$. Since $\Gamma_{G,1}$ is indecomposable (see (5.13)(c)), we have that

$$\mathrm{St}_G \text{ is indecomposable and } \mathrm{hd}(\mathrm{St}_G) \text{ is simple.} \tag{a}$$

Note that St_G is not necessarily isomorphic to the integral version of the Steinberg module as originally defined in [64]. Since we are dealing with the case $s = 1$, the statement in Lemma 5.14 simplifies as follows:

$$\text{For any subset } J \subseteq S, \text{ we have } {}^*R_J^S(\mathrm{St}_G) \cong_{\mathcal{O}L_J} \mathrm{St}_{L_J}. \tag{b}$$

The following result will be crucial in the discussion of (generalized) q-Schur algebras in (6.B); see Corollary 6.16. We only state it here for the Steinberg lattice. For a more general version, see Gruber–Hiss [43, (4.2), (4.3)]. In the case where $G = \mathrm{GL}_n(q)$, it goes back to James [52].

Theorem 5.16. *There exists an injective homomorphism* $\iota\colon \mathrm{St}_G \to \mathrm{Ind}_B^G(\mathcal{O}_B)$, *which is unique up to non-zero scalar multiples. Moreover,* $\overline{\mathrm{St}}_G$ *has a unique composition factor in the Harish-Chandra series* $\mathrm{Irr}_k(G, (\emptyset, k_H))$. *The following conditions are equivalent:*

(a) *The induced map* $\bar{\iota}\colon \overline{\mathrm{St}}_G \to \mathrm{Ind}_B^G(k_B)$ *is injective.*

(b) *The unique composition factor of* $\overline{\mathrm{St}}_G$ *in the series* $\mathrm{Irr}_k(G, (\emptyset, k_H))$ *occurs in the socle of* $\overline{\mathrm{St}}_G$, *and* $\mathrm{soc}(\overline{\mathrm{St}}_G)$ *is simple.*

(c) *For all* $\emptyset \neq J \subseteq S$, $\mathrm{soc}(\overline{\mathrm{St}}_{L_J})$ *does not have any cuspidal simple constituent.*

If (a) *holds, then* St_G *is isomorphic to a pure sublattice of* $\mathrm{Ind}_B^G(\mathcal{O}_B)$; *see* (4.6)(a).

Proof. We set $M_0 := \mathrm{Ind}_B^G(\mathcal{O}_B)$; note that $M_0 = R_\emptyset^S(\mathcal{O}_H)$ and that $\overline{M}_0 = R_\emptyset^S(k_H)$. Using Example 5.15(b) and the adjointness in Lemma 3.6, we have

$$\mathrm{Hom}_{\mathcal{O}G}(\mathrm{St}_G, M_0) \cong_{\mathcal{O}} \mathrm{Hom}_{\mathcal{O}H}({}^*R_\emptyset^S(\mathrm{St}_G), \mathcal{O}_H) \cong_{\mathcal{O}} \mathcal{O}.$$

Hence there exists a non-trivial homomorphism $\iota\colon \mathrm{St}_G \to M_0$, which is unique up to non-zero scalar multiplies. But $\mathrm{St}_{G,K}$ is simple and so ι_K must be injective. Hence ι itself is injective. Thus the first statement is proved. Now ι can certainly be chosen such that $\bar{\iota} \neq 0$. Using the fact that every simple module in $\mathrm{Irr}_k(G, (\emptyset, 1_H))$ occurs in $\mathrm{soc}(\overline{M}_0)$ (see Theorem 3.16), we conclude that $\overline{\mathrm{St}}_G$ must have at least one composition factor in the series $\mathrm{Irr}_k(G, (\emptyset, 1_H))$. In order to prove that there is precisely one such composition factor, we consider the module $P_0 := \mathrm{Ind}_U^G(\mathcal{O}_U)$ which is projective since ℓ does not divide $|U|$. Moreover, since $U \subseteq B$, we have a natural surjective map $P_0 \to M_0$. So, using the fact that every simple module in $\mathrm{Irr}_k(G, (\emptyset, 1_H))$ also occurs in $\mathrm{hd}(\overline{M}_0)$ (see Theorem 3.16), we conclude that P_0 contains a projective cover of every simple module in $\mathrm{Irr}_k(G, (\emptyset, 1_H))$. Hence it is enough to show that $\dim_k \mathrm{Hom}_{kG}(\overline{P}_0, \overline{\mathrm{St}}_G)) \leq 1$, which is clear by Frobenius reciprocity and since $\dim_k \overline{\mathrm{St}}_G = |U|$.

Let us now check the equivalence of (a), (b), (c).

First we show that (a) implies (c). Let $\emptyset \neq J \subseteq S$ and X be a cuspidal simple kL_J-module. Then Lemma 3.14(b) shows that $R_J^S(X)$ has no composition factor in the series $\mathrm{Irr}_k(G, (\emptyset, k_H))$. So there is no non-trivial homomorphism from $R_J^S(X)$ to $\overline{\mathrm{St}}_G$. Hence, by Example 5.15(b) and the adjointness in Lemma 3.6, we have

$$\mathrm{Hom}_{kL_J}(X, \overline{\mathrm{St}}_{L_J}) \cong_k \mathrm{Hom}_{kL_J}(X, {}^*R_J^S(\overline{\mathrm{St}}_G)) \cong_k \mathrm{Hom}_{kG}(R_J^S(X), \overline{\mathrm{St}}_G) = \{0\},$$

as desired. Next we show that (c) implies (b). To see this, let $Z \subseteq \overline{\mathrm{St}}_G$ be any simple submodule. Let $J \subseteq S$ be minimal such that ${}^*R_J^S(Z) \neq \{0\}$. Since Harish-Chandra restriction is exact, every submodule of ${}^*R_J^S(Z)$ is isomorphic to a submodule of ${}^*R_J^S(\overline{\mathrm{St}}_G) \cong_{kL_J} \overline{\mathrm{St}}_{L_J}$ where the latter isomorphism holds by Example 5.15(b). But, by Lemma 3.12, any such submodule of ${}^*R_J^S(Z)$ is cuspidal. So the condition in (c) implies that we must have $J = \emptyset$. Thus every simple submodule of $\overline{\mathrm{St}}_G$ lies in the series $\mathrm{Irr}_k(G, (\emptyset, 1_H))$. Moreover, we already know that $\overline{\mathrm{St}}_G$ only has one composition factor in that series. So (b) holds. Finally, we prove that (b) implies (a). We may assume that $\bar{\iota} \neq 0$. Then $\overline{\mathrm{St}}_G/\ker(\bar{\iota})$ is isomorphic to a submodule of $\overline{M}_0$. Using Theorem 3.16 once more, we see that $\overline{\mathrm{St}}_G/\ker(\bar{\iota})$ has a simple submodule in the series $\mathrm{Irr}_k(G, (\emptyset, k_H))$. Since $\overline{\mathrm{St}}_G$ only has one composition factor in that series, we must have $\ker(\bar{\iota}) = \{0\}$. □

Remark 5.17. A subtle and non-trivial point in the whole theory is to determine whether the three conditions (a), (b), (c) in Theorem 5.16 are satisfied or not. For $G = \mathrm{GL}_n(q)$, this will be shown to be the case in Corollary 6.16. The following example shows that the above conditions are not always satisfied.

Let $G = \mathrm{GU}_3(q)$ and assume that $2, 3 \neq \ell \mid q+1$. Let $M_0 := \mathrm{Ind}_B^G(\mathcal{O}_B)$ and set $V := \mathrm{hd}(\overline{\mathrm{St}}_G)$. In [27, (5.2)], it is shown that there exists a submodule $\overline{D} \subseteq \overline{M}_0$ such that

$$\dim_k \overline{D} \leq q(q^2 - q + 1) < q^3 = \dim_k \overline{\mathrm{St}}_G \quad \text{and} \quad \mathrm{hd}(\overline{D}) \cong_{kG} V.$$

But $\overline{M}_0$ has a simple head (isomorphic to k_G) and $\dim_k \overline{M}_0 = [G : B] = q^3+1$. So, if $\overline{\mathrm{St}}_G$ were isomorphic to a submodule of $\overline{M}_0$, then we would have $\overline{D} \subseteq \mathrm{rad}(\overline{M}_0) \cong_{kG} \overline{\mathrm{St}}_G$. Since $\overline{\mathrm{St}}_G$ has a simple head and $\dim_k \overline{D} < \dim_k \overline{\mathrm{St}}_G$, we could now deduce that V has multiplicity at least 2 in a composition series for $\overline{\mathrm{St}}_G$. This is impossible by the decomposition matrix in [27, (4.3)].

5.C. Iwahori–Hecke algebras

We now consider a setting similar to that in (3.C), but where we work over the ground ring $\mathcal{O}$. Let us fix a subset $J \subseteq S$ and an $\mathcal{O}L_J$-lattice X_0 such that $X := KX_0 \in KL_J$-mod is cuspidal and simple. We set

$$H_0 = \mathrm{End}_{\mathcal{O}G}(M_0)^\circ \quad \text{where } M_0 = R_J^S(X_0),$$
$$H = \mathrm{End}_{\mathcal{O}G}(M)^\circ \quad \text{where } M = R_J^S(X).$$

Note that Harish-Chandra induction and Hom are compatible with scalar extension from $\mathcal{O}$ to K, so we can identify $KM_0 = M$ and $KH_0 = H$. Thus, all the results in (3.C) apply to KH_0. Our aim is to show that, under suitable assumptions, this is sufficient also to determine the structure of the $\mathcal{O}$-algebra H_0. We will also show that Lusztig's strengthening of Theorem 3.28, obtained in [59, Chapter 8], carries over to H_0.

5.18. Let $\rho_0 \colon \mathcal{O}L_J \to \mathrm{End}_{\mathcal{O}}(X_0)$ be the representation afforded by X_0. Then $\rho = (\rho_0)_K \colon KL_J \to \mathrm{End}_K(X)$ is the representation afforded by X. (As usual, we identify X_0 with an $\mathcal{O}$-submodule of X.) We can certainly choose a projective representation $\tilde{\rho}$ extending ρ as in (3.25) such that $\tilde{\rho}(\dot{w}) \in \mathrm{End}_{\mathcal{O}}(X_0)$ and $\overline{\tilde{\rho}(\dot{w})} \neq 0$ for all $w \in W_{(J,X)}$. Following [28, §2.6], we shall now make the following assumption:

(∗) The endomorphism $\tilde{\rho}(\dot{w}) \in \mathrm{End}_{\mathcal{O}}(X_0)$ is invertible over $\mathcal{O}$ for all $w \in W_{(J,X)}$.

Consequently, the corresponding cocycle λ has values in $\mathcal{O}^\times$, and we can also assume that it is normalized as in (3.25) (by the same proof as in [7, (10.3.4)]). Now, identifying $M_0 = \mathrm{Hom}_{\mathcal{O}P_J}(\mathcal{O}G, \tilde{X}_0)$ as in (3.3)(b), we define endomorphisms $B_w \in \mathrm{End}_{\mathcal{O}}(M_0)$ for all $w \in W_{(J,X)}$ by exactly the same formula as in (3.26). We have the following result.

Lemma 5.19 (See [28, (2.6.5)]). *In the above setting, $\{B_w \mid w \in W_{(J,X)}\}$ is an $\mathcal{O}$-basis of H_0. Moreover, $\{(B_w)_K\}$ is a K-basis of H as in Theorem 3.28.*

Proof. The second statement is clear by construction of the elements B_w. This already shows that each B_w commutes with the action of G and that the set $\{B_w\}$ is linearly independent over $\mathcal{O}$. So it remains to show that any $B \in H_0$ is an $\mathcal{O}$-linear combination of $\{B_w\}$. We already know that we can write $B = \sum_w \xi_w B_w$ with $\xi_w \in K$. (In order to abbreviate notation, we omit the subscript K to indicate scalar extension from $\mathcal{O}$ to K.) Now take any $0 \neq v \in X_0$ and define f_v as in (3.26)(a). Then (3.26)(b) shows that

$$(Bf_v)(\dot{w}') = \sum_w \xi_w (B_w f_v)(\dot{w}') = \xi_{w'} |U_J \cap {}^{w'}U_J| \tilde{\rho}(\dot{w}')(v)$$

for all $w' \in W_{(J,X)}$. Since the left hand side lies in X_0 and $|U_J \cap {}^{w'}U_J|$ is invertible in $\mathcal{O}$, we conclude that $\xi_{w'}\tilde{\rho}(\dot{w}')(v) \in X_0$ for all $w' \in W_{(J,X)}$.

Now let $\{v_i\}$ be an $\mathcal{O}$-basis of X_0 and set $u_i = \tilde{\rho}(\dot{w}')(v_i)$ for all i. By condition $(*)$ in (5.18), $\tilde{\rho}(\dot{w}')$ is invertible over $\mathcal{O}$. So the elements $\{u_i\}$ also form an $\mathcal{O}$-basis of X_0. Thus, we have $\xi_{w'} u_i \in X_0$ for all i and so $\xi_{w'} \in \mathcal{O}$, as desired. □

We shall now give criteria for the condition $(*)$ to hold.

Lemma 5.20. *Condition* $(*)$ *in* (5.18) *is satisfied if one of the following holds.*

(a) *The kL_J-module $\overline{X}_0$ is simple (and hence also cuspidal).*

(b) *The cocycle λ can be chosen to be trivial.*

Proof. (a) Let $w \in W_{(J,X)}$. By definition, $\tilde{\rho}(\dot{w})$ is an $\mathcal{O}L_J$-homomorphism between X_0 and $\mathrm{ad}(\dot{w})(X_0)$. Then the induced kL_J-homomorphism between $\overline{X}_0$ and $\mathrm{ad}(\dot{w})(\overline{X}_0)$ is non-zero since $\overline{\tilde{\rho}(\dot{w})} \neq 0$. So, by Schur's Lemma, that homomorphism must be an isomorphism. This certainly implies that $\tilde{\rho}(\dot{w})$ itself is invertible over $\mathcal{O}$. The fact that $\overline{X}_0$ is also cuspidal follows from the more general fact that Harish-Chandra restriction is compatible with extension of scalars. If (b) holds, then the projective representation $\tilde{\rho}$ can be chosen to be an actual representation, so that $\tilde{\rho}(\dot{w})$ is invertible over $\mathcal{O}$ since $\dot{w}$ is invertible. □

Part (a) of the following result was obtained by Geck [28, (2.6.9), (2.6.10)] and also by Dipper–Fleischmann [14, (2.12)].

Proposition 5.21. *Assume that $\overline{X}_0$ is simple and that $W_{(J,X)} = W_{(J,\overline{X}_0)}$.*

(a) *Then Dipper's hypothesis* (D) *holds for M_0. By Lemma* 4.7(b), *this also implies that $\overline{H}_0 = \mathrm{End}_{kG}(\overline{M}_0)$; i.e., M_0 is* reduction stable *in the sense of Dipper–Fleischmann* [14].

(b) *Let $\tau_0 \colon H_0 \to \mathcal{O}$ be defined as in* (3.30). *Then τ_0 is a central function and the determinant of the Gram matrix of the associated bilinear form is invertible in $\mathcal{O}$. In particular, H_0 is symmetric over $\mathcal{O}$.*

Proof. First note that condition $(*)$ in (5.18) is satisfied by Lemma 5.20(a).

(a) Let $\beta_{X_0} \colon P_{X_0} \to X_0$ be a projective cover and $\beta_0 \colon P_0 \to M_0$ be the induced map where $P_0 = R_J^S(P_{X_0})$. By Lemma 2.6 it is enough to show that the injective map

$$H_0 = \mathrm{End}_{\mathcal{O}G}(M_0) \to \mathrm{Hom}_{\mathcal{O}G}(P_0, M_0), \quad \varphi \mapsto \varphi \circ \beta_0,$$

is surjective. For this purpose, it is certainly enough to prove that the induced map $\overline{H}_0 \to \overline{\mathrm{Hom}_{\mathcal{O}G}(P_0, M_0)}$ is surjective. Since that induced map is also injective (note that $\bar{\beta}_0$ is still surjective), it suffices to show that

$$\dim_k \overline{H}_0 = \dim_k \overline{\mathrm{Hom}_{\mathcal{O}G}(P_0, M_0)}.$$

This is proved as follows. Since $\overline{X}_0$ is simple, the induced map $\bar{\beta}_{X_0} \colon \overline{P}_{X_0} \to \overline{X}_0$ is also a projective cover. Moreover, since reduction modulo π and induction commute with each other, we have $\overline{P}_0 = R_J^S(\overline{P}_{X_0})$ and $\overline{M}_0 = R_J^S(\overline{X}_0)$. Now $\overline{X}_0$ is cuspidal; so we can apply Theorem 3.20 and conclude that hypothesis (D) holds for $\bar{\beta}_0 \colon \overline{P}_0 \to \overline{M}_0$. This yields

$$\begin{aligned}\dim_k \operatorname{End}_{kG}(\overline{M}_0) &= \dim_k \operatorname{Hom}_{kG}(\overline{P}_0, \overline{M}_0) && \text{by Lemma 2.6,}\\ &= \dim_k \overline{\operatorname{Hom}_{\mathcal{O}G}(P_0, M_0)} && \text{since } P_0 \text{ is projective.}\end{aligned}$$

Hence it remains to show that $\dim_k \overline{H}_0 = \dim_k \operatorname{End}_{kG}(\overline{M}_0)$. But H_0 is free of rank $|W_{(J,X)}|$, by Lemma 5.19. So we also have $\dim_k \overline{H}_0 = |W_{(J,X)}|$. On the other hand, we know by the results in (3.C) that $\dim_k \operatorname{End}_{kG}(\overline{M}_0) = |W_{(J,\overline{X}_0)}|$. It remains to use the assumption that $W_{(J,X)} = W_{(J,\overline{X}_0)}$.

(b) We know that $(\tau_0)_K \colon KH \to K$ is a central function. Hence τ_0 is a central function. It remains to show the assertion about the associated bilinear form. To see this, we argue as follows. We already know by (a) that M_0 is reduction stable. Hence, using Lemma 5.18, $\{\bar{B}_w \mid w \in W_{(J,X)}\}$ is a k-basis of $\overline{H}_0$. But it is clear that $\bar{B}_w$ is exactly the linear map defined directly in terms of the cuspidal simple module $\overline{X}_0$ by the equation in (3.26). Hence (3.30) shows that $\bar{\tau}_0 \colon \overline{H}_0 \to k$ is a central function which gives $\overline{H}_0$ the structure of a symmetric algebra. Thus the Gram matrix of the associated bilinear form is invertible over k. But that Gram matrix is the reduction modulo π of the Gram matrix of the bilinear form associated with τ_0. So the determinant of the latter matrix is invertible in $\mathcal{O}$. □

Lemma 5.22 (See [30, (6.6)]). *Assume that the center of* $\mathbf{G}$ *is connected and that* ℓ *is good for* $\mathbf{G}$ *(see (5.7)). Furthermore, assume that* $X \in \operatorname{Irr}_K(L_J, s)$*, where* s *has order prime to* ℓ*, and that* $\overline{X}_0$ *is simple. Then we have* $W_{(J,X)} = W_{(J,\overline{X}_0)}$.

Proof. Let $w \in C_J$. Then we have $\dot{w} \in N_G(L_J) \cap N$ (see (3.24)) and so $X_0' := \operatorname{ad}(\dot{w})(X_0)$ is an $\mathcal{O}L_J$-lattice such that $X' := KX_0'$ and $\overline{X}_0'$ are also simple. Note that the following certainly holds.

(1) We have $X' \cong_{KL_J} \operatorname{ad}(\dot{w})(X)$ and $\overline{X}_0' \cong_{kL_J} \operatorname{ad}(\dot{w})(\overline{X}_0)$.

Now let us first check that we have $W_{(J,X)} \subset W_{(J,\overline{X}_0)}$. Assume that $w \in W_{(J,X)}$. Then, using (1), we have $X \cong_{KL_J} X'$ and so, by a classical result of Brauer on decomposition numbers (see [23, (I.17.7)]), $\overline{X}_0$ and $\overline{X}_0'$ have the same sets of composition factors (with multiplicities). Since $\overline{X}_0$ is simple, we must have $\overline{X}_0 \cong_{kL_J} \overline{X}_0'$. Hence, using (1) again, we conclude that $w \in W_{(J,\overline{X}_0)}$, as desired.

In order to prove the reverse inclusion, we need the following statement.

(2) We have $X' \in \operatorname{Irr}_K(L_J, s')$ where $s' \in L_J^*$ has order prime to ℓ.

Indeed, by assumption, X has non-zero multiplicity in some $R_T^{L_J}(\theta)$ where (T, θ) corresponds to a pair (T^*, s); see (5.1)(a). Now, by [10, (13.11)], we have an isomorphism of abelian groups $T^* \cong \operatorname{Irr}(T)$, and θ, s correspond to each other under that isomorphism. Hence, since s has order prime to ℓ, a similar statement also holds for θ. Now, using (1), X' has non-zero multiplicity in $\operatorname{ad}(\dot{w})(R_T^{L_J}(\theta))$. By [10, (13.22)], the latter virtual module is isomorphic to $R_{T'}^{L_J}(\theta')$ where $T' = {}^{\dot{w}}T$ and $\theta' = \mathrm{ad}(\dot{w})(\theta)$. A similar argument as above shows that (T', θ') corresponds to a pair (T'^*, s') where s' has order prime to ℓ. Thus (2) is proved.

Now assume that $w \in W_{(J,\overline{X}_0)}$. Then, by (2), X' is a simple KL_J-module which lies in $\mathrm{Irr}_K(L_J, s')$, where $s' \in L_J^*$ has order prime to ℓ. Hence both X and X' belong to the basic sets of L_J given by Theorem 5.8. But, by assumption (and using (1) once more), we have $\overline{X}_0 \cong_{kL_J} \overline{X}'_0$. By the definition of basic sets this can only happen if $X \cong_{KL_J} X'$ and so $w \in W_{(J,X)}$, as desired. □

Remark 5.23. In [28, (2.6.11)] (see also [30, (6.6)]), it is conjectured that the condition that $\overline{X}_0$ is simple is actually a consequence of the other assumptions of Lemma 5.22.

5.24. Assume that the center of $\boldsymbol{G}$ is connected. Then Lusztig has shown in [59, (8.6)] that the fact that L_J has a cuspidal simple module over K (namely, X) has the following two consequences on the structure of H.

(a) The group $W_{(J,X)}$ is a Coxeter group with a set of generators $S_{(J,X)}$ given in terms of the Weyl group of $C_{\boldsymbol{G}^*}(s)$ as in [59, (8.5.8)], where $s \in L_J^*$ is a semisimple element such that $X \in \mathrm{Irr}_K(L_J, s)$.

(b) We can take $C_{(J,X)} = \{1\}$ in Theorem 3.28, and $\tilde{\rho}$ in (3.25) can be chosen such that the cocycle λ is trivial.

The above two results already show that the structure of H is completely determined by the Coxeter relations in $W_{(J,X)}$ and the quadratic relations in Theorem 3.28(c). The latter relations can also be described more explicitly. For this purpose, we consider a renormalization of the basis elements $\{B_w\}$ as in [7, (10.8.3)]. For each $w \in W_{(J,X)}$, there exists some $\eta_w \in K$ (which has the property that some power of η_w is a power of p) such that, if we set $T_w := \eta_w B_w$, the following holds.

(c) We have $T_w = T_{s_1} \circ \cdots \circ T_{s_t}$ whenever $w = s_1 \ldots s_t$ with $s_i \in S_{(J,X)}$ is a reduced expression.

(d) We have $T_s^2 = q^{c(s)} T_1 + (q^{c(s)} - 1) T_s$ for $s \in S_{(J,X)}$, where $c(s)$ is a positive integer given by [59, (8.2.3)].

Hence, under the above conditions, H is completely determined by the relations in the Coxeter group $W_{(J,X)}$ and the function $c\colon S_{(J,X)} \to \mathbb{Z}$.

Definition 5.25. Assume that the center of $\boldsymbol{G}$ is connected. Let $S_{(J,X)} \subseteq W_{(J,X)}$ and $c\colon S_{(J,X)} \to \mathbb{Z}$ be as in (5.24). Let $A = \mathbb{Z}[u^{\pm 1}]$ be the ring of Laurent polynomials in an indeterminate u. Then the corresponding *generic Iwahori–Hecke algebra*

$$\boldsymbol{H} = \boldsymbol{H}_A(W_{(J,X)}, S_{(J,X)}, c)$$

is an A-algebra with a basis $\{\boldsymbol{T}_w \mid w \in W_{(J,X)}\}$ such that $\boldsymbol{T}_w = \boldsymbol{T}_{s_1} \cdots \boldsymbol{T}_{s_t}$ whenever $w = s_1 \cdots s_t$ with $s_i \in S_{(J,X)}$ is a reduced expression, and such that $\boldsymbol{T}_s^2 = u^{c(s)}\boldsymbol{T}_1 + (u^{c(s)} - 1)\boldsymbol{T}_s$ for all $s \in S_{(J,X)}$

Such an algebra exists by a construction due to Bourbaki; see [9, §68A]. If $A \to R$ is a homomorphism of commutative rings, we may consider R as an A-algebra and obtain a corresponding *specialized algebra* $\boldsymbol{H}_R = R \otimes_A \boldsymbol{H}$.

Corollary 5.26. *Assume that condition* $(*)$ *in* (5.18) *is satisfied and that the center of* $\boldsymbol{G}$ *is connected. Then* $H_{\mathcal{O}}$ *is isomorphic to the specialized algebra* $\boldsymbol{H}_{\mathcal{O}}$ *via the map* $A \to \mathcal{O}$, $u \mapsto q$. *In particular, this means that* $H_{\mathcal{O}}$ *has a basis* $\{T_w \mid w \in W_{(J,X)}\}$ *which satisfies the multiplication rules in* (5.24)(c) *and* (d).

Recall from Lemma 5.20 that condition $(*)$ holds, for example, if $\overline{X}_{\mathcal{O}}$ is simple.

Proof. This follows from Lemma 5.19 and (5.24). It only remains to check that the constants η_w (used in the renormalization of B_w) are units in $\mathcal{O}$. But this is clear since some power of η_w is a power of p. □

6. (Generalized) q-Schur algebras

We keep the notations and the set-up of the previous section; i.e., $\boldsymbol{G}$ is a connected reductive group defined over the finite field $\mathbb{F}_q$, with corresponding Frobenius map $F \colon \boldsymbol{G} \to \boldsymbol{G}$, and ℓ is a prime not dividing q. Let $\mathcal{O}$ be a discrete valuation ring with field of fractions K (of characteristic 0) and residue field k (of characteristic ℓ). Now, we will combine all the techniques developed so far in order to study the ℓ-modular decomposition matrices of the finite group $G = \boldsymbol{G}^F$. The strategy by which we will do this is that described in (4.C). By the results of Dipper–James [16] and Gruber–Hiss [43], this yields complete answers for the finite general linear groups (with no restriction on ℓ) and the finite classical groups (in the so-called linear prime case). We explain the main steps in obtaining these results for unipotent blocks.

6.A. Linear primes

We begin by defining the ingredients required in the strategy of (4.C). In order to be able to control blocks and basic sets as in (5.A), we assume throughout that the center of $\boldsymbol{G}$ is connected and that ℓ is a good prime for $\boldsymbol{G}$ (see (5.7)).

6.1. Let $\mathcal{S}$ be the set of all triples (J, s, X) where $J \subseteq S$, $s \in L_J^*$ is semisimple of order prime to ℓ, and $X \in \operatorname{Irr}_K(L_J, s)$ is cuspidal. If $\sigma = (J, s, X) \in \mathcal{S}$, we also write $\sigma = (J^\sigma, s^\sigma, X^\sigma)$ and denote

$$H^\sigma = \operatorname{End}_{KG}(M^\sigma) \quad \text{where } M^\sigma = R_{J^\sigma}^S(X^\sigma).$$

Note that, by Theorem 3.16, two modules M^σ and $M^{\sigma'}$ (for $\sigma, \sigma' \in \mathcal{S}$) either have no irreducible constituent in common or they are isomorphic. Thus we may assume that the elements of $\mathcal{S}$ are taken modulo a suitable equivalence relation such that we have $M^\sigma \cong_{KG} M^{\sigma'} \Leftrightarrow \sigma = \sigma'$.

Lemma 6.2. *For each $\sigma \in \mathcal{S}$, we let $\mathfrak{B}^\sigma \subseteq \mathrm{Irr}_K(G)$ be the set of composition factors of M^σ and set $\mathfrak{B} := \bigcup_{\sigma \in \mathcal{S}} \mathfrak{B}^\sigma \subseteq \mathrm{Irr}_K(G)$. Then we have $\mathfrak{B} = \bigcup_s \mathrm{Irr}_K(G, s)$, where the union is over all semisimple elements $s \in G^*$ of order prime to ℓ. In particular, by Theorem 5.8, $\mathfrak{B}$ is an ordinary basic set for G.*

Proof. The compatibility of Harish-Chandra induction with Lusztig series (see (5.2)) implies that each simple module in $\mathfrak{B}$ belongs to a Lusztig series $\mathrm{Irr}_K(G, s)$ where s has order prime to ℓ. On the other hand, consider a simple KG-module $Y \in \mathrm{Irr}_K(G, s)$ where s has order prime to ℓ. By Theorem 3.16, there exists a subset $J \subseteq S$ and a cuspidal simple KL_J-module X such that Y is a constituent of $R_J^S(X)$. Let $s' \in L_J^*$ be semisimple such that $X \in \mathrm{Irr}_K(L_J, s')$. Using (5.2) once more, we see that $Y \in \mathrm{Irr}_K(G, s')$. Hence, by (5.1)(b), s and s' are conjugate in G^*. Consequently, s' also has order prime to ℓ and so we have $(J, s', X) \in \mathcal{S}$. □

The critical point of the whole strategy will be to gain control over the cuspidal simple modules over k and their projective covers. The following definition takes care precisely of this problem. (See [36, §3] for some results and examples which show what can happen in general.)

Definition 6.3. Recall that ℓ is a good prime for $\boldsymbol{G}$ which does not divide q. Let $\sigma = (J, s, X) \in \mathcal{S}$; see (6.1). We say that ℓ is *linear for* σ if the following condition is satisfied.

(L)$_\sigma$ If X_0 is an $\mathcal{O}L_J$-lattice such that $X \cong_{KL_J} KX_0$, then $\overline{X}_0$ is simple and X_0 is the ℓ-regular quotient of a projective module.

(Note that if $\overline{X}_0$ is simple then X_0 is unique up to isomorphism such that $X \cong_{KL_J} KX_0$.) We say that ℓ is *linear for G* if (L)$_\sigma$ holds for all $\sigma \in \mathcal{S}$.

Remark 6.4. The term "linear prime" goes back to Fong and Srinivasan ([24], [25]), where purely numerical conditions on ℓ are described (in the case where G is a unitary or classical group) which *imply* condition (L)$_\sigma$. For example, ℓ is linear for a symplectic or an orthogonal group if ℓ and the multiplicative order of q modulo ℓ are odd; for a unitary group, ℓ is linear if the order of $-q$ modulo ℓ is twice the multiplicative order of q^2 modulo ℓ. This is checked directly in [43, (5.4)]; in fact, if ℓ satisfies the appropriate numerical condition, then any $\sigma = (J, s, X) \in \mathcal{S}$ has the property that X belongs to a block with a central defect group.

Our condition (L)$_\sigma$ appears to be somewhat weaker than the original definition of Fong and Srinivasan. First of all, Example 6.5 below shows that all primes are linear for $G = \mathrm{GL}_n(q)$. Moreover, ℓ turns out to be linear if the Sylow ℓ-subgroups of G are cyclic and if G is a finite classical group (see [26]) or if ℓ is a "Coxeter prime" (in the sense of [48, §5]). However, there are cases where the Sylow ℓ-subgroups are cyclic but ℓ is not linear; see the Brauer tree in [48, (3.1)(7)].

Example 6.5. Let $G = \mathrm{GL}_n(q)$. Then every prime ℓ which does not divide q is linear for G. This is seen as follows. Let $\sigma = (J, s, X) \in \mathcal{S}$. We must show that condition $(\mathrm{L})_\sigma$ holds for X. Now L_J is again a direct product of general linear groups over $\mathbb{F}_q$ and X is the outer tensor product of cuspidal simple modules which again lie in Lusztig series corresponding to semisimple elements of order prime to ℓ in the various factors. It is clear that $(\mathrm{L})_\sigma$ holds for X if $(\mathrm{L})_\sigma$ holds for each of the factors. Thus we are reduced to the case where $G = L_J$. Now we have seen in (5.3)(c) that X lies in a Lusztig series by itself. Then, by Theorem 5.6 and Theorem 5.8, the singleton set consisting of X forms an ordinary basic set with respect to the block b_s. This certainly implies that $(\mathrm{L})_\sigma$ holds.

Lemma 6.6. *Assume that ℓ is linear for $\sigma = (J, s, X) \in \mathcal{S}$. Let X_0 be an $\mathcal{O}L_J$-lattice such that $X \cong_{KL_J} KX_0$ and $X_0 \cong \pi_{\ell'}(P_0)$, where P_0 is projective. Then, setting $M_0^\sigma := R_J^S(X_0)$ and $P_0^\sigma := R_J^S(P_0)$, the following hold.*

(a) *We have $M_0^\sigma \cong_{\mathcal{O}G} \pi_{\ell'}(P_0^\sigma)$. Thus Dipper's hypothesis* (D) *holds for M_0^σ.*

(b) *The algebra $H_0^\sigma := \mathrm{End}_{\mathcal{O}G}(M_0^\sigma)$ is symmetric and reduction stable; i.e., we have $\overline{H}_0^\sigma = \mathrm{End}_{kG}(\overline{M}_0^\sigma)$.*

Proof. (a) follows from (5.10) and Lemma 5.11. To prove (b), note that $X \in \mathrm{Irr}_K(G, s)$ where s has order prime to ℓ. So we can apply Lemma 5.22 and hence all the assumptions of Proposition 5.21 are satisfied. □

6.7. Let us fix a triple $\sigma = (J, s, X) \in \mathcal{S}$ and assume that ℓ is linear for σ. Let X_0 be an $\mathcal{O}L_J$-lattice such that $X \cong_{KL_J} KX_0$, and set

$$H_0^\sigma = \mathrm{End}_{\mathcal{O}G}(M_0^\sigma) \quad \text{where } M_0^\sigma = R_J^S(X_0). \tag{a}$$

By condition $(\mathrm{L})_\sigma$, we know that $\overline{X}_0$ is simple. Since the center of $\mathbf{G}$ is connected, we can also apply Corollary 5.26. Let us denote $W^\sigma = W_{(J,X)}$. Then there exists a subset $S^\sigma \subseteq W^\sigma$ and a function $c^\sigma : S^\sigma \to \mathbb{Z}$ such that (W^σ, S^σ) is a Coxeter system and H_0^σ has a basis $\{T_w \mid w \in W^\sigma\}$ satisfying the relations in (5.24).

We now define a set of elements in H_0^σ from which we will select elements as required in Step 2 of the strategy in (4.C). For this purpose, consider any subset $\lambda \subseteq S^\sigma$ and let $W_\lambda^\sigma \subseteq W^\sigma$ be the corresponding standard parabolic subgroup. Then, as in [15, §3], we set

$$y_\lambda^\sigma := \sum_{w \in W_\lambda^\sigma} (-1)^{l(w)} \mathrm{ind}(T_w)^{-1} T_w \in H_0^\sigma \tag{b}$$

where the representation $\mathrm{ind}\colon H_0^\sigma \to \mathcal{O}$ is given by $\mathrm{ind}(T_s) = q^{c^\sigma(s)}$ for $s \in S^\sigma$. Note that, if λ is the empty subset, then y_λ^σ is the identity in H_0^σ and so $\sqrt{y_\varnothing M_0^\sigma} = M_0^\sigma$.

On the other hand, if $\lambda = S$, then we have

$$\sqrt{y_S^\sigma M_0^\sigma} = \{m \in M_0^\sigma \mid T_w(m) = (-1)^{l(w)} m \text{ for all } w \in W^\sigma\}. \qquad \text{(c)}$$

This is an immediate consequence of the following result.

Lemma 6.8. *With the above notation, we have*

$$T_w y_\lambda^\sigma = (-1)^{l(w)} y_\lambda^\sigma \quad \textit{for all } w \in W_\lambda^\sigma.$$

Moreover, the left ideal $H_0^\sigma y_\lambda^\sigma \subseteq H_0^\sigma$ is $\mathcal{O}$-pure in H_0^σ.

Proof. The expression for $T_w y_\lambda^\sigma$ is easily verified by a direct computation, using the multiplication rules in (5.24). (Note that it is enough to do this for $w \in \lambda$.) For the proof of the second statement, we follow the argument in [12, (6.20)].

Let $H_{0,\lambda}^\sigma \subseteq H_0^\sigma$ be the $\mathcal{O}$-submodule generated by T_w for $w \in W_\lambda^\sigma$. Then $H_{0,\lambda}^\sigma$ is a subalgebra of H_0^σ and, by definition, we have $y_\lambda^\sigma \in H_{0,\lambda}^\sigma$. Now H_0^σ is in fact free as an $H_{0,\lambda}^\sigma$-module, with a basis given by the images of T_d where d runs over the distinguished right coset representatives of W_λ^σ in W_λ (see [9, §64C], [15, (2,4)]). This implies that it is enough to show that the left ideal $H_{0,\lambda}^\sigma y_\lambda^\sigma$ is $\mathcal{O}$-pure in $H_{0,\lambda}^\sigma$. Now the above expression for $T_w y_\lambda^\sigma$ shows that $H_{0,\lambda}^\sigma y_\lambda^\sigma$ is simply the $\mathcal{O}$-span of y_λ^σ. Let $h = \sum_w a_w T_w \in H_{0,\lambda}^\sigma$ (where $a_w \in \mathcal{O}$) and $0 \neq a \in \mathcal{O}$ be such that $ah = by_\lambda^\sigma$ for some $b \in \mathcal{O}$. A comparison of coefficients shows that we must have $aa_1 = b$. It follows that $h = a_1 y_\lambda^\sigma \in \mathcal{O} y_\lambda^\sigma$. So $H_{0,\lambda}^\sigma y_\lambda^\sigma \subseteq H_{0,\lambda}^\sigma$ is pure. □

Definition 6.9. Assume that ℓ is linear for $\sigma \in \mathcal{S}$. Let Λ^σ be a collection of subsets of S^σ such that $\emptyset \in \Lambda^\sigma$ and each of the modules $\sqrt{y_\lambda^\sigma M_0^\sigma}$ is the ℓ-regular quotient of a projective module. Then, with the above notation, the algebra

$$S_0^\sigma(\Lambda^\sigma) := \operatorname{End}_{H_0^\sigma}\Big(\dot{+}_{\lambda \in \Lambda^\sigma} H_0^\sigma y_\lambda^\sigma\Big)$$

is called the *generalized q-Schur algebra* of type Λ^σ. By Lemma 6.6, H_0^σ is symmetric and, by Lemma 5.11, each $\sqrt{y_\lambda^\sigma M_0^\sigma}$ satisfies Dipper's hypothesis (D). Thus, taking also into account Lemma 6.8, we can apply Corollary 4.13. So we have

$$S_0^\sigma(\Lambda^\sigma) \cong_{\mathcal{O}} \operatorname{End}_{\mathcal{O}G}\Big(\dot{+}_{\lambda \in \Lambda^\sigma} \sqrt{y_\lambda^\sigma M_0^\sigma}\Big)$$

and the decomposition matrix of $S_0^\sigma(\Lambda^\sigma)$ is a submatrix of that of G.

The above definition was first given by Dipper and James [16] for $G = \mathrm{GL}_n(q)$. There are alternative ways of generalizing that original definition; see Du–Scott [22] and Dipper–James–Mathas [19], [20] and the references there. For applications to finite groups of Lie type, the above definition seems to be the most suitable one.

Remark 6.10. By Lemma 6.6(a), all the conditions in Definition 6.9 are satisfied for $\Lambda^\sigma = \{\emptyset\}$. In this case, we have $S_0^\sigma(\{\emptyset\}) = H_0^\sigma$. However, since H_0^σ is a symmetric algebra, its decomposition matrix can only be a square matrix of determinant ± 1 if

$\overline{H}_0^\sigma$ is semisimple. (This follows from the semisimplicity criterion [39, 7.4.7] and the Geck–Rouquier result [39, 7.5.3].)

Remark 6.11. Assume now that ℓ is linear for G and let D be the ℓ-modular decomposition matrix of G. Let us state precisely which part of D is controlled by generalized q-Schur algebras as above. Consider the basic set $\mathfrak{B}$ (see Lemma 6.2) and let $\dot{D}$ be the square submatrix of D with rows labelled by the elements of $\mathfrak{B}$. For each $\sigma \in \mathcal{S}$, let D^σ be the decomposition matrix of $S_0^\sigma(\Lambda^\sigma)$ for a suitable choice of Λ^σ. Let us enumerate the elements of $\mathcal{S}$ by $\sigma_1, \sigma_2, \ldots, \sigma_m$ (for some $m \geq 1$). Now note that, by Lemma 6.2, Lemma 6.6 and Lemma 6.8, all the requirements in Step 1 and Step 2 of the strategy in (4.C) are satisfied. So, as in Lemma 4.16, we see that $\dot{D}$ has the following shape:

$$\dot{D} = \left[\begin{array}{cccc|c} D^{\sigma_1} & 0 & \ldots & 0 & \\ 0 & D^{\sigma_2} & & \vdots & \\ \vdots & & \ddots & 0 & * \\ 0 & \ldots & 0 & D^{\sigma_m} & \end{array}\right].$$

The remaining problem is to find suitable collections Λ^σ which satisfy the conditions in Definition 6.9 and condition ($\square$) in Lemma 4.16. That problem is solved by Dipper–James [16] and Gruber–Hiss [43] for general linear and classical groups. We will say more about this point in (6.B) below.

Example 6.12. The case of "strongly" linear primes. Assume that ℓ is a good prime which divides $q-1$ but does not divide the order of the Weyl group $\boldsymbol{W}$ nor the order of the automorphism of $\boldsymbol{W}$ induced by F. (This case is studied in detail by Hiss [45, Kapitel 6] by a slightly different approach; see also Puig [62].) The conditions on ℓ imply the following two facts.

(a) Every cuspidal simple KL_J-module (for any $J \subseteq S$) belongs to a block with a central defect group, and so ℓ is linear.

(b) For each $\sigma \in \mathcal{S}$, the decomposition matrix of H_0^σ is the identity matrix.

Thus we can just take $\Lambda^\sigma = \{\emptyset\}$ ($\sigma \in \mathcal{S}$) in Remark 6.11 and then the whole matrix $\dot{D}$ is the identity matrix. So the decomposition matrix of G is known.

The statements in (a) and (b) are proved as follows. Let $X \in \operatorname{Irr}_K(L_J)$ be cuspidal and X_0 be an $\mathcal{O}L_J$-lattice such that $X \cong_{KL_J} KX_0$. By [37, (3.2)] and [29, (3.1)], the conditions on ℓ imply that the ℓ-part of X equals the ℓ-part of the order of $L_J/Z(L_J)$. Hence X belongs to a block with a central defect group. But then $\overline{X}_0$ is simple and, if P_0 denotes the projective cover of $\overline{X}_0$, the ordinary character of KP_0 is the sum of irreducible characters of L_J which all have the same restriction to the ℓ-regular elements of L_J (see [23, (V.4.6)]). So KP_0 can only have one constituent which belongs to an ordinary basic set for L_J. This implies that condition $(\mathrm{L})_\sigma$ in Definition 6.3 holds. To prove (b), first note that, by Lemma 6.6, H_0^σ is reduction stable. Now, using the condition $\ell \mid q-1$ and the multiplication rules in (5.24), we see that $\overline{H_0^\sigma}$ is isomorphic to the group algebra of W^σ over k. Hence, since ℓ does not divide the order W^σ, the algebra $\overline{H_0^\sigma}$ is semisimple. So, applying Tits' Deformation Theorem in [9, §68A], we obtain (b).

6.B. Unipotent blocks

We will now explain the main results of Dipper–James (for general linear groups) and Gruber–Hiss (for classical groups) in an important special case where we restrict ourselves to the study of the unipotent blocks only.

Recall that the simple modules in the Lusztig series $\mathrm{Irr}_K(G, 1)$ are called *unipotent*. Similarly, the blocks b of G such that $bb_1 \neq 0$ (with b_1 as in Theorem 5.6) are called *unipotent blocks* and the simple kG-modules belonging to b_1 will also be called *unipotent*. Since we are assuming that ℓ is a good prime for G and that $\mathbf{G}$ has a connected center, the simple modules in $\mathrm{Irr}_K(G, 1)$ form an ordinary basic set with respect to b_1. We denote by D_{uni} the decomposition matrix of the simple modules belonging to b_1 and by $\dot{D}_{\mathrm{uni}}$ the square submatrix of D_{uni} with rows labelled by the simple modules in $\mathrm{Irr}_K(G, 1)$.

We begin by discussing some special features of unipotent simple modules which lead to some technical simplifications.

6.13. For the study of unipotent blocks, we only have to consider the subset

$$\mathcal{S}_{\mathrm{uni}} := \{\sigma = (J, s, X) \in \mathcal{S} \mid s = 1\}.$$

(This follows from (5.2)(b).) Assume that ℓ is linear for all $\sigma \in \mathcal{S}_{\mathrm{uni}}$.

(a) Let $\sigma = (J, 1, X) \in \mathcal{S}_{\mathrm{uni}}$. The first step is to describe the ramification group $W^\sigma = W_{(J,X)}$; see (3.24). By Lusztig [59, Chapter 8], X is invariant under all of $N_G(L_J) \cap N$ and J is self-opposed in the sense of [7, (10.10)]. This means that W^σ is a Coxeter group with a set of generators $S^\sigma = \{\bar{s} \mid s \in S \setminus J\}$ where we set

$$\bar{s} = (w_0)_{J \cup \{s\}}(w_0)_J \quad \text{for any } s \in S \setminus J.$$

Here, for any subset $I \subseteq S$, we denote the longest element in the parabolic subgroup $W_I \subseteq W$ by $(w_0)_I$.

(b) Next let $\sigma = (J, 1, X) \in \mathcal{S}_{\mathrm{uni}}$ and let X_0 be an $\mathcal{O}L_J$-lattice such that $X \cong_{KL_J} KX_0$. We consider the endomorphism algebra H_0^σ as in (6.7). Then the type of W^σ and the function $c^\sigma : S^\sigma \to \mathbb{Z}$ are explicitly given by the table in [7, p. 464]. Thus we have a complete description of H_0^σ.

Before we can treat the general case, we need some preparations concerning the general linear groups.

6.14. We consider the unipotent blocks of $G = \mathrm{GL}_n(q)$. What can we already say about $\dot{D}_{\mathrm{uni}}$ at this stage? First recall from Example 6.5 that every ℓ is linear for G.

By (5.3)(b), $\mathrm{Irr}_K(G, 1)$ consists of the simple modules in the Harish-Chandra series $\mathrm{Irr}_K(G, (\emptyset, K_H))$. So $\mathcal{S}_{\mathrm{uni}}$ is the singleton set consisting of $\sigma_0 := (\emptyset, 1, K_H)$. By definition, $W_{(\emptyset, K_H)}$ is the Weyl group $W = \mathfrak{S}_n$ of G. The construction in [10, (15.8)] shows in fact that we have a natural bijection $\mathrm{Irr}_K(\mathfrak{S}_n) \leftrightarrow \mathrm{Irr}_K(G, 1)$. Now, as is well-known, $\mathrm{Irr}_K(\mathfrak{S}_n)$ has a natural parametrization by the partitions of n (see, for example, [39, 5.4.5]). If $\lambda \vdash n$, we denote by M_λ the corresponding $K\mathfrak{S}_n$-module and write $Y_\lambda := Y_{M_\lambda}$. For example, we have $Y_{(n)} = K_G$ and $Y_{(1^n)} = \mathrm{St}_{G,K}$. Thus the rows of $\dot{D}_{\mathrm{uni}}$ have a natural labelling by the partitions of n.

We will now place ourselves in the setting of (6.A). To simplify the notation, we omit the superscript σ_0 if there is no danger of confusion. As before, we set

$$M_0 = \mathrm{Ind}_B^G(\mathcal{O}_B) \quad \text{and} \quad H_0 := \mathrm{End}_{\mathcal{O}}(M_0). \tag{a}$$

By [7, p. 464], the function c is identically 1. Hence H_0 is the classical Hecke algebra of type A_{n-1}, as studied by Dipper and James [15]. By Remark 6.10, the decomposition matrix of H_0 is a submatrix of $\dot{D}_{\mathrm{uni}}$. In fact we obtain complete columns of $\dot{D}_{\mathrm{uni}}$ but we will only obtain all columns when $\overline{H}_0$ is semisimple. Thus, in general, we will have to consider a (generalized) q-Schur algebra corresponding to a suitable set Λ. For this purpose, we proceed with the following constructions.

With every partition $\mu \vdash n$, we can also associate a Levi subgroup of G as follows. Recall from Example 3.1 that the Coxeter generators of $\mathfrak{S}_n$ are the basic transpositions $s_i = (i, i+1)$ for $1 \leq i \leq n-1$. Then, if μ has parts $\mu_1 \geq \cdots \geq \mu_r$, we let $J_\mu \subseteq S = \{s_1, \ldots, s_{n-1}\}$ be the subset obtained by deleting the generators indexed by the numbers $\mu_1 + \cdots + \mu_i$ for $1 \leq i < r$. Consider the corresponding Levi subgroup L_{J_μ} and let Γ_μ denote its Gelfand–Graev lattice. We set

$$Q_\mu := R_{J_{\mu'}}^S(\Gamma_{\mu,1}) \quad \text{with } \Gamma_{\mu,1} \text{ as in (5.13)(a).} \tag{b}$$

Then Q_μ is a projective $\mathcal{O}G$-lattice since Γ_μ is projective. Moreover, all simple KG-modules occurring in KQ_μ belong to b_1, by the compatibility of Lusztig series and Harish-Chandra series and the definition of b_1 in Theorem 5.6.

Lemma 6.15. *For $\lambda, \mu \vdash n$, we define*

$$\kappa_{\lambda,\mu} := [KQ_{\mu'} : Y_\lambda] = [R_{J_{\mu'}}^S(K\Gamma_{\mu'}) : Y_\lambda]$$

where μ' denotes the dual partition. Then the matrix $(\kappa_{\lambda,\mu})$ is lower triangular with 1 *along the diagonal if the partitions of n are ordered downward lexicographically.* (*I.e., in that ordering, K_G comes first and* $\mathrm{St}_{G,K}$ *is the last.*)

Proof. By the compatibility of Lusztig series and Harish-Chandra series, we have

$$\kappa_{\lambda,\mu} = [R_{J_{\mu'}}^S(\mathrm{St}_{L_{\mu'},K}) : Y_\lambda] \quad \text{for all } \lambda, \mu \vdash n.$$

Now the Steinberg module can be obtained from the trivial module by applying the Alvis–Curtis–Kawanaka duality operation (see [9, §71]). Since that operation commutes with Harish-Chandra induction and is a self-adjoint involution, we conclude that

$$\kappa_{\lambda,\mu} = [R^S_{J_{\mu'}}(K_{L_{\mu'}}) : D_G(Y_\lambda)] \quad \text{for all } \lambda, \mu \vdash n.$$

Furthermore, by [9, (71.14)], we have $D_G(Y_\lambda) \cong_{KG} Y_{\varepsilon \otimes M_\lambda}$ where ε is the sign representation of $\mathfrak{S}_n$. By [39, 5.4.9], we have $\varepsilon \otimes M_\lambda \cong_{K\mathfrak{S}_n} M_{\lambda'}$ and so

$$\kappa_{\lambda,\mu} = [R^S_{J_{\mu'}}(K_{L_{\mu'}}) : Y_{\lambda'}] = [\mathrm{Ind}_{\mathfrak{S}_{\mu'}}^{\mathfrak{S}_n}(K_{\mathfrak{S}_{\mu'}}) : M_{\lambda'}] \quad \text{for all } \lambda, \mu \vdash n,$$

where the second equality holds by [9, (68.24)]. Now, by [39, 5.4.7], the multiplicities on the right hand side form a matrix which is lower triangular with 1 along the diagonal for the given ordering of the partitions of n. □

Part (b) of the following result is already contained in James [52]; for part (a), see also Hiss [47, (6.3)].

Corollary 6.16. *Assume that G is a group of type* GL, *as in* (5.4). *Then the following hold.*

(a) *If $X \in \mathrm{Irr}_k(G, 1)$ is cuspidal, then $X \cong_{kG} \mathrm{hd}(\mathrm{St}_G)$.*

(b) *We have $\mathrm{St}_G \cong_{\mathcal{O}G} \sqrt{y_S M_0}$, with y_S as in* (6.7)(b).

Note that, using (6.7)(c), we now obtain the following explicit description of St_G as a submodule of M_0: we have

$$\mathrm{St}_G \cong_{\mathcal{O}G} \{m \in M_0 \mid T_w(m) = (-1)^{l(w)} m \ \text{ for all } \ w \in \mathfrak{S}_n\}.$$

Proof. (a) First consider the case where $G = \mathrm{GL}_n(q)$. Since $\mathrm{Irr}_K(G, 1)$ is an ordinary basic set with respect to b_1, the simple kG-modules belonging to b_1 are also parametrized by the partitions of n. Let $\{P_\nu \mid \nu \vdash n\}$ be a set of projective indecomposable modules such that $\{\mathrm{hd}(P_\nu) \mid \nu \vdash n\}$ are the simple unipotent kG-modules (up to isomorphism). Then, by definition, we have

$$\dot{D}_{\mathrm{uni}} = (d_{\lambda,\nu}) \quad \text{where } d_{\lambda,\nu} = [KP_\nu : Y_\lambda].$$

Now, for $\mu \vdash n$, all simple KG-modules occurring in $Q_{\mu'}$ belong to b_1. So we can write $Q_{\mu'} = \sum_{\nu \vdash n} t_{\nu,\mu} P_\nu$ where $t_{\nu,\mu}$ are non-negative integers. Then we have

$$\kappa_{\lambda,\mu} = [KQ_{\mu'} : Y_\lambda] = \sum_{\nu \vdash n} t_{\mu,\nu}[KP_\nu : Y_\lambda] = \sum_{\nu \vdash n} d_{\lambda,\nu} t_{\nu,\mu}.$$

Thus the matrix $(\kappa_{\lambda,\mu})$ can be expressed as a matrix product where one factor is $\dot{D}_{\mathrm{uni}}$ and the second factor is a square matrix of non-negative integers. Since, by Lemma 6.15, the product has a lower triangular shape with 1 along the diagonal, the same must hold for the two factors, up to possible permutations of rows and columns. It follows that the labelling of the P_μ can be chosen such that

each P_μ is a direct summand of $Q_{\mu'}$ (with multiplicity 1).

By construction, each $Q_{\mu'}$ is obtained by Harish-Chandra induction from a Levi subgroup of G. Hence Remark 3.11 implies that $\mathrm{hd}(P_\mu)$ is non-cuspidal if that Levi subgroup is properly contained in G. Thus, if $\mathrm{hd}(P_\mu)$ is cuspidal, then we must have $\mu' = (n)$ and so $\mathrm{hd}(P_\mu)$ is a quotient of St_G.

Thus we have settled the special case where $G = \mathrm{GL}_n(q)$. Using (5.4)(b), we can easily reduce the general case to that special case. Note that we are only dealing with unipotent modules and these modules always have the center in their kernel (see [7, p. 380]).

(b) First we show that St_G is isomorphic to an $\mathcal{O}$-pure sublattice of M_0. This will follow if we can verify one of the equivalent conditions in Theorem 5.16. We will check that condition (c) holds. Since every Levi subgroup L_J of G is again a group of type GL, it is enough to consider the case where $J = S$. Assume now that $\mathrm{soc}(\overline{\mathrm{St}}_G)$ has a cuspidal simple constituent. We must show that this implies $G = H$. Now that cuspidal simple constituent is unipotent and so, by (a), it also occurs in $\mathrm{hd}(\overline{\mathrm{St}}_G)$. Since the simple kG-module in the head of $\overline{\mathrm{St}}_G$ has multiplicity 1 in $\overline{\mathrm{St}}_G$ and since $\overline{\mathrm{St}}_G$ is indecomposable, we conclude that $\overline{\mathrm{St}}_G$ is simple and cuspidal. But this is impossible if $G \neq H$. Indeed, since $\mathrm{St}_{G,K}$ lies in the Harish-Chandra series $\mathrm{Irr}_K(G, (\emptyset, K_H))$, we have $\mathrm{Fix}_U(\mathrm{St}_{G,K}) \neq \{0\}$ and so also $\mathrm{Fix}_U(\overline{\mathrm{St}}_G) \neq \{0\}$ since the order of U is not divisible by ℓ.

Thus St_G is isomorphic to a pure sublattice of M_0. Since $\mathrm{St}_{G,K}$ has multiplicity 1 in KM_0 (see [7, (6.2.3)]), that pure sublattice is uniquely determined. So it is enough to show that $\mathrm{St}_{G,K} \cong_{KG} y_S KM_0$. To see this, note that the formula in Lemma 6.8 implies that y_S is a scalar multiple of an idempotent. So, by (2.1)(b), we have $F_{KM_0}(y_S KM_0) \cong_K Hy_S$ as H-modules. In fact, Hy_S has dimension 1 and so $y_S KM_0$ must be simple. Moreover, the character of Hy_S is given by $T_w \mapsto (-1)^{l(w)}$. Hence the desired assertion follows from [9, (67.10)]. □

Remark 6.17. The above result raises the question of determining precisely when $\mathrm{hd}(\mathrm{St}_G)$ is a cuspidal simple kG-module. A general criterion is given in [37, §4]. In the case where $G = \mathrm{GL}_n(q)$, this yields that $\mathrm{hd}(\mathrm{St}_G)$ is cuspidal if and only if we have $n = d\ell^i$ for some $i \geq 0$, where d denotes the multiplicative order of q modulo ℓ (see [37, (7.6)]). In the case where $G = \mathrm{GU}_n(q)$, the exact conditions are given in [37, (8.6)]. For general G, the answer seems to be open.

We now return to the general case. The following result provides the basis for choosing a suitable set Λ^σ as required in Definition 6.9. We only state this result for unipotent modules but note that it is true in general for general linear groups (see James [53]) and finite classical groups (see Gruber–Hiss [43]). We also remark that its proof proceeds by reduction to the case of general linear groups, where the required reduction arguments are already contained in [36, §4].

Theorem 6.18. *Assume that ℓ is linear for $\sigma = (J, 1, X) \in \mathcal{S}_{\mathrm{uni}}$. Let $\lambda \subseteq S^\sigma$ in the notation of* (6.13) *and assume that the following conditions hold.*

(a) *Every element of $J(\lambda)$ commutes with every element of J, where we set $J(\lambda) := \{s \in S \setminus J \mid \bar{s} \in \lambda\}$.*

(b) *The Levi subgroup $L_{J(\lambda)}$ is a group of type* GL, *as in* (5.4).

Then $\sqrt{y_\lambda^\sigma M_0^\sigma}$ is the ℓ-regular quotient of a projective module.

Proof. We reduce the problem of showing that $\sqrt{y_\lambda^\sigma M_0^\sigma}$ is the ℓ-regular quotient of a projective module in several steps to the case where $G = \mathrm{GL}_n(q)$.

Step 1. The following arguments provide a reduction to the case where $S = J \cup J(\lambda)$. We set

$$M_{0,\lambda}^\sigma := R_J^{J\cup J(\lambda)}(X_0) \quad \text{and} \quad H_{0,\lambda}^\sigma := \mathrm{End}_{\mathcal{O}L_{J\cup J(\lambda)}}(M_{0,\lambda}^\sigma). \tag{1}$$

Then Corollary 5.26 also applies to $H_{0,\lambda}^\sigma$, and it follows that $H_{0,\lambda}^\sigma$ has a basis $\{T_w \mid w \in W_\lambda^\sigma\}$ satisfying the multiplication rules in (5.24). Now, as in (3.22), we have a natural map

$$\iota_{J\cup J(\lambda)}\colon H_{0,\lambda}^\sigma \to H_0^\sigma, \quad \varphi \mapsto R_{J\cup J(\lambda)}^S(\varphi).$$

In fact, it is shown in [43, (2.5)] that this map sends the standard basis element T_w ($w \in W_\lambda^\sigma$) to the corresponding standard basis element, also denoted by T_w, in H_0^σ. Thus $H_{0,\lambda}^\sigma$ can be identified with a standard parabolic subalgebra of H_0^σ. The defining formula (6.7)(b) shows that, under this identification, we can regard y_λ^σ as an element of $H_{0,\lambda}^\sigma$. With a slight abuse of notation, this means that

$$R_{J\cup J(\lambda)}^S(y_\lambda^\sigma M_{0,\lambda}^\sigma) = y_\lambda^\sigma R_{J\cup J(\lambda)}^S(M_{0,\lambda}^\sigma). \tag{2}$$

Combining the above statements, we now obtain

$$\begin{aligned} \sqrt{y_\lambda^\sigma M_\lambda^\sigma} &\cong_{\mathcal{O}G} \sqrt{y_\lambda^\sigma R_{J\cup J(\lambda)}^S(M_{0,\lambda}^\sigma)} && \text{by (1) and transitivity,} \\ &\cong_{\mathcal{O}G} \sqrt{R_{J\cup J(\lambda)}^S(y_\lambda^\sigma M_{0,\lambda}^\sigma)} && \text{using (2)} \\ &\cong_{\mathcal{O}G} R_{J\cup J(\lambda)}^S\Big(\sqrt{y_\lambda^\sigma M_{0,\lambda}^\sigma}\Big), \end{aligned}$$

where the last isomorphism follows from the fact that Harish-Chandra induction is exact and preserves $\mathcal{O}$-torsion modules. Thus, using (5.10)(c), we are reduced to showing that $\sqrt{y_\lambda^\sigma M_{0,\lambda}^\sigma}$ is the ℓ-regular quotient of a projective module.

Step 2. Assume now that we have $S = J \cup J(\lambda)$. The following arguments provide a reduction to the case where $J(\lambda) = S$.

By Lemma 5.12, we may also assume that $\boldsymbol{G}$ is semisimple of adjoint type. Then it known that $\boldsymbol{G}$ is in fact the direct product of simple groups of adjoint type. Consequently we also have a decomposition $G = G_J \times G'$, where G_J has Weyl group

W_J and G' has Weyl group $W_{J(\lambda)}$. More precisely, let Φ be the root system of G and $X_\alpha \subseteq G$ be the root subgroup corresponding to $\alpha \in \Phi$; see [7, §2.5]. Then, setting $H_J := H \cap G_J$ and $H' := H \cap G'$, we have

$$G_J = \langle H_J, X_\alpha \mid \alpha \in \Phi_J \rangle \quad \text{and} \quad G' = \langle H', X_\alpha \mid \alpha \in \Phi_{J(\lambda)} \rangle.$$

Moreover, the description of parabolic subgroups in [7, §2.6] shows that

$$L_J = G_J \times H' \quad \text{and} \quad P_J = G_J \times B'$$

where $B' \subseteq G'$ denotes the standard Borel subgroup of G'.

Since X_0 is unipotent, $H' \subseteq Z(L_J)$ acts trivially on X_0. So we may regard X_0 as a module for G_J. It follows that

$$M_0^\sigma = R_J^S(X_0) = \operatorname{Ind}_{P_J}^G(\tilde{X}_0) = \operatorname{Ind}_{G_J \times B'}^{G_J \times G'}(X_0 \boxtimes \mathcal{O}_{B'}) \cong_{\mathcal{O}G} X_0 \boxtimes \operatorname{Ind}_{B'}^{G'}(\mathcal{O}_{B'}).$$

Setting $M_0' := \operatorname{Ind}_{B'}^{G'}(\mathcal{O}_{B'})$, we now also see that

$$H_0^\sigma \cong_{\mathcal{O}} \operatorname{End}_{\mathcal{O}G'}(M_0'),$$

and y_λ^σ may be naturally regarded as an endomorphism of M_0'. Hence, in order to verify that $\sqrt{y_\lambda^\sigma M_0^\sigma}$ is the ℓ-regular quotient of a projective module for G, it is enough to verify that $\sqrt{y_\lambda^\sigma M_0'}$ is the ℓ-regular quotient of a projective module for G'. Note that X_0 is the ℓ-regular quotient of a projective module for L_J by the condition that ℓ is linear.

Step 3. Finally, assume that $S = J(\lambda)$. Then G itself is a group of type GL and the desired conclusion follows from Corollary 6.16(b); note that, by definition, the Steinberg lattice is the ℓ-regular quotient of a projective module. □

6.19. The Dipper–James q-Schur algebra. Let $G = \mathrm{GL}_n(q)$. We place ourselves in the setting of (6.14). Recall that $W = \mathfrak{S}_n$, with Coxeter generators S given by the basic transposition $(i, i+1)$ for $1 \le i \le n-1$. For any subset $\lambda \subseteq S$, the conditions in Theorem 6.18 are trivially satisfied. So, in Definition 6.9, we can take for Λ the set of all subsets of the generators of W. The resulting algebra $S_0(\Lambda)$ is the original q-Schur algebra introduced by Dipper and James [16].

The decomposition matrix of $S_0(\Lambda)$ is known to be a square uni-triangular matrix (for a suitable ordering of the rows and columns). This has been shown by Dipper–James [16, (4.19)]. In the present context, this follows from the criterion in (4.17) and Lemma 6.15; using the reduction argument in Step 1 of the proof of Theorem 6.18, we see that

$$\sqrt{y_\lambda M_0} \cong_{\mathcal{O}G} R_{J(\lambda)}^S(\mathrm{St}_{L_{J(\lambda)}}) \quad \text{for all subsets } \lambda \subseteq S,$$

and these modules are precisely the ℓ-regular quotients of the projective modules Q_μ considered in Lemma 6.15. Thus, the discussion in Remark 6.11 shows that $\dot{D}_{\mathrm{uni}}$ is completely determined by the decomposition matrix of $S_0(\Lambda)$.

For example, let $G = \mathrm{GL}_2(q)$ and assume that $\ell \mid q+1$. There are two unipotent simple KG-modules, the trivial module K_G and the Steinberg module $\mathrm{St}_{G,K}$. Then $\dot{D}_{\mathrm{uni}}$ is given by

K_G	1	.
$\mathrm{St}_{G,K}$	1	1

The first column is the decomposition matrix of H_0 and the second column is given by $\Gamma_{G,1}$. The simple kG-module labelling the second column is cuspidal.

More examples of decomposition matrices for $\mathrm{GL}_n(q)$ have been computed by James [53]. For further reading on modular representations of $\mathrm{GL}_n(q)$, including a description of cuspidal modules and Harish-Chandra series, see James [52] and Dipper's survey in [12].

6.20. The finite classical groups. Let G be a finite classical group and assume that ℓ is linear. For $\sigma \in \mathcal{S}_{\mathrm{uni}}$, let Λ^σ be the collection of all subsets of S^σ such that conditions (a) and (b) in Theorem 6.18 hold. Note that (b) holds if and only if W_λ itself is a direct product of Weyl groups of type A. Then we have a corresponding (generalized) q-Schur algebra $S_0^\sigma(\Lambda^\sigma)$ as in Definition 6.9.

These algebras were first defined in [36, §4] and then studied in general by Gruber and Hiss [43]. The fact that the decomposition matrix of $S_0^\sigma(\Lambda^\sigma)$ is a square matrix is non-trivial: its proof depends on the results of Dipper and James [18] on modular representations of Hecke algebras of type B, and extensions of these results to type D; see [43, §7]. For unitary, symplectic and odd-dimensional orthogonal groups, an independent argument (for the case of unipotent blocks) is given in [36, (4.7)]. This part of the construction can be somewhat simplified using the more recent results on generalized q-Schur algebras in [19] and [22].

Examples of decomposition matrices for finite unitary groups in the linear prime case have been computed in [37, (9.7)].

6.C. Concluding remarks

In this final paragraph, we briefly discuss James' conjecture on the "generic" behaviour of decomposition numbers in the linear prime case and give an example in the non-linear prime case.

6.21. Assume we are in a setting as in (6.A). Fix $\sigma \in \mathcal{S}$ and let $S_0^\sigma(\Lambda^\sigma)$ be a (generalized) q-Schur algebra as in Definition 6.9. Recall that $S_0^\sigma(\Lambda^\sigma)$ is described purely in terms of the Hecke algebra H_0^σ (defined as in (6.7)).

As in Definition 5.25, we now consider the generic Iwahori–Hecke algebra $\boldsymbol{H}^\sigma = \boldsymbol{H}_A(W^\sigma, S^\sigma, c^\sigma)$. Then, by Corollary 5.26, H_0^σ is isomorphic to the specialized algebra $\mathcal{O} \otimes_A \boldsymbol{H}^\sigma$ via the map $u \mapsto q$. For $\lambda \subseteq S^\sigma$, we set

$$\mathbf{y}_\lambda^\sigma := \sum_{w \in W_\lambda^\sigma} (-1)^{l(w)} \operatorname{ind}(\boldsymbol{T}_w)^{-1} \boldsymbol{T}_w \in \boldsymbol{H}^\sigma,$$

where the homomorphism ind: $\boldsymbol{H}^\sigma \to A$ is given by $\operatorname{ind}(\boldsymbol{T}_s) = u^{c^\sigma(s)}$ for $s \in S$. Then the *generic generalized q-Schur algebra* associated with $(W^\sigma, S^\sigma, c^\sigma)$ and of type Λ^σ is defined as

$$\boldsymbol{S}^\sigma(\Lambda^\sigma) := \operatorname{End}_{\boldsymbol{H}^\sigma}\Big(\dot{+}_{\lambda\in\Lambda^\sigma} \boldsymbol{H}^\sigma \mathbf{y}_\lambda^\sigma\Big).$$

By the same argument as in [17, (1.3)], we see that $\boldsymbol{S}^\sigma(\Lambda^\sigma)$ is free over A of finite rank. Moreover, we have $S_0^\sigma(\Lambda^\sigma) \cong_{\mathcal{O}} \mathcal{O} \otimes_A \boldsymbol{S}^\sigma(\Lambda^\sigma)$ via the map $A \to \mathcal{O}$, $u \mapsto q$.

The fact that $S_0^\sigma(\Lambda^\sigma)$ can be obtained via specialization from a generic algebra implies that the decomposition matrix of $S_0^\sigma(\Lambda^\sigma)$ admits a factorization into a "generic" part and an "adjustment" part. This is made more precise as follows.

6.22. Let $\boldsymbol{K}$ be a sufficiently large extension of the field of fractions of A, such that $\boldsymbol{K}\boldsymbol{S}^\sigma(\Lambda^\sigma)$ is a split algebra. We consider an extension of $A \to K$, $u \mapsto q$, to the ring of algebraic integers in $\boldsymbol{K}$. Since $KS_0^\sigma(\Lambda^\sigma)$ is split semisimple, Tits' Deformation Theorem [9, §68A] implies that the specialization $u \mapsto q$ (extended as above) induces a bijection

$$\operatorname{Irr}_{\boldsymbol{K}}(\boldsymbol{S}^\sigma(\Lambda^\sigma)) \longleftrightarrow \operatorname{Irr}_K(S_0^\sigma(\Lambda^\sigma)). \qquad \text{(a)}$$

Thus, as far as the simple modules over K are concerned, we can work with the generic algebra. Now let $d \geq 1$ be the multiplicative order of q modulo ℓ. Let $\zeta_d \in \mathcal{O}$ be a primitive d-th root of unity such that ζ_d and q have the same image in k under the canonical map $\mathcal{O} \to k$. Then the specialization $u \mapsto \zeta_d$ gives rise to a decomposition matrix of the generic algebra $\boldsymbol{S}^\sigma(\Lambda^\sigma)$; see [33, §2]. Denote that decomposition matrix by $\mathbf{D}_{\zeta_d}$. Using (a), we can identify the simple modules labelling the rows of $\mathbf{D}_{\zeta_d}$ with the simple modules labelling the rows of the decomposition matrix D^σ of $S_0^\sigma(\Lambda^\sigma)$. Then a completely general argument (due to Geck–Rouquier; see [33, §2]) implies that we have

$$D^\sigma = \mathbf{D}_{\zeta_d} D', \qquad \text{(b)}$$

where D' is a certain "adjustment matrix" whose entries are non-negative integers. The term "adjustment matrix" is due to James [53], where examples are computed for type A. The above factorization provides the framework for formulating the following conjecture.

Conjecture 6.23 (See James [53, §4] for type A). *Assume that ℓ is coprime to the order of W^σ and to $c^\sigma(s)$ for all $s \in S^\sigma$. Then the adjustment matrix D' in* (6.22)(b) *is the identity matrix. In other words, the decomposition matrix of $S_0^\sigma(\Lambda^\sigma)$ only depends on the multiplicative order of q modulo ℓ, but not on the particular values of q or ℓ.*

See also [34] for an equivalent formulation of James' conjecture in terms of Kazhdan–Lusztig cells.

Actually, James proposes a slightly weaker condition on ℓ. The argument in [33, (2.7)] implies the conjecture holds if ℓ is large enough. So the conjecture is really about specifying the correct lower bound for ℓ. In any case, we can already conclude that if ℓ is large enough and if we are in a situation where the decomposition matrix $\dot{D}$ is completely given by (generalized) q-Schur algebras as in Remark 6.11 (i.e., if there are no columns represented by "$*$"), then $\dot{D}$ only depends on the multiplicative order of q modulo ℓ.

Our last example is meant to serve as an illustration of the problems that can arise when we consider non-linear primes.

Example 6.24. Let $G = \mathrm{GU}_3(q)$, the full unitary group on a 3-dimensional space. There are 3 unipotent simple KG-modules: the trivial module K_G, a cuspidal simple module Y_0, and the Steinberg module St_G. Their degrees are 1, $q(q-1)$ and q^3, respectively. Assume now that $\ell \neq 2, 3$ and ℓ divides $q+1$. Let ℓ^d be the maximal power of ℓ dividing $q+1$. Then, by [27], the submatrix of the decomposition matrix of G with rows labelled by the above 3 modules is given by

$$\begin{array}{r|ccc} K_G & 1 & . & . \\ Y_0 & . & 1 & . \\ \mathrm{St}_{G,K} & 1 & \alpha & 1 \end{array} \qquad \text{where } 2 \leq \alpha \leq \frac{1}{3}(\ell^d + 1).$$

The first column is determined by the decomposition matrix of the Hecke algebra of the permutation module on the cosets of B and the third column is given by $\Gamma_{G,1}$. The two simple kG-modules labelling the second and the third column are cuspidal. Hence, in whatever way we define generalized q-Schur algebras, we can never expect to cover the second column of the above matrix! Note also that St_G does not satisfy the conditions in Theorem 5.16; see Remark 5.17.

The problem of determining α is still open. It is conjectured that we always have $\alpha = 2$. The full decomposition matrix in [27, (4.3)] shows that the conjecture is true if and only if the simple kG-module labelling the third column of the above matrix can be lifted to characteristic 0.

A similar problem about decomposition numbers also occurs for $\mathrm{Sp}_4(q)$. This has now been solved by Okuyama and Waki [61]. In their paper they make essential use of the fact that $\mathrm{Sp}_4(q)$ has a BN-pair of rank 2. So it does not seem to be possible to apply their methods to $\mathrm{GU}_3(q)$. However, it is likely that some of the open questions on $G_2(q)$ (see Hiss [44]) can be solved in this way.

Acknowledgements. This paper is an expanded version of a series of talks given at the "Symposium on Modular Representation Theory of Finite Groups" held at Charlottesville, VA, in May 1998. I wish to thank M. Collins, B. Parshall and L. Scott for the invitation and the University of Charlottesville for its hospitality.

I am indebted to G. Hiss for many helpful discussions. Finally, I wish to thank the referee for a careful reading of this paper.

References

[1] S. Ariki, On the decomposition numbers of the Hecke algebra of $G(m, 1, n)$, J. Math. Kyoto Univ. 36 (1996), 789–808.

[2] R. E. Borcherds, Coxeter groups, Lorentzian lattices, and K3 surfaces, Internat. Math. Res. Notices 19 (1998), 1011–1031.

[3] M. Broué and J. Michel, Blocs et séries de Lusztig dans un groupe réductif fini, J. Reine Angew. Math. 395 (1989), 56–67.

[4] M. Broué and G. Malle, Generalized Harish-Chandra theory, in: Representations of reductive groups (R. W. Carter and M. Geck, eds.), Cambridge University Press, 1988, 85–103.

[5] M. Cabanes, A criterion for complete reducibility and some applications, in: Représentations linéaires de groupes de finis, Luminy, 16–21 mai 1988, Astérisque 181–182 (1990), 93–112.

[6] M. Cabanes, Algèbres de Hecke comme algèbres symétriques et théorème de Dipper, C. R. Acad. Sci. Paris Sér. I Math. 327 (1998), 531–536.

[7] R. W. Carter, Finite groups of Lie type: Conjugacy classes and complex characters, Wiley, New York 1985.

[8] C. W. Curtis and I. Reiner, Representation theory of finite groups and associative algebras, Wiley, New York 1962.

[9] C. W. Curtis and I. Reiner, Methods of representation theory Vol. I and II, Wiley, New York 1981 and 1987.

[10] F. Digne and J. Michel, Representations of finite groups of Lie type, London Math. Soc. Stud. Texts 21, Cambridge University Press, 1991.

[11] R. Dipper, On quotients of Hom-functors and representations of finite general linear groups I, J. Algebra 130 (1990), 235–259.

[12] R. Dipper, On quotients of Hom-functors and representations of finite general linear groups II, J. Algebra 209 (1998), 199–269.

[13] R. Dipper and J. Du, Harish-Chandra vertices, J. Reine Angew. Math. 437 (1993), 101–130.

[14] R. Dipper and P. Fleischmann, Modular Harish-Chandra theory I, Math. Z. 211 (1992), 49–71; II, Arch. Math. 62 (1994), 26–32.

[15] R. Dipper and G. D. James, Representations of Hecke algebras of general linear groups, Proc. London Math. Soc. 52 (1986), 20–52.

[16] R. Dipper and G. D. James, The q-Schur algebra, Proc. London Math. Soc. 59 (1989), 23–50.

[17] R. Dipper and G. D. James, q-Tensor space and q-Weyl modules, Trans. Amer. Math. Soc. 327 (1991), 251–282.

[18] R. Dipper and G. D. James, Representations of Hecke algebras of type B, J. Algebra 146 (1992), 454–481.

[19] R. Dipper, G. D. James and A. Mathas, The (Q, q)-Schur algebra, Proc. London Math. Soc. 77 (1998), 327–361.

[20] R. Dipper, G. D. James and A. Mathas, Cyclotomic q-Schur algebras, Math. Z. 229 (1998), 385–416.

[21] S. Donkin, The q-Schur algebra, London Math. Soc. Lecture Note Ser. 253, Cambridge University Press, 1998.

[22] J. Du and L. Scott, The q-Schur^2 algebra, Trans. Amer. Math. Soc. 352 (2000), 4325–4353.

[23] W. Feit, The representation theory of finite groups, North–Holland Publishing Company, Amsterdam, New York, Oxford 1982.

[24] P. Fong and B. Srinivasan, The blocks of finite general linear and unitary groups, Invent. Math. 69 (1982), 109–153.

[25] P. Fong and B. Srinivasan, The blocks of finite classical groups, J. Reine Angew. Math. 396 (1989), 122–191.

[26] P. Fong and B. Srinivasan, Brauer trees in classical groups, J. Algebra 131 (1990), 179–225.

[27] M. Geck, Irreducible Brauer characters of the 3-dimensional special unitary groups in non-defining characteristic, Comm. Algebra 18 (1990), 563–584.

[28] M. Geck, Verallgemeinerte Gelfand-Graev Charaktere und Zerlegungszahlen endlicher Gruppen vom Lie-Typ, Dissertation, RWTH Aachen, 1990.

[29] M. Geck, On the classification of ℓ-blocks of finite groups of Lie type, J. Algebra 151 (1992), 180–191.

[30] M. Geck, Brauer trees of Hecke algebras, Comm. Algebra 20 (1992), 2937–2973.

[31] M. Geck, A note on Harish–Chandra induction, Manuscripta Math. 80 (1993), 393–401.

[32] M. Geck, Basic sets of Brauer characters of finite groups of Lie type, II, J. London Math. Soc. 47 (1993), 255–268; III, Manuscripta Math. 85 (1994), 195–216.

[33] M. Geck, Representations of Hecke algebras at roots of unity, Séminaire Bourbaki, année 1997-98, Astérisque No. 252 (1998), Exp. 836, 33–55.

[34] M. Geck, Kazhdan–Lusztig cells, q-Schur algebras, and James' conjecture, J. London Math. Soc. 63 (2001), 336–352.

[35] M. Geck and G. Hiss, Basic sets of Brauer characters of finite groups of Lie type, J. Reine Angew. Math. 418 (1991), 173–188.

[36] M. Geck and G. Hiss, Modular representations of finite groups of Lie type in non-defining characteristic, in: Finite reductive groups: Related structures and representations (M. Cabanes, ed.), Progr. Math. 141, Birkhäuser, Basel 1997, 195–249.

[37] M. Geck, G. Hiss, and G. Malle, Cuspidal unipotent Brauer characters, J. Algebra 168 (1994), 182–220.

[38] M. Geck, G. Hiss, and G. Malle, Towards a classification of the irreducible representations in non-defining characteristic of a finite group of Lie type, Math. Z. 221 (1996), 353–386.

[39] M. Geck and G. Pfeiffer, Characters of finite Coxeter groups and Iwahori–Hecke algebras, London Math. Soc. Monogr. (N.S.) 21, Clarendon Press, Oxford 2000.

[40] F. M. Goodman and H. Wenzl, Crystal bases of quantum affine algebras and affine Kazhdan-Lusztig polynomials, Internat. Math. Res. Notices 5 (1999), 251–275.

[41] J. A. Green, On a theorem of Sawada, J. London Math. Soc. 18 (1978), 247–252.

[42] J. A. Green, Polynomial representations of GL_n, Lecture Notes in Math. 830, Springer-Verlag, Berlin 1980.

[43] J. Gruber and G. Hiss, Decomposition numbers of finite classical groups for linear primes, J. Reine Angew. Math. 485 (1997), 55–91.

[44] G. Hiss, On the decomposition numbers of $G_2(q)$, J. Algebra 120 (1989), 339–360.

[45] G. Hiss, Zerlegungszahlen endlicher Gruppen vom Lie-Typ in nicht-definierender Charakteristik, Habilitationsschrift, RWTH Aachen, 1990.

[46] G. Hiss, Regular and semisimple blocks of finite reductive groups, J. London Math. Soc. 41 (1990), 63–68.

[47] G. Hiss, Harish-Chandra series of Brauer characters in a finite group with a split BN-pair, J. London Math. Soc. 48 (1993), 219–228.

[48] G. Hiss, F. Lübeck and G. Malle, The Brauer trees of the exceptional Chevalley groups of type E_6, Manuscripta Math. 87 (1995), 131–144.

[49] R. B. Howlett, Normalizers of parabolic subgroups of reflection groups, J. London Math. Soc. 21 (1980), 62–80.

[50] R. B. Howlett and G. I. Lehrer, Induced cuspidal representations and generalized Hecke rings, Invent. Math. 58 (1980), 37–64.

[51] R. B. Howlett and G. I. Lehrer, On Harish-Chandra induction for modules of Levi subgroups, J. Algebra 165 (1994), 172–183.

[52] G. D. James, The irreducible representations of the finite general linear groups, Proc. London Math. Soc. 52 (1986), 236–268.

[53] G. D. James, The decomposition matrices of $\mathrm{GL}_n(q)$ for $n \leq 10$, Proc. London Math. Soc. 60 (1990), 225–265.

[54] A. Lascoux, B. Leclerc and J. Y. Thibon, Hecke algebras at roots of unity and crystal bases of quantum affine algebras, Comm. Math. Phys. 181 (1996), 205–263.

[55] B. Leclerc, Decomposition numbers and canonical bases, Algebr. Represent. Theory 3 (2000), 277–287.

[56] B. Leclerc and J. Y. Thibon, Canonical bases of q-deformed Fock spaces, Internat. Math. Res. Notices 9 (1996), 447–456.

[57] M. Linckelmann, Letter to M. Geck and G. Hiss, April 25th, 1994.

[58] G. Lusztig, On the finiteness of the number of unipotent classes, Invent. Math. 34 (1976) 201–213

[59] G. Lusztig, Characters of reductive groups over a finite field, Ann. of Math. Stud. 107, Princeton University Press, 1984.

[60] G. Lusztig, On the representations of reductive groups with disconnected centre, Astérisque 168 (1988), 157–166.

[61] T. Okuyama and K. Waki, Decomposition numbers of Sp$(4, q)$, J. Algebra 199 (1998), 544–555.

[62] L. Puig, Algèbres de source de certains blocs des groupes de Chevalley, Astérisque 181–182 (1990), 221–236.

[63] W. Soergel, Conjectures de Lusztig, Séminaire Bourbaki, année 1994–95, Astérisque No. 237 (1996), Exp. No. 793, 75–85.

[64] R. Steinberg, Prime power representations of finite linear groups, II, Canad. J. Math. 9 (1957), 347–351.

[65] M. Varagnolo and E. Vasserot, On the decomposition matrices of the quantized Schur algebra, Duke J. Math. 100 (1999), 267–297.

[66] M. F. Vigneras, Induced representations of reductive p-adic groups in characteristic $\ell \neq p$, Selecta Math. (N.S) 4 (1998), 549–623.

UFR de Mathématiques and UMR 7586, Université Paris 7, 2 Place Jussieu, 75251 Paris, France

Current address: Institut Girard Desargues, Université Lyon 1, 21 av. Claude Bernard, 69622 Villeurbanne cedex, France

E-mail: geck@desargues.univ-lyon1.fr

Tensor products and restrictions in type A

Jonathan Brundan and Alexander S. Kleshchev

Abstract. The goal of this article is to give an exposition of some recent results on tensor products and restrictions for rational representations of the general linear group in positive characteristic. The exposition is based on our papers [13, 14, 15]. We also outline the relations with the LLT algorithm and the ideal structure of the group algebra of the finitary symmetric group.

Contents

Introduction 67

Notation 69

1 General results 70

2 Translation functors in type A 73

3 Connections with the LLT algorithm 78

4 Relating tensor products and restrictions 81

5 The symmetric group 85

6 Group algebra of the finitary symmetric group 93

References 95

Introduction

We begin with some motivating discussion about *translation functors*, following Jantzen [40]. Let G be a reductive algebraic group over an algebraically closed field $\mathbb{F}$ of characteristic $p > 0$, and let $\mathcal{C}$ denote the category of all rational $\mathbb{F}G$-modules. Writing X (resp. X^+) for the set of integral weights (resp. dominant integral weights) corresponding to the root system of G, we have for each $\lambda \in X^+$ the modules $L(\lambda)$, $\Delta(\lambda)$, $\nabla(\lambda)$ and $T(\lambda)$ which are the irreducible, standard (Weyl), costandard (induced) and indecomposable tilting modules of highest weight λ respectively. We would of course like to describe the inverse decomposition numbers $[L(\lambda) : \Delta(\mu)]$, allowing us to compute the formal characters of irreducibles as linear combinations of Weyl characters, and also the ∇-filtration multiplicities $[T(\lambda) : \nabla(\mu)]_\nabla$. Jantzen's translation functors have played a key role in attacking (and in the analogous quantum problems, solving) these questions.

We recall briefly the definition of the translation functor T_λ^μ, for $\lambda, \mu \in X$ lying in the closure of the same alcove. Let $\mathcal{C}(\mu)$ denote the *linkage class* corresponding to μ, that is, the full subcategory of $\mathcal{C}$ consisting of all modules with composition factors of the form $\{L(w \cdot \mu) \mid w \in W_p\}$ where W_p denotes the affine Weyl group acting on X by the usual dot action. There is an exact projection functor

$$\mathrm{pr}_\mu : \mathcal{C} \to \mathcal{C}(\mu)$$

given on objects by taking the largest submodule belonging to $\mathcal{C}(\mu)$. Then the translation functor

$$T_\lambda^\mu : \mathcal{C}(\lambda) \to \mathcal{C}(\mu)$$

is the functor $\mathrm{pr}_\mu \circ (? \otimes \nabla(\nu))$ where ν is the unique dominant weight conjugate under the Weyl group to $(\mu - \lambda)$. When λ and μ belong to the interior of the same alcove (or, more generally, the same facet) the functor T_λ^μ is an equivalence of categories.

The key situation to consider is when $\lambda \in X^+$ lies in some alcove A and μ lies on the interior of a wall in the upper closure of A. Let λs denote the reflection of λ in the wall containing μ (s can viewed as a simple reflection in W_p). Then $T_\lambda^\mu L(\lambda) \cong L(\mu)$, but the module $T_\mu^\lambda L(\mu)$ is far more complicated. It is known that $T_\mu^\lambda L(\mu)$ has simple head and socle isomorphic to $L(\lambda)$, and also that $T_\mu^\lambda L(\mu)$ contains $L(\lambda s)$ as a composition factor with multiplicity 1. In particular, this implies that $T_\mu^\lambda L(\mu)$ has Loewy length at least 3. Moreover, the Lusztig conjecture is equivalent to the statement that $T_\mu^\lambda L(\mu)$ has length exactly 3 for λ within a certain region. This equivalence was proved by Andersen [2, 2.16] and Cline–Parshall–Scott [18]. So understanding the structure of $T_\mu^\lambda L(\mu)$ even for such special configurations of λ and μ is a fundamental problem.

Question. What does $T_\mu^\lambda L(\mu)$ look like in general, for $\mu, \lambda \in X^+$ lying in the closure of the same alcove but neither lying in the closure of the facet containing the other? For example, when is it non-zero? When is it irreducible or indecomposable? Can one give a lower bound on its Loewy length?

In this article, we will describe some recent results [14] which answer these questions in special cases for $G = \mathrm{GL}_n(\mathbb{F})$. For instance, we will see *in our special cases* that the Loewy length of $T_\mu^\lambda L(\mu)$ has a natural lower bound equal to

$$2 \dim \mathrm{End}_G(T_\mu^\lambda L(\mu)) - 1.$$

This lower bound can take any odd value, for suitable choices of λ, μ and sufficiently large n. The lower bound 3 mentioned above is then a special case of our results.

Actually, we will not work with the functor T_μ^λ in type A, but instead introduce functors F_α and E_α for $\alpha \in \mathbb{Z}/p\mathbb{Z}$ which roughly speaking are given by tensoring with the natural $\mathrm{GL}_n(\mathbb{F})$-module V or its dual V^*, then projecting onto certain linkage classes determined by α. Our results are in keeping with the philosophy behind the algorithm of Lascoux, Leclerc and Thibon [6, 53, 54, 67]: in type A, one should always expect to be able to generalize results involving the affine Weyl group W_p, with associated alcove combinatorics, to results involving the Fock space of the affine Kac–Moody algebra $\hat{\mathfrak{gl}}_p$, with associated Young diagram combinatorics. Just as, at the level of representation theory, the 'wall crossing functor' $\Theta_s = T_\mu^\lambda \circ T_\lambda^\mu$ plays the role of the simple reflection $s \in W_p$, our functors E_α and F_α play the role (in a way we make precise later) of the simple root generators $e_\alpha, f_\alpha \in \hat{\mathfrak{gl}}_p$.

The remainder of the article is organized as follows. We begin in Section 1 with some quite general results from [15] about the structure of tensor products of the form $M \otimes \nabla(\nu)$ in characteristic p, remembering that for a G-module M, $T_\lambda^\mu M$ is by definition a certain linkage class of $M \otimes \nabla(\nu)$ for suitable ν. Then, for the remainder of the paper, we specialize to $G = \mathrm{GL}(n)$, when there are very close connections between the tensor product $M \otimes \nabla(\nu)$ for special ν and the restriction of M to the subgroup $\mathrm{GL}(n-1)$. In Section 2, we state in detail the definitions of the functors E_α and F_α and our main results from [14]. The proof of these depended on first reformulating the results in terms of the branching problems from $\mathrm{GL}(n)$ to $\mathrm{GL}(n-1)$ studied in our earlier work [45, 46, 47, 49, 50, 10, 11, 16]. In Section 3, we explain the connection between the functors F_α and the LLT algorithm [54, 53], basing our account on the recent results of Varagnolo and Vasserot [67]. In Section 4 we discuss some of the other connections between tensor products and restrictions in $\mathrm{GL}(n)$ obtained in [13], in particular the relationship with tilting modules and the work of Mathieu and Papadopoulou [56]. As applications of these techniques, Section 5 contains the corollaries of our main results for the symmetric group, while the relations between tensor ideals [5, 30, 31] and ideals of group algebras of the finitary symmetric group [8] are outlined in Section 6.

The material described here was presented in lectures at the Symposium on modular representation theory at the University of Virginia. The second author would like to thank the organizers M. Collins, B. Parshall and L. Scott for hospitality and for the possibility to publish this exposition in the proceedings of the conference.

Notation

General conventions: If G and H are two groups, L is an $\mathbb{F}G$-module and M is an $\mathbb{F}H$-module, we write $L \boxtimes M$ for the *outer* tensor product of L and M (which is a module over $G \times H$). If N is another $\mathbb{F}G$-module we write $L \otimes N$ for the *inner* tensor product of L and N (which is a G-module). If L is irreducible, I is indecomposable, and M is an arbitrary $\mathbb{F}G$-module, then $[M : L]$ stands for the multiplicity of L as a composition factor of M, and $(M : I)$ stands for the multiplicity of I as an indecomposable summand of M. If G is an algebraic group, a G-module will always mean a rational $\mathbb{F}G$-module, unless otherwise stated.

Notation in arbitrary type: If G is an arbitrary reductive algebraic group over $\mathbb{F}$, we will follow Jantzen [40] for notation. In particular, R denotes the root system of G with respect to a fixed maximal torus T, $R^+ \subset R$ denotes the set of positive roots determined by a choice of Borel subgroup B^+ containing T, and $\{\alpha_1, \dots, \alpha_\ell\} \subset R^+$ is the corresponding base for R. We write $X(T)$ for the character group $\mathrm{Hom}(T, \mathbb{F}^\times)$, $Y(T)$ for the cocharacter group $\mathrm{Hom}(\mathbb{F}^\times, T)$, and let $\langle \cdot, \cdot \rangle$ be the natural pairing $X(T) \times Y(T) \to \mathbb{Z}$. For $\alpha \in R$, $\alpha^\vee$ denotes the corresponding coroot in $Y(T)$, and $X^+(T) = \{\lambda \in X(T) \mid \langle \lambda, \alpha_i^\vee \rangle \geq 0, i = 1, \dots, \ell\}$ denotes the set of dominant weights. Given a weight $\nu \in X(T)$ and a T-module M, M_ν will denote the ν-weight space of M, and the formal character of M is $\mathrm{ch}\, M$.

Notation in type A: In addition, if $G = \mathrm{GL}(n) = \mathrm{GL}_n(\mathbb{F})$, we will make the following choices. We always take T to be all diagonal matrices in $\mathrm{GL}(n)$ and B^+ to be all upper triangular matrices. We identify the weight lattice $X(T)$ with the set $X(n)$ of all n-tuples $\lambda = (\lambda_1, \lambda_2, \dots, \lambda_n)$ of integers, λ corresponding to the character $\mathrm{diag}(t_1, \dots, t_n) \mapsto t_1^{\lambda_1} \dots t_n^{\lambda_n}$, and $X^+(T)$ with the set $X^+(n) = \{\lambda \in X(n) \mid \lambda_1 \geq \dots \geq \lambda_n\}$. We also write ϵ_i for the weight $(0, \dots, 0, 1, 0, \dots, 0)$ with 1 in the ith position. The natural n-dimensional $\mathrm{GL}(n)$-module with highest weight ϵ_1 will be denoted by V. Its dual V^* has highest weight $-\epsilon_n$. Whenever we need to clarify the group $\mathrm{GL}(n)$ that we are referring to, we will add a subscript n to our notation, giving us $\mathrm{GL}(n)$-modules V_n, $L_n(\lambda)$, $\Delta_n(\lambda)$, $\nabla_n(\lambda)$ and $T_n(\lambda)$.

1. General results

Let G denote an arbitrary reductive algebraic group over $\mathbb{F}$. Throughout the article, we consider two types of problems. Firstly, we are interested in tensor products of irreducible (standard, costandard, tilting) G-modules. Secondly, we study the restrictions of irreducible (standard, costandard, tilting) modules from G to its Levi subgroups. We also want to reveal various connections between the two types of problems.

Some of the problems have 'characteristic-free' answers but we try to consider those which do depend on the characteristic. For example, assume for a moment that the ground field $\mathbb{F}$ has characteristic 0. Let $c_{\mu\nu}^\lambda$ be the multiplicity of $L(\lambda)$ in $L(\mu) \otimes L(\nu)$. If $G = \mathrm{GL}(n)$, the constant $c_{\mu\nu}^\lambda$ is a Littlewood–Richardson coefficient. Then, in *any* characteristic, the tensor product $\nabla(\mu) \otimes \nabla(\nu)$ has a ∇-filtration, with $\nabla(\lambda)$ appearing $c_{\mu\nu}^\lambda$ times. Indeed, the fact that a ∇-filtration exists follows from the fundamental Donkin–Mathieu theorem on good filtrations (see [66, 20, 55]), and the multiplicities do not depend on the characteristic as the formal characters of costandard modules do not (they are given by Weyl's character formula).

Moreover, by a theorem of Cline, Parshall, Scott and van der Kallen [19], [40, 4.13], $\mathrm{Ext}^1_G(\Delta(\lambda), \nabla(\gamma)) = 0$ for any λ and γ. Also, it is well known (see for example [40, II.4.13]) that $\dim \mathrm{Hom}_G(\Delta(\lambda), \nabla(\gamma)) = \delta_{\lambda\gamma}$ where $\delta_{\lambda\gamma}$ is the Kronecker delta. It follows that

$$\dim \mathrm{Hom}_G \big(\Delta(\lambda), \nabla(\mu) \otimes \nabla(\nu)\big) = c^{\lambda}_{\mu\nu}. \tag{1}$$

We say a vector $v \in M$ is *primitive* if it is invariant with respect to the unipotent radical U^+ of B^+. By the universal property of standard modules [40, II.2.13 b)], we see from (1) that the space of the primitive vectors of weight λ in $\nabla(\mu) \otimes \nabla(\nu)$ has dimension $c^{\lambda}_{\mu\nu}$, just like in characteristic 0.

However, we do get a 'modular problem' if we want to understand which of these primitive vectors generate simple modules. Equivalently, we are interested in the *socle* of $\nabla(\mu) \otimes \nabla(\nu)$ or in the dimensions of the spaces

$$\mathrm{Hom}_G \big(L(\lambda), \nabla(\mu) \otimes \nabla(\nu)\big) \tag{2}$$

for all triples (λ, μ, ν). These dimensions can be thought of as *modular Littlewood–Richardson coefficients* (in type A). By dualizing, the space (2) is isomorphic to $\mathrm{Hom}_G(\Delta(-w_0\mu), L(-w_0\lambda) \otimes \nabla(\nu))$, where w_0 is the longest element of the Weyl group $W = N_G(T)/T$. So, as λ, μ, ν are arbitrary, our problem is equivalent to describing the spaces

$$\mathrm{Hom}_G \big(\Delta(\lambda), L(\mu) \otimes \nabla(\nu)\big), \tag{3}$$

for arbitrary triples (λ, μ, ν). Equivalently, we want to understand the primitive vectors in tensor products of the form $L(\mu) \otimes \nabla(\nu)$.

Our first result on the space (3) generalizes a well known fact in characteristic 0. Let $\mathrm{Dist}(G)$ be the algebra of distributions of G as in [40, I.7], which is generated by $\mathrm{Dist}(T)$ and the 'divided power' root generators $X^{(n)}_\alpha, Y^{(n)}_\alpha$ for $\alpha \in R^+, n \geq 1$. Write $X^{(n)}_i = X^{(n)}_{\alpha_i}, Y_i = Y^{(n)}_{\alpha_i}$ for $i = 1, \ldots, \ell$. If G is semisimple and simply connected, $\mathrm{Dist}(G)$ coincides with the hyperalgebra of G arising from the Chevalley construction. We note that any G-module is a $\mathrm{Dist}(G)$-module in a natural way; see [40, I.7.11, II.1.20]. For any G-module M, a dominant weight $\nu \in X(T)^+$ and any weight $\gamma \in X(T)$, we define

$$M^\nu := \{v \in M \mid X^{(b_i)}_i v = 0 \text{ for all } b_i > \langle \nu, \alpha_i^\vee \rangle \text{ and } i = 1, 2, \ldots, \ell\}$$

and let $M^\nu_\gamma := M^\nu \cap M_\gamma$ denote its γ-weight space.

Theorem 1.1. *Let $\lambda, \nu \in X^+(T)$, and M be any (rational) G-module. Then*

$$\dim \mathrm{Hom}_G \big(\Delta(\lambda), M \otimes \nabla(\nu)\big) = \dim M^\nu_{\lambda-\nu}.$$

To explain our interest in the theorem, suppose that $M = L(\mu)$ is an irreducible module for some fixed $\mu \in X^+(T)$. Then, for ν large relative to μ, we have $M^\nu_{\lambda-\nu} =$

$M_{\lambda-\nu}$ and, by the theorem, $\dim L(\mu)_{\lambda-\nu} = \dim \mathrm{Hom}_G(\Delta(\lambda), L(\mu) \otimes \nabla(\nu))$. So to compute the formal character of $L(\mu)$ it suffices to describe the Hom-space in 1.1 for $M = L(\mu)$ and λ, ν large. In view of the universality of standard modules, this is equivalent to describing the primitive vectors of weight λ in $L(\mu) \otimes \nabla(\nu)$.

A complete proof of 1.1 can be found in [15, Theorem A]. The main idea is to use the well-known presentation of standard modules by generators and relations: for $\mu \in X^+(T)$, we have that

$$\Delta(\mu) \cong U/I(\mu),$$

where $I(\mu)$ is the left ideal of $\mathrm{Dist}(G)$ generated by $\{X_\alpha^{(b_\alpha)} \mid \alpha \in R^+, b_\alpha \geq 1\} \cup \{H - \mu(H) \mid H \in \mathrm{Dist}(T)\} \cup \{Y_i^{(b_i)} \mid 1 \leq i \leq \ell, b_i > \langle \mu, \alpha_i^\vee \rangle\}$.

Now suppose that $G = \mathrm{GL}(n)$ and embed $\mathrm{GL}(n-1)$ into the top left hand corner of $\mathrm{GL}(n)$. Observe that if we take $\mu = -\ell\epsilon_n$ for $\ell \geq 0$, then the space $M^\mu_{\lambda-\mu}$ appearing in 1.1 is precisely the space of vectors in $M_{\lambda-\mu}$ which are primitive with respect to the subgroup $\mathrm{GL}(n-1)$, and which satisfy in addition $X_{n-1}^{(b)} v = 0$ for any $b > \ell$. Recalling that $\nabla_n(-\ell\epsilon_n)$ is precisely the ℓth symmetric power $S^\ell(V_n^*)$, one obtains a connection between $\mathrm{GL}(n)$-primitive vectors in the tensor product $M \otimes S^\ell(V_n^*)$ and $\mathrm{GL}(n-1)$-primitive vectors in the restriction $M \downarrow_{\mathrm{GL}(n-1)}$. We obtained the following extension in one important special case [15, Theorem C]:

Theorem 1.2. *Fix $\lambda, \mu \in X^+(n)$ with $\lambda_n = \mu_n$ and set $\ell = \sum_{i=1}^n (\lambda_i - \mu_i)$. Then, for any submodule M of $\nabla_n(\lambda)$,*

$$\mathrm{Hom}_{\mathrm{GL}(n)}\left(L_n(\mu), M \otimes S^\ell(V_n^*)\right) \cong \mathrm{Hom}_{\mathrm{GL}(n-1)}(L_{n-1}(\bar{\mu}), M \downarrow_{\mathrm{GL}(n-1)})$$

where $\bar{\mu} = (\mu_1, \ldots, \mu_{n-1})$ denotes the restriction of μ to $T \cap \mathrm{GL}(n-1)$.

We believe it is an important problem to find the socle of $L_n(\lambda) \downarrow_{\mathrm{GL}(n-1)}$ for any $\lambda \in X^+(n)$. This socle is described by the right hand side of the equation in 1.2, taking $M = L_n(\lambda)$. We refer to this problem as the *modular branching problem for the general linear group*. A complete answer only exists in the special case that $L_n(\lambda) \downarrow_{\mathrm{GL}(n-1)}$ is completely reducible; see [16].

If we take $\ell = 1$ in 1.2, we see that the problems of computing the socle of $L_n(\lambda) \otimes V_n^*$ and the socle of part of the restriction $L_n(\lambda) \downarrow_{\mathrm{GL}(n-1)}$, known as the *first level*, are equivalent. The first level of such restrictions has been studied extensively in our earlier work, especially [47, 10]. This connection was exploited in [14]; we will discuss the results in detail in the next section.

Finally, we describe one other general result from [15] about the structure of tensor products of the form $M \otimes \nabla(\nu)$, again valid in arbitrary type. A dominant weight λ is called *p^r-restricted* if $\langle \lambda, \alpha_i^\vee \rangle < p^r$ for all $i = 1, 2, \ldots, \ell$. A semisimple module is called p^r-restricted if all of its composition factors have p^r-restricted highest weights. Going back to the space (2), we would like to understand when the socle of $\nabla(\mu) \otimes \nabla(\nu)$ is p-restricted. A necessary condition is that μ and ν are both p-restricted. Even though the converse is not quite true (see [15, Remark 2.10] for a counterexample), we have the following result from [15, Theorem B]:

Theorem 1.3. *Let $\mu, \nu \in X^+(T)$ and $\alpha_0 \in R$ be the highest root. Suppose that μ is p^r-restricted and $\langle \nu, \alpha_0^\vee \rangle < p^r$. Then the socle of $\nabla(\mu) \otimes \nabla(\nu)$ is p^r-restricted. In particular, the socle of $L(\mu) \otimes L(\nu)$ is p^r-restricted.*

We can of course take ν to be a miniscule weight in 1.3 so:

Corollary 1.4. *Let μ be a dominant p^r-restricted weight, and ν be any miniscule weight. If M is any submodule of $\nabla(\mu)$, then the socle of $M \otimes \nabla(\nu)$ is p^r-restricted. In particular, the socle of $L(\mu) \otimes \nabla(\nu)$ is p^r-restricted.*

Remark 1.5. (I) We note that even under the assumptions of 1.3, the tensor product $\nabla(\mu) \otimes \nabla(\nu)$ can in general have many non-p^r-restricted composition factors. The fact that they never appear in the socle is somewhat miraculous.

(II) The result 1.3 can also be interpreted in terms of translation functors as in the introduction. Suppose that $\lambda, \mu \in X(T)$ lie in the closure of the same alcove and that $-p < \langle \lambda - \mu, \alpha^\vee \rangle < p$ for all roots α. Then for any $w \in W_p$ such that $w \cdot \mu \in X^+(T)$ is p^r-restricted, the socle of $T_\mu^\lambda \nabla(w \cdot \mu)$ is also p^r-restricted.

2. Translation functors in type A

For the remainder of the article, we specialize to the case $G = \mathrm{GL}(n)$. Given $(a, b) \in \mathbb{Z} \times \mathbb{Z}$, define the corresponding *p-residue* $\mathrm{res}(a, b)$ to be $(b - a)$ regarded as an element of the ring $\mathbb{Z}/p\mathbb{Z}$. For $\alpha \in \mathbb{Z}/p\mathbb{Z}$ and $\lambda \in X(n)$, define the *α-content* of λ to be the integer:

$$\mathrm{cont}_\alpha(\lambda) = \left| \left\{ (a, b) \;\middle|\; \begin{array}{c} 1 \le a \le n, 0 < b \le \lambda_a, \\ \mathrm{res}(a, b) = \alpha \end{array} \right\} \right| - \left| \left\{ (a, b) \;\middle|\; \begin{array}{c} 1 \le a \le n, \lambda_a \le b < 0, \\ \mathrm{res}(a, b) = \alpha \end{array} \right\} \right|.$$

Say $\lambda, \mu \in X(n)$ are *linked*, written $\lambda \sim \mu$, if $\mathrm{cont}_\alpha(\lambda) = \mathrm{cont}_\alpha(\mu)$ for all $\alpha \in \mathbb{Z}/p\mathbb{Z}$. The *linkage principle* proved in [17] implies that if $\mathrm{Ext}^1_{\mathrm{GL}(n)}(L(\lambda), L(\mu)) \neq 0$, for $\lambda, \mu \in X^+(n)$, then $\lambda \sim \mu$.

Let $\mathcal{C}$ denote the category of all rational $\mathrm{GL}(n)$-modules. For any $\lambda \in X(n)$, let $\mathcal{C}(\lambda)$ denote the full subcategory of $\mathcal{C}$ consisting of all $M \in \mathcal{C}$ such that all composition factors of M are of the form $L(\mu)$ for $\mu \sim \lambda$. By the linkage principle, any module $M \in \mathcal{C}$ can be written uniquely as

$$M \cong \bigoplus_\lambda \mathrm{pr}_\lambda M$$

where λ runs over a set of $\sim$-equivalence class representatives in $X(n)$, and $\mathrm{pr}_\lambda M$ denotes the largest submodule of M belonging to $\mathcal{C}(\lambda)$.

Fix a residue $\alpha \in \mathbb{Z}/p\mathbb{Z}$. We can now define the functors

$$E_\alpha : \mathcal{C} \to \mathcal{C} \quad \text{and} \quad F_\alpha : \mathcal{C} \to \mathcal{C}.$$

We will first define their restrictions to $\mathcal{C}(\lambda)$ for any $\lambda \in X(n)$, and then extend additively to obtain the functors on the whole category $\mathcal{C}$. Given $M \in \mathcal{C}(\lambda)$, we let $F_\alpha M$ (resp. $E_\alpha M$) denote the largest submodule of $M \otimes V$ (resp. $M \otimes V^*$) all of whose composition factors are of the form $L(\mu)$ with

$$\mathrm{cont}_\alpha(\mu) = \mathrm{cont}_\alpha(\lambda) + 1 \quad (\text{resp. } \mathrm{cont}_\alpha(\mu) = \mathrm{cont}_\alpha(\lambda) - 1)$$

and $\mathrm{cont}_\beta(\mu) = \mathrm{cont}_\beta(\lambda)$ for all $\alpha \neq \beta \in \mathbb{Z}/p\mathbb{Z}$. Given a morphism $\theta : M \to N$, $F_\alpha\theta$ is just the restriction to $F_\alpha M$ of the natural map $\theta \otimes 1 : M \otimes V \to N \otimes V$, and similarly for E_α. We have

$$M \otimes V \cong \bigoplus_{\alpha \in \mathbb{Z}/p\mathbb{Z}} F_\alpha M \quad \text{and} \quad M \otimes V^* \cong \bigoplus_{\alpha \in \mathbb{Z}/p\mathbb{Z}} E_\alpha M$$

for any $M \in \mathcal{C}$.

On any fixed linkage class $\mathcal{C}(\lambda)$, the functor F_α (resp. E_α), for a suitable choice of α, coincides with the translation functor T_λ^μ defined in [40, II.7.6] for any weight $\mu \in X(n)$ such that the dominant conjugate of $(\mu - \lambda)$ is equal to the highest weight of V (resp. V^*). We note initially that the argument of [40, II.7.6] shows easily that the functors F_α and E_α are (left and right) adjoint to one another, and both are exact.

In the next combinatorial definitions, the notions of normal and good first appeared in [47], while the dual notions of conormal and cogood were introduced in [14]. (The reader may be more familiar with normal and good *nodes* – we reserve this terminology for the symmetric group setting when definitions are 'transposed'). In the definitions, we call a map ψ from a set $M \subseteq \mathbb{Z}$ to a set $N \subseteq \mathbb{Z}$ *increasing* (resp. *decreasing*) if $\psi(m) > m$ (resp. $\psi(m) < m$) for all $m \in M$.

Fix $\lambda \in X^+(n)$ and $1 \leq i \leq n$. We say i is *λ-removable* if *either* $i = n$ *or* $1 \leq i < n$ and $\lambda_i > \lambda_{i+1}$; equivalently, i is λ-removable if $\lambda - \epsilon_i \in X^+(n)$. We say i is *λ-addable* if *either* $i = 1$ *or* $1 < i \leq n$ and $\lambda_i < \lambda_{i-1}$; equivalently, i is λ-addable if $\lambda + \epsilon_i \in X^+(n)$.

Say i is *normal for* λ if i is λ-removable and there is a decreasing injection from the set of

$$\lambda\text{-addable } j \text{ with } i < j \leq n \text{ and } \mathrm{res}(i, \lambda_i) = \mathrm{res}(j, \lambda_j + 1)$$

into the set of

$$\lambda\text{-removable } j' \text{ with } i < j' < n \text{ and } \mathrm{res}(i, \lambda_i) = \mathrm{res}(j', \lambda_{j'}).$$

Say i is *good for* λ if i is normal for λ and there is no j that is normal for λ with $1 \leq j < i$ and $\mathrm{res}(j, \lambda_j) = \mathrm{res}(i, \lambda_i)$.

Say i is *conormal for* λ if i is λ-addable and there is an increasing injection from the set of

$$\lambda\text{-removable } j \text{ with } 1 \le j < i \text{ and } \operatorname{res}(j, \lambda_j) = \operatorname{res}(i, \lambda_i + 1)$$

into the set of

$$\lambda\text{-addable } j' \text{ with } 1 < j' < i \text{ and } \operatorname{res}(j', \lambda_{j'} + 1) = \operatorname{res}(i, \lambda_i + 1).$$

Say i is *cogood for* λ if i is conormal for λ and there is no j that is conormal for λ with $i < j \le n$ and $\operatorname{res}(j, \lambda_j + 1) = \operatorname{res}(i, \lambda_i + 1)$.

Example 2.1. We pause to give an example illustrating the definitions. Consider $n = 4$, $\lambda = (6, 5, 2, 0)$ and $p = 3$. The 3-residues of the addable and removable nodes are:

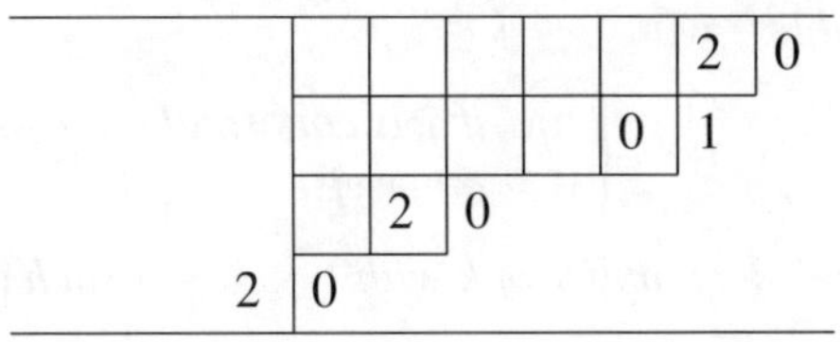

For $i = 1, 2, 3$ or 4, i is normal for λ if $i = 1, 3, 4$ (when $\operatorname{res}(i, \lambda_i) = 2$) but not if $i = 2$. So 1 is good for λ. Similarly, for $i = 1, 2, 3$ or 4, i is conormal for λ if $i = 1, 4$ (when $\operatorname{res}(i, \lambda_i + 1) = 0$) or if $i = 2$ (when $\operatorname{res}(i, \lambda_i + 1) = 1$), but not if $i = 3$. So 2 and 4 are cogood for λ.

In general, for a fixed $\alpha \in \mathbb{Z}/p\mathbb{Z}$, there is at most one good i for λ such that $\operatorname{res}(i, \lambda_i) = \alpha$. Moreover there is exactly one such i if and only if there is at least one normal j for λ with $\operatorname{res}(j, \lambda_j) = \alpha$. A similar result is true for conormal and cogood.

Our first result [14, Theorem A] describes the effect of F_α on standard modules (the analogous result for costandard modules follows easily since F_α commutes with contravariant duality):

Theorem 2.2. *Fix $\lambda \in X^+(n)$ and a residue $\alpha \in \mathbb{Z}/p\mathbb{Z}$. Then $F_\alpha \Delta(\lambda)$ is zero unless there is at least one λ-addable i with $1 \le i \le n$ and* $\operatorname{res}(i, \lambda_i + 1) = \alpha$. *In that case,*

(i) *$F_\alpha \Delta(\lambda)$ has a filtration with factors $\Delta(\lambda + \epsilon_j)$ for all λ-addable j with $1 \le j \le n$ and* $\operatorname{res}(j, \lambda_j + 1) = \alpha$, *each appearing with multiplicity one;*

(ii) *the head of $F_\alpha \Delta(\lambda)$ is $\bigoplus_j L(\lambda + \epsilon_j)$ where the sum is over all j with $1 \le j \le n$ such that j is normal for $\lambda + \epsilon_j$ and* $\operatorname{res}(j, \lambda_j + 1) = \alpha$;

(iii) *if λ is p^r-restricted, every element of the head of $F_\alpha \Delta(\lambda)$ is also p^r-restricted.*

By contravariant duality, (iii) is a special case of 1.4.

To illustrate the theorem, take λ as in 2.1. Then $F_0 \Delta(\lambda)$ has a Δ-filtration with factors (from the bottom up) $\Delta(\lambda + \epsilon_1)$, $\Delta(\lambda + \epsilon_3)$ and $\Delta(\lambda + \epsilon_4)$. Moreover, $F_0 \Delta(\lambda)$ has simple head $L(\lambda + \epsilon_4)$ and so is certainly indecomposable.

Next we consider F_α applied to an irreducible module. This is [14, Theorem B]:

Theorem 2.3. *Fix $\lambda \in X^+(n)$ and a residue $\alpha \in \mathbb{Z}/p\mathbb{Z}$. Then $F_\alpha L(\lambda)$ is zero unless there is at least one j conormal for λ and such that* $\mathrm{res}(j, \lambda_j + 1) = \alpha$. *In that case,*

(i) *$F_\alpha L(\lambda)$ is an indecomposable, contravariantly self-dual module, with simple socle and head isomorphic to $L(\lambda + \epsilon_i)$ where i is the cogood with* $\mathrm{res}(i, \lambda_i + 1) = \alpha$*;*

(ii) *for any $\mu \in X^+(n)$,*

$$\mathrm{Hom}_{\mathrm{GL}(n)}\left(\Delta(\mu), F_\alpha L(\lambda)\right) = \begin{cases}]\mathbb{F} & \textit{if } \mu = \lambda + \epsilon_j \textit{ for some } j \textit{ conormal for } \lambda, \\ & \textit{such that } \mathrm{res}(j, \lambda_j + 1) = \alpha; \\ 0 & \textit{otherwise;} \end{cases}$$

(iii) *for any λ-addable j with $1 \leq j \leq n$,*

$$[F_\alpha L(\lambda) : L(\lambda + \epsilon_j)] = \begin{cases} b_j & \textit{if } j \textit{ is conormal for } \lambda \textit{ and } \mathrm{res}(j, \lambda_j + 1) = \alpha, \\ 0 & \textit{otherwise,} \end{cases}$$

where b_j denotes the number of k with $1 \leq k \leq j$ such that k is conormal for λ and $\mathrm{res}(k, \lambda_k + 1) = \alpha$*;*

(iv) *the endomorphism ring* $\mathrm{End}_{\mathrm{GL}(n)}(F_\alpha L(\lambda))$ *is isomorphic to the truncated polynomial ring $\mathbb{F}[T]/(T^b)$, of dimension b, where b is the number of j with $1 \leq j \leq n$ such that j is conormal for λ and* $\mathrm{res}(j, \lambda_j + 1) = \alpha$.

Again, we illustrate the theorem using the setup of 2.1. Then $F_0 L(\lambda)$ has simple head and socle $L(\lambda + \epsilon_4)$, and $L(\lambda + \epsilon_1)$ appears in $F_0 L(\lambda)$ with multiplicity 1. So we see at once that the Loewy length of $F_0 L(\lambda)$ is at least 3. In fact, this example is similar to the length 3 case of translation functors mentioned in the introduction: λ lies in the upper closure of the facet containing $\lambda + \epsilon_4$ and $\lambda + \epsilon_1$ is the reflection of $\lambda + \epsilon_4$ across a wall containing λ.

Now suppose quite generally that N is a G-module (for any reductive group G) such that

(1) N is a submodule of a *finite dimensional* G-module with a ∇-filtration;

(2) $N \cong N^\tau$ where N^τ denotes the contravariant dual of N as in [40, p.205].

Then, according to [14, Proposition 4.7]:

Lemma 2.4. *The G-module N is completely reducible if and only if*

$$\mathrm{Hom}_G(L(\lambda), N) \cong \mathrm{Hom}_G(\Delta(\lambda), N)$$

for all $\lambda \in X^+(T)$.

We apply 2.4 to the module $N = F_\alpha L(\lambda)$, observing that this certainly satisfies (1) and (2). The results in 2.3 then allow us to obtain the following *necessary and sufficient* condition for $F_\alpha L(\lambda)$ to be irreducible:

Corollary 2.5. *With the notation of* 2.3, $F_\alpha L(\lambda)$ *is irreducible if and only if there is a unique* j *such that* j *is conormal for* λ *and* $\operatorname{res}(j, \lambda_j + 1) = \alpha$. *Hence* $L(\lambda) \otimes V$ *is completely irreducible if and only if every conormal* i *is cogood.*

In fact it is well-known that $T_\mu^\lambda L(\mu)$ is irreducible whenever λ lies in the upper closure of the facet containing μ (see [40, Proposition 7.15]), as is the case in our example for $F_1 L(\lambda) \cong L(\lambda + \epsilon_2)$. However, the corollary shows that there are many other more general circumstances when $F_\alpha L(\lambda)$ is irreducible. The significance of this is the following, which follows immediately from 2.3 by exactness of F_α:

Corollary 2.6. *Fix* $\lambda \in X^+(n)$ *and* $\alpha \in \mathbb{Z}/p\mathbb{Z}$ *such that* λ *has a unique conormal* i *with* $\operatorname{res}(i, \lambda_i + 1) = \alpha$. *Then the inverse decomposition number* $[L(\lambda + \epsilon_i) : \Delta(\mu)]$ *is equal to* $\sum_j [L(\lambda) : \Delta(\mu - \epsilon_j)]$ *summed over all* μ*-removable* j.

The final result [14, Theorem C] gives further information about the structure of $F_\alpha L(\lambda)$:

Theorem 2.7. *Fix* $\lambda \in X^+(n)$ *and a residue* $\alpha \in \mathbb{Z}/p\mathbb{Z}$. *Let* $1 = s_1 < s_2 < \cdots < s_b$ *denote the set of all* j *with* $1 \le j \le n$ *such that* j *is conormal for* λ *and* $\operatorname{res}(j, \lambda_j + 1) = \alpha$. *Then* $N := F_\alpha L(\lambda)$ *has a filtration* $0 = N_0 < N_1 < \cdots < N_b = N$ *such that:*

(i) *for* $1 \le i \le b$, N_i/N_{i-1} *is a non-zero quotient of* $\Delta(\lambda + \epsilon_{s_i})$;

(ii) *for* $1 \le i \le j \le b$,
$$\dim \operatorname{Hom}_{\mathrm{GL}(n)}(N_j/N_{j-1}, N_i/N_{i-1}) = [N_i/N_{i-1} : L(\lambda + \epsilon_{s_j})] = 1;$$

(iii) *for* $1 \le i < b$, *the extension* $0 \to N_i/N_{i-1} \to N_{i+1}/N_{i-1} \to N_{i+1}/N_i \to 0$ *does not split;*

(iv) *the Loewy length of* N_i/N_{i-1} *is at least* $b - i + 1$;

(v) *the Loewy length of* N *is at least* $2b - 1$.

It is easy to construct examples of λ so that the number b from 2.3(iv), 2.7 is arbitrarily large. Thus we see by 2.7(v) that the Loewy length of $F_\alpha L(\lambda)$ can be arbitrarily large.

We have not mentioned the functor E_α adjoint to F_α yet. In fact, there are entirely analogous statements to 2.2–2.7 in this case, all of which follow directly from the above and some combinatorial arguments. Roughly, one needs to swap 'addable' and 'removable', 'normal' and 'conormal', 'good' and 'cogood' in the statements, but there are some other differences too; for precise statements we refer the reader to [14, Theorems A′, B′, C′] (where the functor E_α is denoted Tr_α).

3. Connections with the LLT algorithm

We need to switch to working with *polynomial representations*. Let $\Lambda(n) \subset X(n)$ denote all n-tuples $(\lambda_1, \ldots, \lambda_n)$ satisfying $\lambda_i \geq 0$ for $i = 1, \ldots, n$, and $\Lambda^+(n) := \Lambda(n) \cap X^+(n)$. Let $\Lambda(n, r)$ denote all n-tuples $(\lambda_1, \ldots, \lambda_n) \in \Lambda(n)$ satisfying $|\lambda| := \lambda_1 + \cdots + \lambda_n = r$, and $\Lambda^+(n, r) := \Lambda(n, r) \cap X^+(n)$. We call elements of $\Lambda(n, r)$ *compositions* of r (with at most n non-zero parts), and elements of $\Lambda^+(n, r)$ *partitions* of r (with at most n non-zero parts).

If λ is any partition we denote by λ^t the *transpose* partition, i.e. the partition whose Young diagram is the transpose of the Young diagram of λ. The following result of Donkin from [22] will be important:

Theorem 3.1. *For $\lambda, \mu \in \Lambda^+(n, r)$, $[\Delta(\lambda) : L(\mu)] = [T(\mu^t) : \nabla(\lambda^t)]_\nabla$.*

We note that $\lambda \in \Lambda^+(n, r)$ does *not* necessarily imply that $\lambda^t \in \Lambda^+(n, r)$, but we can always find some $m > n$ so that $\lambda^t \in \Lambda^+(m, r)$. The expression $[T(\mu^t) : \nabla(\lambda^t)]_\nabla$ in the theorem needs to be interpreted as a ∇-filtration multiplicity inside the group $\mathrm{GL}(m)$ for this larger m. Since such multiplicities remain stable as m increases, there is no ambiguity.

Let q be an indeterminate and $U = U_q(\hat{\mathfrak{sl}}_p)$ denote Lusztig's integral form for the quantized enveloping algebra of the affine Lie algebra $\hat{\mathfrak{sl}}_p$. The negative part U^- of this algebra is generated by certain elements $\{f_\alpha \,|\, \alpha \in \mathbb{Z}/p\mathbb{Z}\}$ together with all of their *quantized divided powers*, subject to various well-known relations. The *Fock space* $\mathcal{F}$ is a certain integrable representation of U with basis given by *all* partitions. A beautiful account of this representation (or rather its unquantized analogue) is given in the book [42]. For its extension to the quantum case, we refer the reader to [58, 43, 54].

Fix n and let $\mathcal{F}_n$ denote the free $\mathbb{Z}[q, q^{-1}]$-module with basis $\{\Delta_\lambda \mid \lambda \in \Lambda^+(n)\}$. This can be regarded as a quotient of $\mathcal{F}$ as a U^--module in an obvious way, so that [43, 54] give us a well-defined action of U^- on $\mathcal{F}_n$ satisfying, for each $\alpha \in \mathbb{Z}/p\mathbb{Z}$,

$$f_\alpha \Delta_\lambda = \sum_i q^{a_{<i} - r_{<i}} \Delta_{\lambda + \epsilon_i}$$

where the sum is over all λ-addable i such that $\mathrm{res}(i, \lambda_i + 1) = \alpha$, and $a_{<i}$ denotes the number of λ-addable $j < i$ with $\mathrm{res}(j, \lambda_j + 1) = \alpha$ and $r_{<i}$ denotes the number of λ-removable $k < i$ with $\mathrm{res}(k, \lambda_k) = \alpha$.

Now assume for simplicity that $p \geq n$. Everything we are saying generalizes to the case $p < n$ too, but the definitions become considerably more technical. Then it is quite easy to see (because $p \geq n$) that $\mathcal{F}_n$ is generated as a U^--module by the vectors $\{\Delta_{i\delta}\}_{i \geq 0}$ where $i\delta$ denotes the weight $(i, i, \ldots, i) \in \Lambda^+(n)$ corresponding to the ith power of determinant. Moreover, as follows directly from [54] or [53] on passing to the quotient $\mathcal{F}_n$ of the space $\mathcal{F}$ there, there is a unique ring homomorphism

$$\bar{\ }: \mathcal{F}_n \to \mathcal{F}_n$$

such that $\bar{q} = q^{-1}$, $\bar{\Delta}_{i\delta} = \Delta_{i\delta}$ for all $i \geq 0$ and commuting with the action of the f_α for $\alpha \in \mathbb{Z}/p\mathbb{Z}$.

Theorem 3.2. *There exist unique bases $\{L_\lambda \mid \lambda \in \Lambda^+(n)\}$ and $\{T_\lambda \mid \lambda \in \Lambda^+(n)\}$ for $\mathcal{F}_n$ which are $^-$-invariant and satisfy*

$$L_\lambda = \Delta_\lambda + \sum_{\mu < \lambda} e_{\lambda,\mu}(q)\Delta_\mu, \quad T_\lambda = \Delta_\lambda + \sum_{\mu < \lambda} d_{\lambda,\mu}(q)\Delta_\mu$$

for polynomials $e_{\lambda,\mu} \in q^{-1}\mathbb{Z}[q^{-1}]$ and $d_{\lambda,\mu} \in q\mathbb{Z}[q]$.

The basis $\{T_\lambda\}_{\lambda\in\Lambda^+(n)}$ is called the *lower global crystal basis* or *canonical basis* of $\mathcal{F}_n$, and $\{L_\lambda\}_{\lambda\in\Lambda^+(n)}$ is the *upper global crystal basis*. The (purely combinatorial) construction of these bases follows directly from the more general [53, Theorem 4.1] on passing to the quotient $\mathcal{F}_n$ of $\mathcal{F}$. In fact, the Leclerc–Thibon construction gives similar bases of $\mathcal{F}_n$ even for $n < p$. The strategy in the general case is to work not just with $U_q(\hat{\mathfrak{sl}}_p)$ but with the algebra $U_q(\hat{\mathfrak{gl}}_p)$, which is a sum of $U_q(\hat{\mathfrak{sl}}_p)$ and a Heisenberg algebra, and its action on $\mathcal{F}$ constructed in [43]. There is also a more geometric construction of the canonical basis of $\mathcal{F}_n$ in terms of the Hall algebra associated to the cyclic quiver of type $\hat{A}_p$ originating in [32] and exploited by Varagnolo and Vasserot [67].

To state the main result of [67], we let ξ be a primitive pth root of unity in $\mathbb{C}$. For $\lambda \in \Lambda^+(n)$, write $L_\xi(\lambda)$, $\Delta_\xi(\lambda)$, $\nabla_\xi(\lambda)$ and $T_\xi(\lambda)$ for the irreducible, standard, costandard and indecomposable tilting modules for "quantum GL_n" over $\mathbb{C}$ at the root of unity ξ. Let $\mathcal{G}_\xi$ denote the Grothendieck group of the category of polynomial representations of quantum GL_n at root of unity ξ. Then there are three natural bases for $\mathcal{G}_\xi$, namely, $\{[\Delta_\xi(\lambda)] = [\nabla_\xi(\lambda)]\}_{\lambda\in\Lambda^+(n)}$, $\{[L_\xi(\lambda)]\}_{\lambda\in\Lambda^+(n)}$ and $\{[T_\xi(\lambda)]\}_{\lambda\in\Lambda^+(n)}$ corresponding to the standard, irreducible and tilting modules respectively. Then the main results of [67] can be stated as:

Theorem 3.3. *Suppose that $p \geq n$. Identify the specialization $\mathcal{F}_n \otimes_{\mathbb{Z}[q,q^{-1}]} \mathbb{Z}$ at $q = 1$ with the Grothendieck group $\mathcal{G}_n$ so that*

$$\Delta_\lambda \otimes 1 = [\Delta_\xi(\lambda)]$$

for all $\lambda \in \Lambda^+(n)$. Then, for all $\lambda \in \Lambda^+(n)$,

$$L_\lambda \otimes 1 = [L_\xi(\lambda)], \quad T_\lambda \otimes 1 = [T_\xi(\lambda)].$$

In particular, this implies that the inverse decomposition numbers $[L_\xi(\lambda) : \Delta_\xi(\mu)]$ and the ∇-filtration multiplicities $[T_\xi(\lambda) : \nabla_\xi(\mu)]_\nabla$ can be computed from knowledge of the lower and upper global crystal bases of $\mathcal{F}_n$:

Corollary 3.4. *For $\lambda, \mu \in \Lambda^+(n)$,*

$$[L_\xi(\mu) : \Delta_\xi(\mu)] = e_{\lambda,\mu}(1), \quad [T_\xi(\lambda) : \nabla_\xi(\mu)]_\nabla = d_{\lambda,\mu}(1).$$

Remark 3.5. (I) We have stated 3.3 somewhat differently from [67]. The result concerning the basis $\{T_\lambda\}_{\lambda\in\Lambda^+(n)}$ stated here follows from [67, Corollary 11.2] by the definition of the decomposition matrix of the q-Schur algebra adopted in [67, 53], together with the quantum analogue of 3.1 (proved e.g. in [23]). The result concerning L_λ stated here is precisely [67, Theorem 12].

(II) In fact, for $p \geq n$, the result 3.3 in the case of the basis $\{T_\lambda\}_{\lambda \in \Lambda^+(n)}$ can also be deduced directly from Ariki's earlier theorem proving the original LLT conjecture [6, 54]. However, it is *not* immediately clear how to deduce the result about the basis $\{L_\lambda\}_{\lambda \in \Lambda^+(n)}$ from this using tricks like 3.1, the problem being that the necessary restriction $p \geq n$ does not allow one to transpose partitions.

(III) In [67, Proposition 9.3(j)], it is shown that the coefficients $e_{\lambda,\mu}(q)$ coincide with the signed sums of Kazhdan–Lusztig polynomials appearing in the statement of the (quantum) Lusztig conjecture (for λ, μ lying in interiors of alcoves). Thus [67] gives an alternative proof of the Lusztig conjecture for quantum groups of type A.

(IV) In [67, Proposition 9.3(k)], it is shown that the coefficients $d_{\lambda,\mu}(q)$ coincide with certain Kazhdan–Lusztig polynomials for the affine Hecke algebra of type A. In particular, this shows that the coefficients of the polynomials $d_{\lambda,\mu}(q)$ are all nonnegative integers, as conjectured in [53]. There is a seemingly quite different approach due to Soergel [64, 65] to computing the multiplicities $[T_\xi(\lambda) : \nabla_\xi(\mu)]_\nabla$ (in arbitrary type!) for λ, μ lying in interiors of alcoves, again involving affine Kazhdan–Lusztig polynomials. The Kazhdan–Lusztig polynomials appearing in both approaches are the same. This has also been verified combinatorially by Goodman and Wenzl [33].

(V) The bases $\{L_\lambda\}_{\lambda \in \Lambda^+(n)}$ and $\{T_\lambda\}_{\lambda \in \Lambda^+(n)}$ were constructed more generally for $p < n$ in [53], as we have said. The analogue of 3.3 for $p < n$ was also conjectured by Leclerc and Thibon. This was not quite proved by Vasserot and Varagnolo: the problem left was to show that the canonical basis constructed there geometrically using the Hall algebra coincided with the combinatorial construction of Leclerc and Thibon. This identification, completing the proof of the Leclerc–Thibon conjecture, has recently been made by Schiffmann [63].

Now we wish to discuss how the results in Section 2 relate to 3.3 in the quantum case. We will assume that *all* the earlier results described in Section 2 in the classical setting have analogues for quantum groups. We have no doubt that this is true, even in the quantum mixed case, with precisely the same statements: there are no modifications necessary resulting from the different Steinberg tensor product theorems in the two settings. However, to date, full proofs in the quantum case have only been given in roughly half of the results, see [10]; the remaining quantizations have not been carried out yet. We remark that it is not obvious how to deduce all the results from Section 2 in the quantum case directly from 3.3: the latter at present only gives information about multiplicities, not about submodule structure.

First, we observe that at $q = 1$, $f_\alpha \Delta_\lambda = \sum_i \Delta_{\lambda+\epsilon_i}$, summed over all λ-addable i with $\mathrm{res}(i, \lambda_i + 1) = \alpha$, which is precisely the same as the effect of the (quantum analogue of the) functor F_α on the basis $[\Delta_\xi(\lambda)]$ of $\mathcal{G}_\xi$. Consequently, we can identify the operator f_α and the functor F_α in their actions on the Grothendieck group. So we can calculate the composition multiplicities (resp. the tilting module multiplicities) in $F_\alpha L_\xi(\lambda)$ (resp. $F_\alpha T_\xi(\lambda)$) algorithmically by first computing $f_\alpha L_\lambda$ (resp. $f_\alpha T_\lambda$) using the known action of f_α on the Δ_λ's, then rewriting the resulting expression in terms of the L_λ's (resp. the T_λ's). This observation was first made in [27] in a slightly different setting.

Let us give a very simple example. Suppose that $n = 4$, $p = 3$ and $\lambda = (4, 2, 0)$. This is the highest weight of the Steinberg module, so

$$L_{(4,2,0)} = \Delta_{(4,2,0)}.$$

Now apply f_1 to deduce that

$$f_1 L_{(4,2,0)} = \Delta_{(5,2,0)} + q\Delta_{(4,3,0)} + q^2\Delta_{(4,2,1)}$$

which on rewriting in terms of the L basis gives

$$f_1 L_{(4,2,0)} = L_{(5,2,0)} + (q + q^{-1})L_{(4,3,0)} + (q^2 + 1 + q^{-2})L_{(4,2,1)}.$$

Now, 2.3(iii) tells us that $F_1 L_\xi((4, 2, 0))$ has Loewy length at least 5 and contains the composition factors $L_\xi((5, 2, 0))$, $L_\xi((4, 3, 0))$ and $L_\xi((4, 2, 1))$ with multiplicities $1, 2$ and 3 respectively.

In fact, it is easy to see using the positivity 3.5(IV) that, quite generally, one can always write

$$f_\alpha L_\lambda = \sum_\mu b_{\lambda,\mu}(q) L_\mu$$

for Laurent polynomials $b_{\lambda,\mu}(q)$ which are always positive linear combinations of quantum integers. Evaluating at $q = 1$, $b_{\lambda,\mu}(1)$ computes the composition multiplicity $[F_\alpha L_\xi(\lambda) : L_\xi(\mu)]$. It seems reasonable to expect that, in the quantum case, the Loewy length of $F_\alpha L_\xi(\lambda)$ is exactly as predicted in 2.7(v), and moreover that the powers of q in the polynomial $b_{\lambda,\mu}(q)$ indicate the Loewy levels that the irreducible $L_\xi(\mu)$ appears in $F_\alpha L_\xi(\lambda)$.

One consequence of these remarks is that there should be an alternative approach to proving the quantum analogue of 2.5, very similar to the result of [27]. In the language of crystal bases, 2.5 is equivalent to describing precisely when the operator f_α sends an upper global crystal basis element to a single upper global crystal basis element. It would be interesting to determine in a similar way when f_α sends a lower global crystal basis element to a single lower global crystal basis element, that is, when $f_\alpha T_\lambda = T_\mu$ for some μ. In terms of representation theory, this is:

Question. For $\lambda \in \Lambda^+(n)$, when is $F_\alpha T(\lambda)$ an indecomposable tilting module?

4. Relating tensor products and restrictions

We have already mentioned in Section 1 one source of connections between tensor products and restrictions to Levi subgroups. We will now describe two more ways such connections arise. The results in this section were all obtained in [13].

Fix $a \geq 1$ and $\nu = (n_1, \dots, n_a) \in \Lambda(a, n)$ (a composition of n with a non-zero parts). Let $\mathrm{GL}(\nu) = \mathrm{GL}(n_1) \times \cdots \times \mathrm{GL}(n_a)$ denote the *standard Levi subgroup* of $\mathrm{GL}(n)$ consisting of all invertible block diagonal matrices with block sizes $n_1, \dots, n_a$. Of course, if $\nu = (n)$ then $\mathrm{GL}(\nu) = \mathrm{GL}(n)$ while, at the other extreme, if $\nu = (1, \dots, 1)$ then $\mathrm{GL}(\nu)$ is the maximal torus $T < \mathrm{GL}(n)$. The following theorem is [13, Theorem 2.8].

Theorem 4.1. *Let $\nu \in \Lambda(a, n)$ and $\mu^{(1)}, \dots, \mu^{(a)} \in \Lambda^+(n)$ be partitions such that $\mu^{(i)}$ has at most n_i non-zero rows for each i. Let $\bar{\mu}^{(i)} = (\mu_1^{(i)}, \dots, \mu_{n_i}^{(i)}) \in \Lambda^+(n_i)$. For any polynomial $\mathrm{GL}(n)$-module M,*

$$\begin{aligned}\mathrm{Hom}_{\mathrm{GL}(n)}\big(M, \nabla_n(\mu^{(1)}) \otimes \cdots \otimes \nabla_n(\mu^{(a)})\big) \\ \cong \mathrm{Hom}_{\mathrm{GL}(\nu)}\big(M{\downarrow}_{\mathrm{GL}(\nu)}, \nabla_{n_1}(\bar{\mu}^{(1)}) \boxtimes \cdots \boxtimes \nabla_{n_a}(\bar{\mu}^{(a)})\big).\end{aligned}$$

The following corollary of 4.1 (with $a = 2$) should be compared with 1.2:

Corollary 4.2. *Fix $\lambda, \mu \in \Lambda^+(n)$ with $\mu_n = 0$. Then*

$$\mathrm{Hom}_{\mathrm{GL}(n)}\big(L_n(\lambda), \nabla_n(\mu) \otimes S^\ell(V)\big) \cong \mathrm{Hom}_{\mathrm{GL}(n-1)}(\Delta_{n-1}(\bar{\mu}), L_n(\lambda) \downarrow_{\mathrm{GL}(n-1)})$$

where $\ell = |\lambda| - |\mu|$.

The main tool used in the proof of 4.1 is a *polynomial induction functor* from Levi subgroups. This notion goes back to [21] (see also [25]). However we prove a new property of this functor (see 4.3 below), which is crucial for 4.1.

Let $M_{\mathbb{F}}(n)$ (resp. $M_{\mathbb{F}}(n, r)$) be the category of polynomial $\mathrm{GL}(n)$-modules (resp. of degree r) and, analogously, let $M_{\mathbb{F}}(\nu)$ (resp. $M_{\mathbb{F}}(\nu, r)$) be the category of polynomial $\mathrm{GL}(\nu)$-modules (resp. of degree r). The restriction of a polynomial module (resp. a polynomial module of degree r) from $\mathrm{GL}(n)$ to $\mathrm{GL}(\nu)$ is again a polynomial module (resp. a polynomial module of degree r). So we have the exact restriction functor

$$R^n_\nu : M_{\mathbb{F}}(n) \to M_{\mathbb{F}}(\nu).$$

We now describe how to construct a functor which is right adjoint to R^n_ν. Let $A(n)$ denote the subalgebra of the algebra of regular functions $\mathbb{F}[\mathrm{GL}(n)]$ generated by the functions $\{c_{ij} \mid 1 \leq i, j \leq n\}$, where c_{ij} picks out the ij-entry of a matrix $g \in \mathrm{GL}(n)$. There are two commuting *left* actions of $\mathrm{GL}(n)$ on $A(n)$, the *left regular* and *right regular* actions, which we define for $g, g' \in \mathrm{GL}(n)$, $f \in A(n)$ by $(g \cdot_l f)(g') = f(g^{-1}g')$ and $(g \cdot_r f)(g') = f(g'g)$ respectively.

For $M \in M_{\mathbb{F}}(\nu)$, we define the $\mathrm{GL}(n)$-module

$$(M \otimes A(n))^{\mathrm{GL}(\nu)}$$

where the $\mathrm{GL}(n)$-action on the induced module comes from the right regular action of $\mathrm{GL}(n)$ on $A(n)$ and the trivial action on M, and the action of $\mathrm{GL}(\nu)$ on $M \otimes A(n)$ under which we are taking fixed points comes from the given action on M and the left regular action on $A(n)$.

Define the *polynomial induction functor*

$$I_\nu^n : M_{\mathbb{F}}(\nu) \to M_{\mathbb{F}}(n)$$

by letting $I_\nu^n M := (M \otimes A(n))^{\mathrm{GL}(\nu)}$, with the obvious definition on morphisms. For $M \in M_{\mathbb{F}}(\nu)$, $I_\nu^n M$ can be described alternatively as the largest polynomial submodule of $\mathrm{Ind}_{\mathrm{GL}(\nu)}^{\mathrm{GL}(n)} M$, where $\mathrm{Ind}_{\mathrm{GL}(\nu)}^{\mathrm{GL}(n)}$ denotes the usual induction functor in the sense of algebraic groups [40, I.3.3]. One easily checks that I_ν^n is right adjoint to R_ν^n, so is left exact and sends injectives in $M_{\mathbb{F}}(\nu)$ to injectives in $M_{\mathbb{F}}(n)$. Moreover, for $M \in M_{\mathbb{F}}(n)$ and any $N \in M_{\mathbb{F}}(\nu)$ with a ∇-filtration,

$$\mathrm{Ext}^i_{\mathrm{GL}(\nu)}(R_\nu^n M, N) \cong \mathrm{Ext}^i_{\mathrm{GL}(n)}(M, I_\nu^n N)$$

for all $i \geq 0$. As observed in [13], this implies, using the well-known analogous result about restrictions of modules with ∇-filtrations to Levi subgroups, that I_ν^n sends modules with ∇-filtrations to modules with ∇-filtrations.

Now fix $m \leq n$, and let $\nu = (m, 1, \dots, 1)$, a composition of n. Then $\mathrm{GL}(m)$ is a normal subgroup of the Levi subgroup $\mathrm{GL}(\nu) \leq \mathrm{GL}(n)$, and we have an exact inflation functor $\mathrm{Infl}_m^\nu : M_{\mathbb{F}}(m) \to M_{\mathbb{F}}(\nu)$ which sends each $M_{\mathbb{F}}(m, r)$ into $M_{\mathbb{F}}(\nu, r)$. Let $H(m) \leq \mathrm{GL}(\nu)$ be the $(n-m)$-dimensional torus such that $\mathrm{GL}(\nu) = \mathrm{GL}(m) \times H(m)$. Then the fixed point functor $M \mapsto M^{H(m)}$ is right adjoint to Infl_ν^m (see e.g. [40, I.6.4]). Now define the functors

$$I_m^n : M_{\mathbb{F}}(m) \to M_{\mathbb{F}}(n) \quad \text{and} \quad R_m^n : M_{\mathbb{F}}(n) \to M_{\mathbb{F}}(m)$$

by the compositions

$$I_m^n := I_\nu^n \circ \mathrm{Infl}_m^\nu, \quad R_m^n := (-)^{H(m)} \circ R_\nu^n.$$

We note that I_m^n is right adjoint to the exact functor R_m^n, hence is left exact and sends injectives to injectives.

Our main motivation for introducing the functors I_m^n and I_ν^n is the following result [13, Theorem 2.7] describing polynomial induction applied to an outer tensor product:

Theorem 4.3. *Let $\nu \in \Lambda(a, n)$. Take any modules $M_1, \dots, M_a$ with $M_i \in M_{\mathbb{F}}(n_i)$, so that $M_1 \boxtimes \cdots \boxtimes M_a \in M_{\mathbb{F}}(\nu)$. Then*

$$I_\nu^n(M_1 \boxtimes \cdots \boxtimes M_a) \cong (I_{n_1}^n M_1) \otimes \cdots \otimes (I_{n_a}^n M_a).$$

Finally, we observe that, for $\mu \in \Lambda^+(n)$ and $m \leq n$ with $\mu_{m+1} = \cdots = \mu_n = 0$, we showed in [13] that

$$I_m^n \nabla_m(\bar{\mu}) \cong \nabla_n(\mu) \quad \text{and} \quad I_m^n L_m(\bar{\mu}) \cong L_n(\mu)$$

where $\bar{\mu} = (\mu_1, \dots, \mu_m) \in \Lambda^+(m)$. Combining this with 4.3, our original result 4.1 follows immediately from 'Frobenius reciprocity'.

Remark 4.4. There is also an analogous result to 4.3 for quantum GL_n. The details appear in [12].

Now we discuss our second, quite different, relation between tensor products and restrictions to Levi subgroups. This comes from Donkin's characteristic free version of Howe duality (see [13, section 3] for more details). Denote

$$\Lambda^+(n \times m) := \{\lambda = (\lambda_1, \lambda_2, \dots) \in \Lambda^+(n) \mid \lambda_1 \leq m\}.$$

The following theorem is [13, Theorem 3.5]:

Theorem 4.5. *Fix $a, n, m \geq 1$, and choose $\nu \in \Lambda(a, n)$. Let $\mathrm{GL}(\nu)$ denote the standard Levi subgroup of $\mathrm{GL}(n)$. Choose $\lambda \in \Lambda^+(n \times m)$ and $\mu^{(i)} \in \Lambda^+(n_i \times m)$ for $i = 1, \dots, a$. Then*

(i) $$(T_n(\lambda) \downarrow_{\mathrm{GL}(\nu)} : T_{n_1}(\mu^{(1)}) \boxtimes \cdots \boxtimes T_{n_a}(\mu^{(a)})) = \left[L_m((\mu^{(1)})^t) \otimes \cdots \otimes L_m((\mu^{(a)})^t) : L_m(\lambda^t)\right];$$

(ii) $$\left[L_n(\lambda) \downarrow_{\mathrm{GL}(\nu)} : L_{n_1}(\mu^{(1)}) \boxtimes \cdots \boxtimes L_{n_a}(\mu^{(a)}))\right] = \left(T_m((\mu^{(1)})^t) \otimes \cdots \otimes T_m((\mu^{(a)})^t) : T_m(\lambda^t)\right).$$

Remark 4.6. There is a generalization of 4.5 involving the Andersen and Jantzen filtrations [40, 3, 4]. We illustrate this in a special case. For $\lambda \in \Lambda^+(n \times m)$, $\mu \in \Lambda^+(n \times m_1)$ and $\nu \in \Lambda^+(n \times m_2)$ with $m = m_1 + m_2$, the space

$$E_\lambda\big(T_n(\mu) \otimes T_n(\nu)\big) = \mathrm{Hom}_{\mathrm{GL}(n)}\big(T_n(\mu) \otimes T_n(\nu), \nabla_n(\lambda)\big)$$

has an *Andersen filtration* $\mathfrak{A}^i E_\lambda(T_n(\mu) \otimes T_n(\nu))$ $(i \geq 0)$ (taking notation from [57]). The space $\Delta_m(\lambda^t)$ has a *Jantzen filtration* $\mathfrak{J}^i \Delta_m(\lambda^t)$ $(i \geq 0)$. One can prove that the dimension of $\mathfrak{A}^i E_\lambda(T_n(\mu) \otimes T_n(\nu))$ is precisely the composition multiplicity $[\mathfrak{J}^i \Delta_m(\lambda^t) \downarrow_{\mathrm{GL}(m_1) \times \mathrm{GL}(m_2)} : L_{m_1}(\mu^t) \boxtimes L_{m_2}(\nu^t)]$. The result 4.5(i) follows directly since the top factor in the Andersen filtration computes $(T_n(\mu) \otimes T_n(\nu) : T_n(\lambda))$ and the top factor in the Jantzen filtration computes the composition multiplicity $[L_m(\lambda^t) \downarrow_{\mathrm{GL}(m_1) \times \mathrm{GL}(m_2)} : L_{m_1}(\mu^t) \boxtimes L_{m_2}(\nu^t)]$.

A special case of 4.5(ii) gives (compare with 1.2 and 4.2):

Corollary 4.7. *For $\lambda \in \Lambda^+(n \times m)$ and $\mu \in \Lambda^+(n \times (m-1))$ with $|\lambda| \geq |\mu|$, put $\ell = |\lambda| - |\mu|$. Then*

$$\big(T_n(\mu) \otimes \textstyle\bigwedge^\ell(V) : T_n(\lambda)\big) = \big[L_m(\lambda^t) \downarrow_{\mathrm{GL}(m-1)} : L_{m-1}(\mu^t)\big].$$

The significance of this corollary is that it explains the connection between the work of Mathieu and Papadopoulou [56] and the paper [16]. Both papers obtained a general character formula for $L_n(\lambda)$ for a special class of highest weights λ, which we called the *completely splittable weights* in [16]. The approach in [16] depended on first understanding precisely the restriction $L_n(\lambda) \downarrow_{\mathrm{GL}(n-1)}$ in the special cases that it is completely reducible, that is, the right hand side of 4.7 for special λ. On the other hand, Mathieu and Papadopoulou exploited the results of [30, 31] to determine the structure of $T_n(\lambda) \otimes \bigwedge^i(V)$ for special λ, modulo a certain 'tilting ideal'. We will discuss further relations between branching rules and tilting ideals in Section 6.

Finally, we note that taking $\nu = (1, 1, \dots, 1)$ in 4.1 and 4.5, we obtain the following character formulas:

Corollary 4.8. *Let* $\lambda \in \Lambda^+(n \times m)$, $\mu = (\mu_1, \dots, \mu_n) \in \Lambda(n \times m)$. *Let* V *and* W *be the natural* $\mathrm{GL}(n)$- *and* $\mathrm{GL}(m)$-*modules respectively. Then,*

(i) $\dim L_n(\lambda)_\mu = (S^{\mu_1}(V) \otimes \cdots \otimes S^{\mu_n}(V) : Q_n(\lambda))$, *where* $Q_n(\lambda)$ *is the injective hull of* $L_n(\lambda)$ *in the category of polynomial* $\mathrm{GL}(n)$-*modules;*

(ii) $\dim L_n(\lambda)_\mu = (\bigwedge^{\mu_1}(W) \otimes \cdots \otimes \bigwedge^{\mu_n}(W) : T_m(\lambda^t))$;

(iii) $\dim T_n(\lambda)_\mu = [\bigwedge^{\mu_1}(W) \otimes \cdots \otimes \bigwedge^{\mu_n}(W) : L_m(\lambda^t)]$.

Of these, (i) is a result of Donkin [22, Lemma 3.4(i)] and (ii) is due to Mathieu and Papadopoulou [56].

5. The symmetric group

The results in Section 2 on translation functors can be translated into analogous results about symmetric groups using the techniques of the Schur functors. In this section we describe these results on the symmetric groups, most but not all of which can be found in [14], and shortly discuss the 'translation' techniques.

Let Σ_r be the symmetric group on r letters. If λ is a partition of r we write $\lambda \vdash r$. We denote by S^λ (resp. Y^λ, M^λ) the Specht (resp. Young, permutation) module over $\mathbb{F}\Sigma_r$ corresponding to a partition $\lambda \vdash r$, and by D^λ the irreducible $\mathbb{F}\Sigma_r$-module corresponding to a *p-regular* partition $\lambda \vdash r$. The reader is referred to [37], [38] or [39] for these and other standard notions of the representation theory of symmetric groups. We denote by sgn the 1-dimensional sign representation of Σ_r.

Fix a partition $\lambda = (\lambda_1, \lambda_2, \dots) \vdash r$. We identify λ with its *Young diagram*

$$\lambda = \{(i, j) \in \mathbb{N} \times \mathbb{N} \mid j \leq \lambda_i\}.$$

The elements of $\mathbb{N} \times \mathbb{N}$ are called *nodes*. A node of the form (i, λ_i) is called a *removable node* (of λ) if $\lambda_i > \lambda_{i+1}$; a node of the form $(i, \lambda_i + 1)$ is called an *addable node* (for λ) if $i = 1$ or $i > 1$ and $\lambda_i < \lambda_{i-1}$. If $A = (i, \lambda_i)$ is a removable node, we denote by

$$\lambda_A = \lambda \setminus \{A\} = (\lambda_1, \dots, \lambda_{i-1}, \lambda_i - 1, \lambda_{i+1}, \dots)$$

the partition of $r - 1$ obtained by removing A from λ. If $B = (i, \lambda_i + 1)$ is an addable node, we denote by

$$\lambda^B = \lambda \cup \{B\} = (\lambda_1, \dots, \lambda_{i-1}, \lambda_i + 1, \lambda_{i+1}, \dots)$$

the partition of $r+1$ obtained by adding B to λ. The p-residue of a node $A = (i, j)$ is defined as in Section 2: $\operatorname{res} A = (j - i) \pmod p$.

The next definitions are the same as the analogous ones in Section 2 except that we have transposed partitions.

A removable node A (of λ) is called *normal* if for every addable node B to the right of A with $\operatorname{res} B = \operatorname{res} A$ there exists a removable node $C(B)$ strictly between A and B with $\operatorname{res} C(B) = \operatorname{res} A$, and $B \neq B'$ implies $C(B) \neq C(B')$. A removable node is called *good* if it is the leftmost among the normal nodes of a fixed residue.

An addable node B (for λ) is called *conormal* if for every removable node A to the left of B with $\operatorname{res} A = \operatorname{res} B$ there exists an addable node $C(A)$ strictly between B and A with $\operatorname{res} C(A) = \operatorname{res} B$, and $A \neq A'$ implies $C(A) \neq C(A')$. An addable node is called *cogood* if it is the rightmost among the conormal nodes of a fixed residue.

For $\alpha \in \mathbb{Z}/p\mathbb{Z}$ and a partition λ, define the *α-content* of λ to be the integer $\operatorname{cont}_\alpha(\lambda) := |\{A \in \lambda \mid \operatorname{res} A = \alpha\}|$, which is again a special case of the corresponding definition in Section 2. For two partitions λ and μ we write $\lambda \sim \mu$ if $\operatorname{cont}_\alpha(\lambda) = \operatorname{cont}_\alpha(\mu)$ for all $\alpha \in \mathbb{Z}/p\mathbb{Z}$. The 'Nakayama Conjecture' (see e.g. [39]) claims that $\mathbb{F}\Sigma_r$-modules D^λ and D^μ belong to the same block if and only if $\lambda \sim \mu$.

Fix a residue $\alpha \in \mathbb{Z}/p\mathbb{Z}$. We define the functors

$$\operatorname{Ind}^\alpha : \mathbb{F}\Sigma_r\text{-mod} \to \mathbb{F}\Sigma_{r+1}\text{-mod} \quad \text{and} \quad \operatorname{Res}_\alpha : \mathbb{F}\Sigma_r\text{-mod} \to \mathbb{F}\Sigma_{r-1}\text{-mod}$$

by defining them first on a module M in an (arbitrary fixed) block and then extending additively to the whole of $\mathbb{F}\Sigma_r$-mod. Assume M belongs to the block corresponding to the residue contents $c_0, c_1, \dots, c_{p-1}$ - this means that for any irreducible module D^λ in this block, $\operatorname{cont}_\beta(\lambda) = c_\beta$ for all $\beta \in \mathbb{Z}/p\mathbb{Z}$. We now let $\operatorname{Ind}^\alpha M$ (resp. $\operatorname{Res}_\alpha M$) denote the largest submodule of $M \uparrow^{\Sigma_{r+1}}$ (resp. $M \downarrow_{\Sigma_{r-1}}$) all of whose composition factors are of the form D^μ with

$$\operatorname{cont}_\alpha(\mu) = c_\alpha + 1 \quad (\text{resp. } \operatorname{cont}_\alpha(\mu) = c_\alpha - 1),$$

and

$$\operatorname{cont}_\beta(\mu) = c_\beta \quad \text{for all } \alpha \neq \beta \in \mathbb{Z}/p\mathbb{Z}.$$

Given a morphism $\theta : M \to N$, $\operatorname{Ind}^\alpha \theta$ is just the restriction to $\operatorname{Ind}^\alpha M$ of the natural map $\hat{\theta} : M \uparrow^{\Sigma_{r+1}} \to N \uparrow^{\Sigma_{r+1}}$, induced by θ, and similarly for $\operatorname{Res}_\alpha$. We have

$$M \uparrow^{\Sigma_{r+1}} \cong \bigoplus_{\alpha \in \mathbb{Z}/p\mathbb{Z}} \operatorname{Ind}^\alpha M \quad \text{and} \quad M \downarrow_{\Sigma_{r-1}} \cong \bigoplus_{\alpha \in \mathbb{Z}/p\mathbb{Z}} \operatorname{Res}_\alpha M.$$

The functors just defined are called *Robinson's α-induction* and *α-restriction functors* (cf. [39, 6.3.16]).

We collect the following known results about the effect of $\operatorname{Ind}^\alpha$ applied to Specht modules. Of these, part (i) is very well-known, see for example [37, 17.14], part (ii) is [14, Theorem D], while part (iii) follows easily from [47, Theorem 0.4] by Frobenius reciprocity.

Theorem 5.1. *Fix a partition $\lambda \vdash r$ and a residue $\alpha \in \mathbb{Z}/p\mathbb{Z}$. Then $\mathrm{Ind}^\alpha S^\lambda$ is zero unless there is at least one addable node B with $\mathrm{res}\, B = \alpha$. In that case:*

(i) *There is a filtration $(0) = S_0 \subset S_1 \subset \cdots \subset S_k = \mathrm{Ind}^\alpha S^\lambda$ with $S_i/S_{i-1} \cong S^{\lambda^{B_i}}$, $i = 1, 2, \ldots, k$, where $B_1, B_2, \ldots, B_k$ are the addable nodes (for λ) of residue α counted from left to right.*

(ii) *Assume that λ^t is p-regular, and $\mu \vdash r+1$ is a partition such that μ^t is p-regular. Then*

$$\mathrm{Hom}_{\Sigma_{r+1}}((S^\mu)^*, \mathrm{Ind}^\alpha S^\lambda) = \begin{cases} \mathbb{F} & \text{if } \mu = \lambda^B \text{ for some addable } B \text{ of residue } \alpha, \\ 0 & \text{otherwise.} \end{cases}$$

(iii) *Assume that λ is p-regular. Then the head of $\mathrm{Ind}^\alpha S^\lambda$ is $\bigoplus_B D^{\lambda^B}$ where the sum is taken over all addable nodes B of residue α such that B is normal for λ^B (and all such λ^B are p-regular).*

Next we consider Ind^α applied to an irreducible module.

Theorem 5.2. *Fix a p-regular partition $\lambda \vdash r$ and a residue $\alpha \in \mathbb{Z}/p\mathbb{Z}$. Then $\mathrm{Ind}^\alpha D^\lambda$ is zero unless λ has at least one conormal node of residue α. In that case,*

(i) *$\mathrm{Ind}^\alpha D^\lambda$ is an indecomposable, self-dual module, with simple socle and head isomorphic to D^{λ^B} where B is the cogood node of residue α;*

(ii) *$\mathrm{Ind}^\alpha D^\lambda$ is irreducible if and only if there is a unique conormal node of residue α;*

(iii) *for any p-regular $\mu \vdash r+1$,*

$$\mathrm{Hom}_{\Sigma_{r+1}}(S^\mu, \mathrm{Ind}^\alpha D^\lambda) = \begin{cases} \mathbb{F} & \text{if } \mu = \lambda^B \text{ for some conormal } B \\ & \text{with } \mathrm{res}\, B = \alpha, \\ 0 & \text{otherwise;} \end{cases}$$

(iv) *for any addable node B such that λ^B is p-regular,*

$$[\mathrm{Ind}^\alpha D^\lambda : D^{\lambda^B}] = \begin{cases} d_B & \text{if } B \text{ is conormal for } \lambda \text{ and } \mathrm{res}\, B = \alpha, \\ 0 & \text{otherwise} \end{cases}$$

where d_B denotes the number of conormal nodes C to the left of B (counting B itself) such that $\mathrm{res}\, C = \alpha$;

(v) *the endomorphism ring $\mathrm{End}_{\Sigma_{r+1}}(\mathrm{Ind}^\alpha D^\lambda)$ is isomorphic to the truncated polynomial ring $\mathbb{F}[T]/(T^d)$ of dimension d, where d is the number of conormal nodes B with $\mathrm{res}\, B = \alpha$.*

As a consequence of 5.2(ii), we have:

Corollary 5.3. *For p-regular λ, the induced module $D^\lambda \uparrow^{\Sigma_{r+1}}$ is completely reducible if and only if all conormal nodes have different residues.*

We note that a criterion of complete reducibility for $D^\lambda \uparrow^{\Sigma_{r+1}}$ different from the one in 5.3 was found in [49]. Bessenrodt and Olsson informed us that they were able to prove directly that the two combinatorial conditions are equivalent.

Our next result gives further information about the structure of $\mathrm{Ind}^\alpha D^\lambda$:

Theorem 5.4. *Fix a p-regular partition $\lambda \vdash r$ and a residue $\alpha \in \mathbb{Z}/p\mathbb{Z}$. Let $B_1, B_2 \dots B_d$ be all conormal nodes of residue α counted from left to right. Then $N := \mathrm{Ind}^\alpha D^\lambda$ has a filtration $0 = N_0 < N_1 < \cdots < N_d = N$ such that*

(i) *for $1 \le i \le d$, N_i/N_{i-1} is a non-zero quotient of $S^{\lambda^{B_i}}$, and*

(ii) *for $1 \le i \le j \le d$ such that λ^{B_j} is p-regular, $[N_i/N_{i-1} : D^{\lambda^{B_j}}] = 1$.*

(iii) *Let $1 < j < d$. The extension*

$$0 \to N_j/N_{j-1} \to N_{j+1}/N_{j-1} \to N_{j+1}/N_j \to 0$$

does not split.

(iv) *If all λ^{B_j} for $1 \le j \le d$ are p-regular, then the Loewy length of N is at least $2d - 1$.*

We state now the dual results to 5.1, 5.2, 5.4 for the functor Res_α. The results of 5.5(i), (ii), 5.6, 5.7 were obtained in [46]–[50].

Theorem 5.5. *Fix a partition $\lambda \vdash r$ and a residue $\alpha \in \mathbb{Z}/p\mathbb{Z}$. Then $\mathrm{Res}_\alpha S^\lambda$ is zero unless there is at least one removable node A with $\mathrm{res}\, A = \alpha$. In that case:*

(i) *There is a filtration $(0) = S_0 \subset S_1 \subset \cdots \subset S_k = \mathrm{Res}_\alpha S^\lambda$ with $S_i/S_{i-1} \cong S^{\lambda_{A_i}}$, $i = 1, 2, \dots, k$, where $A_1, A_2, \dots, A_k$ are the removable nodes (of λ) of residue α counted from right to left.*

(ii) *Assume that λ is p-regular, and $\mu \vdash r-1$ is another p-regular partition. Then*

$$\mathrm{Hom}_{\Sigma_{r-1}}((S^\mu)^*, \mathrm{Res}_\alpha S^\lambda) = \begin{cases} \mathbb{F} & \text{if } \mu = \lambda^A \text{ for some removable } A \\ & \text{of residue } \alpha, \\ 0 & \text{otherwise.} \end{cases}$$

(iii) *Assume that λ is p-regular. Then the head of $\mathrm{Res}_\alpha S^\lambda$ is $\bigoplus_A D^{\lambda_A}$ where the sum is taken over all removable nodes A of residue α such that A is conormal for λ_A (and every such λ_A is p-regular).*

Theorem 5.6. *Fix a p-regular partition $\lambda \vdash r$ and a residue $\alpha \in \mathbb{Z}/p\mathbb{Z}$. Then $\mathrm{Res}_\alpha D^\lambda$ is zero unless λ has at least one normal node of residue α. In that case,*

(i) *$\mathrm{Res}_\alpha D^\lambda$ is an indecomposable, self-dual module, with simple socle and head isomorphic to D_{λ_A} where A is the good node of residue α;*

(ii) $\mathrm{Res}_\alpha D^\lambda$ *is irreducible if and only if there is a unique normal node of residue* α*;*

(iii) *for any p-regular* $\mu \vdash r-1$,

$$\mathrm{Hom}_{\Sigma_{r-1}}(S^\mu, \mathrm{Res}_\alpha D^\lambda) = \begin{cases} \mathbb{F} & \text{if } \mu = \lambda_A \text{ for some normal } A \text{ with } \mathrm{res}\, A = \alpha, \\ 0 & \text{otherwise;} \end{cases}$$

(iv) *for any removable node A such that* λ_A *is p-regular,*

$$[\mathrm{Res}_\alpha D^\lambda : D^{\lambda_A}] = \begin{cases} d_A & \text{if } A \text{ is normal for } \lambda \text{ and } \mathrm{res}\, A = \alpha, \\ 0 & \text{otherwise,} \end{cases}$$

where d_A *denotes the number of normal nodes D to the right of A (counting A itself) such that* $\mathrm{res}\, D = \alpha$*;*

(v) *the endomorphism ring* $\mathrm{End}_{\Sigma_{r-1}}(\mathrm{Res}_\alpha D^\lambda)$ *is isomorphic to the truncated polynomial ring* $\mathbb{F}[T]/(T^d)$ *of dimension d, where d is the number of normal nodes A with* $\mathrm{res}\, A = \alpha$.

In particular, 5.6(ii) gives a criterion for complete reducibility of $D^\lambda\downarrow_{\Sigma_{r-1}}$: the restriction $D^\lambda\downarrow_{\Sigma_{r-1}}$ is completely reducible if and only if all normal nodes have different residues.

Theorem 5.7. *Fix a p-regular partition* $\lambda \vdash r$ *and a residue* $\alpha \in \mathbb{Z}/p\mathbb{Z}$*. Let* $A_1, A_2 \dots A_d$ *be all normal nodes of residue* α *counted from right to left. Then* $N := \mathrm{Res}_\alpha D^\lambda$ *has a filtration* $0 = N_0 < N_1 < \cdots < N_d = N$ *such that*

(i) *for* $1 \le i \le d$, N_i/N_{i-1} *is a non-zero quotient of* $S^{\lambda_{A_i}}$*, and*

(ii) *for* $1 \le i \le j \le d$ *with* λ_{A_j} *is p-regular,* $[N_i/N_{i-1} : D^{\lambda_{A_j}}] = 1$.

(iii) *Let* $1 < j < d$*. The extension*

$$0 \to N_j/N_{j-1} \to N_{j+1}/N_{j-1} \to N_{j+1}/N_j \to 0$$

does not split.

(iv) *If all* λ_{A_j} *for* $1 \le j \le d$ *are p-regular, then the Loewy length of N is at least* $2d-1$.

Now we want to say a little more about the translation techniques to go from the results on tensor products in Section 2 to the results stated here about Ind^α (and analogously how to deduce the results stated here about Res_α from branching rules from $\mathrm{GL}(n)$ to $\mathrm{GL}(n-1)$).

Fix integers n, r and compositions

$$\nu = (n_1, \dots, n_a) \in \Lambda(a, n) \quad \text{and} \quad \rho = (r_1, \dots, r_a) \in \Lambda(a, r)$$

for some a such that $n \ge r$ and $n_i \ge r_i$ for $i = 1, \dots, a$. We denote by $\Sigma_\rho \cong \Sigma_{r_1} \times \cdots \times \Sigma_{r_a} < \Sigma_r$ the standard Young subgroup of Σ_r corresponding to the composition ρ. Let $\mathrm{GL}(\nu)$ be the corresponding Levi subgroup of $\mathrm{GL}(n)$, and Σ_ρ be the corresponding Young subgroup $\Sigma_{r_1} \times \cdots \times \Sigma_{r_a}$ of Σ_r. We first consider the effect of Schur functors (which amount to taking certain weight spaces) on restrictions from $\mathrm{GL}(n)$ to $\mathrm{GL}(\nu)$.

Let ω be the weight

$$\omega = (\underbrace{\overbrace{1,\dots,1}^{r_1 \text{ entries}},0,\dots,0}_{n_1 \text{ entries}},\underbrace{\overbrace{1,\dots,1}^{r_2 \text{ entries}},0,\dots,0}_{n_2 \text{ entries}},\dots,\underbrace{\overbrace{1,\dots,1}^{r_a \text{ entries}},0,\dots,0}_{n_a \text{ entries}}).$$

Let Perm_r denote the subgroup of $\mathrm{GL}(n)$ consisting of all permutation matrices g with $g_{i,i} = 1$ whenever $\omega_i = 0$. Obviously, Perm_r is isomorphic to Σ_r. We *always* identify Perm_r with the symmetric group Σ_r in the obvious way.

Given any $M \in M_{\mathbb{F}}(n,r)$, the ω-weight space M_ω of M is stable under the action of Perm_r. We now define the *Schur functor*

$$F_{n,r} : M_{\mathbb{F}}(n,r) \to \operatorname{mod} \mathbb{F}\Sigma_r$$

on objects by letting $F_{n,r}M := M_\omega$, and by restriction on morphisms.

The next lemma describe a number of useful properties of $F_{n,r}$. Parts (i)-(iii) are proved in [34, §6], and the rest of 5.8 is proved in (or follows easily from) [22, (3.5), (3.6)].

Lemma 5.8. *Fix $\lambda \in \Lambda^+(n,r)$ and $\mu \in \Lambda(n,r)$.*

(i) $F_{n,r}\nabla_n(\lambda) \cong S^\lambda$;

(ii) $F_{n,r}\Delta_n(\lambda) \cong (S^\lambda)^* \cong S^{\lambda^t} \otimes \mathrm{sgn}$;

(iii) $F_{n,r}L_n(\lambda)$ *is zero unless λ is p-restricted, in which case*

$$F_{n,r}L_n(\lambda) \cong D^{\lambda^t} \otimes \mathrm{sgn};$$

(iv) $F_{n,r}(\bigwedge^{\mu_1}(V) \otimes \cdots \otimes \bigwedge^{\mu_n}(V)) \cong M^\mu \otimes \mathrm{sgn}$;

(v) $F_{n,r}(S^{\mu_1}(V) \otimes \cdots \otimes S^{\mu_n}(V)) \cong M^\mu$;

(vi) $F_{n,r}T_n(\lambda) \cong Y^{\lambda^t} \otimes \mathrm{sgn}$;

(vii) $F_{n,r}P_n(\lambda) \cong F_{n,r}Q_n(\lambda) \cong Y^\lambda$, *where $P_n(\lambda)$ (resp. $Q_n(\lambda)$) is a projective cover (resp. injective hull) of $L_n(\lambda)$ in $M_{\mathbb{F}}(n,r)$.*

Let $\mathrm{Perm}_\rho := \mathrm{Perm}_r \cap \mathrm{GL}(\nu)$. Then Perm_ρ is isomorphic to the Young subgroup $\Sigma_\rho < \Sigma_r$. We can now define a more general Schur functor

$$F_{\nu,\rho} : M_{\mathbb{F}}(\nu,r) \to \operatorname{mod} \mathbb{F}\mathrm{Perm}_\rho$$

given on $M \in M_{\mathbb{F}}(\nu,r)$ by letting $F_{\nu,\rho}M$ denote the weight space M_ω, noting that this is stable under the action of $\mathrm{Perm}_\rho < \mathrm{GL}(\nu)$. The next lemma is well known.

Part (i) follows from the definitions, see [13, 4.8], and part (ii) can be found in [24, Lemma 2.5], [13, 4.13].

Lemma 5.9. (i) *Given a module $M \in M_{\mathbb{F}}(n,r)$,*

$$F_{\nu,\rho}(R^n_\nu M) \cong (F_{n,r}M)\downarrow^{\mathbb{F}\mathrm{Perm}_r}_{\mathbb{F}\mathrm{Perm}_\rho}.$$

(ii) *Given modules $M_i \in M_{\mathbb{F}}(n, r_i)$ for $i = 1, \dots, a$,*

$$F_{n,r}(M_1 \otimes \cdots \otimes M_a) \cong \big(F_{n,r_1}M_1 \boxtimes \cdots \boxtimes F_{n,r_a}M_a\big)\uparrow^{\mathbb{F}\Sigma_r}_{\mathbb{F}\Sigma_\rho}.$$

Using these two lemmas, it is quite straightforward to reformulate 4.5 as the following result about the symmetric group (see [13, section 4]):

Theorem 5.10. *Fix partitions $\lambda \vdash r$ and $\mu^{(i)} \vdash r_i$ for $i = 1, \dots, a$. Regard λ^t as an element of $\Lambda^+(n,r)$ and each $(\mu^{(i)})^t$ as an element of $\Lambda^+(n_i, r_i)$ or $\Lambda^+(n, r_i)$. Then*

(i) $(Y^\lambda\downarrow_{\Sigma_\rho} : Y^{\mu^{(1)}} \boxtimes \cdots \boxtimes Y^{\mu^{(a)}}) = [L_n(\mu^{(1)}) \otimes \cdots \otimes L_n(\mu^{(a)}) : L_n(\lambda)]$;

(ii) $((Y^{\mu^{(1)}} \boxtimes \cdots \boxtimes Y^{\mu^{(a)}})\uparrow^{\Sigma_r} : Y^\lambda) = [L_n(\lambda)\downarrow_{\mathrm{GL}(\nu)} : L_{n_1}(\mu^{(1)}) \boxtimes \cdots \boxtimes L_{n_a}(\mu^{(a)})]$.

Moreover, if all the partitions are p-regular, then

(iii) $[D^\lambda\downarrow_{\Sigma_\rho} : D^{\mu^{(1)}} \boxtimes \cdots \boxtimes D^{\mu^{(a)}}] = (T_n(\mu^{(1)}) \otimes \cdots \otimes T_n(\mu^{(a)}) : T_n(\lambda))$;

(iv) $[(D^{\mu^{(1)}} \boxtimes \cdots \boxtimes D^{\mu^{(a)}})\uparrow^{\Sigma_r} : D^\lambda] = (T_n(\lambda)\downarrow_{\mathrm{GL}(\nu)} : T_{n_1}(\mu^{(1)}) \boxtimes \cdots \boxtimes T_{n_a}(\mu^{(a)}))$.

The deduction of the results on Ind^α from the results in Section 2 is also quite straightforward.

Finally, note that parts (iii) and (iv) of 5.4 and 5.7 are new so we sketch their proof here in more detail. The proof for Res^α is similar to that for Ind^α and we concentrate on the latter one.

So fix a p-regular partition λ of r, an integer $n > r$, and a residue $\alpha \in \mathbb{Z}/p\mathbb{Z}$. Set

$$I := \mathrm{Ind}^\alpha D^\lambda \quad \text{and} \quad N := F_{-\alpha}L_n(\lambda^t).$$

It follows from 5.8(iii) and 5.9(ii) that $I \otimes \mathrm{sgn} = F_{n,r}N$.

Let $B_1, B_2, \dots, B_d$ be the conormal nodes for λ of residue α counted from left to right. Since the definitions of conormal for $\mathrm{GL}(n)$ and Σ_r are 'transpose to each other', there are exactly d conormal i with $\mathrm{res}(i, \lambda^t_i + 1) = -\alpha$. Let

$$\{s_1 < \cdots < s_d\} = \{s \mid 1 \le s \le n,\ s \text{ is conormal for } \lambda^t \text{ and } \mathrm{res}(s, \lambda_s + 1) = -\alpha\},$$

and let

$$(0) = N_0 < N_1 < \cdots < N_d = N$$

be the filtration from 2.7. We know from 2.7, 2.3(v) and the proof of Theorem 8.14 from [14] that

(a) N_j/N_{j-1} is a non-zero quotient of $\Delta_n(\lambda^t + \epsilon_{s_j})$ $(1 \leq j \leq d)$;

(b) $[N_j/N_{j-1} : L_n(\lambda^t + \epsilon_k)] = 1$ $(1 \leq j \leq k \leq d)$;

(c) there exists $\psi \in \mathrm{End}_{\mathrm{GL}(n)}(N)$ such that $\{\mathrm{id}_N, \psi, \dots, \psi^{d-1}\}$ is a basis of $\mathrm{End}_{\mathrm{GL}(n)}(N)$, $\psi(N_j) \subseteq N_{j-1}$, and the induced homomorphism

$$\overline{\psi^{k-j}} : N_j/N_{j-1} \to N_k/N_{k-1}$$

is a non-zero homomorphism whose image is properly contained in N_k/N_{k-1} $(1 < k < j \leq d)$.

Let $I_j := F_{n,r+1}(N_j) \otimes \mathrm{sgn}$, $j = 0, 1, \dots, d$. Then

$$(0) = I_0 < I_1 < \cdots < I_d = I \tag{4}$$

is a filtration of I. It follows from 5.8(ii), (iii) and (a), (b) above that I_j/I_{j-1} is a non-zero quotient of $S^{\lambda^{B_j}}$, $1 \leq j \leq d$, and $[I_j/I_{j-1} : D^{\lambda^{B_k}}] = 1$ if $1 \leq j \leq k \leq d$ and λ^{B_k} is p-regular. As B_d is good, λ^{B_d} is always p-regular. Note that $D^{\lambda^{B_d}}$ appears in every I_j/I_{j-1} so the filtration (4) is strict. Let $\hat{\chi} = F_{n,r}(\psi)$. Then $\hat{\chi} = \chi \otimes \mathrm{id}_{\mathrm{sgn}}$ for some $\chi \in \mathrm{End}_{\Sigma_{r+1}}(I)$. It is proved in [49, Theorem B] and [14, section 3] that

$$\{\mathrm{id}_I, \chi, \dots, \chi^{d-1}\}$$

is a basis of $\mathrm{End}_{\Sigma_{r+1}}(I)$. Moreover it follows from the corresponding properties of ψ that $\chi(I_j) \subseteq I_{j-1}$ and that the induced homomorphism $\bar{\chi} : I_{j+1}/I_j \to I_j/I_{j-1}$ is non-zero provided $\lambda^{B_{j+1}}$ is p-regular. If, additionally, λ^{B_j} is p-regular then the image of $\bar{\chi}$ is properly contained in I_j/I_{j-1}.

Claim 1. Let $1 < k < j \leq d$. Then the induced map $\overline{\chi^{j-k}} : I_j/I_{j-1} \to I_k/I_{k-1}$ is non-zero.

Proof. Since the image of $\overline{\chi^{j-k}}$ contains that of the map $\overline{\chi^{d-k}} : I_d/I_{d-1} \to I_k/I_{k-1}$, it is enough to prove the result for $j = d$. To prove that the latter map is non-zero, consider the corresponding map for $\mathrm{GL}(n)$, $\overline{\psi^{d-k}} : N_d/N_{d-1} \to N_k/N_{k-1}$. By (c), this map is non-zero, and the head of N_d/N_{d-1} is p-restricted. It follows that $F_{n,r}(\overline{\psi^{d-k}}) : I_d/I_{d-1} \otimes \mathrm{sgn} \to I_k/I_{k-1} \otimes \mathrm{sgn}$ is also non-zero, whence $\overline{\chi^{d-k}}$ is non-zero.

Claim 2. Let $1 < j < d$. The extension

$$0 \to I_j/I_{j-1} \to I_{j+1}/I_{j-1} \to I_{j+1}/I_j \to 0$$

does not split.

Proof. χ induces the map $\bar{\chi} : I_{j+1}/I_{j-1} \to I_{j+1}/I_{j-1}$. We know that this map is non-zero by Claim 1. On the other hand, $\bar{\chi}$ annihilates I_j/I_{j-1} and I_{j+1}/I_j, and the claim follows.

Claim 3. Let $1 \leq j \leq d$. Assume that λ^{B_k} is p-regular for all $k = j, j+1, \dots, d$. Then the Loewy length of I_j is at least $b + 1 - j$.

Proof. Put $L = I_j/I_{j-1}$, and for $k = j, j+1, \dots, d$ pick w_k to be any vector in I_k such that $w_k + I_{k-1}$ is in the head of I_k/I_{k-1}. Note that this head is simple by assumption. Hence $w_k + I_{k-1}$ generates I_k/I_{k-1} as an $F\Sigma_{r+1}$-module. Set

$$v_k := \chi^{k-j} w_k + I_{j-1} \in L \quad (k = j, j+1, \dots, d).$$

By Claim 1, $v_j, v_{j+1}, \dots, v_b$ are non-zero vectors of L. Denote $L_k = F\Sigma_{r+1} v_k$. Then

$$(0) \lneqq L_d \lneqq L_{d-1} \lneqq \cdots \lneqq L_j = L,$$

the filtration being strict since D^{B_k} is a composition factor of L_k but not of L_{k-1}. Let $(0) = S_0 < S_1 < \dots$ denote the socle series of L. We show that $v_k \notin S_{d-k}$ by downward induction on $k = d, d-1, \dots, j$. In particular this will show that $v_j \notin S_{d-j}$; hence the length of the socle series is at least $d - j + 1$. The base of induction is clear as v_d is non-zero. So let us take $k < d$, and assume that $v_{k+1} \notin S_{d-(k+1)}$. Suppose for a contradiction that $v_k \in S_{d-k}$. Let $\bar{v}_k = v_k + S_{d-(k+1)} \in S_{d-k}/S_{d-(k+1)}$. Since $v_{k+1} \in F\Sigma_{r+1} v_k$ we also have $v_{k+1} \in S_{d-k}$, and we get two non-zero vectors $\bar{v}_k$ and $\bar{v}_{k+1} := v_{k+1} + S_{d-(k+1)}$. We note that the submodule $F\Sigma_{r+1}\bar{v}_k \subseteq S_{d-k}/S_{d-(k+1)}$ is a quotient of L_k, so it has a simple head. But $S_{d-k}/S_{d-(k+1)}$ is semisimple, so this submodule must be simple and equal to $D^{\lambda^{B_k}}$. But $\bar{v}_{k+1}$ is non-zero vector of this submodule, and so it generates $D^{\lambda^{B_{k+1}}}$ by the same reason. Since $D^{\lambda^{B_k}} \not\cong D^{\lambda^{B_{k+1}}}$, we get a contradiction.

Claim 4. Assume that λ^{B_j} is p-regular for all $j = 1, 2, \dots, d$. Then the Loewy length of I is at least $2d - 1$.

Proof. The proof is similar to that of Corollary 8.15 of [14] but uses Claim 3 instead of Theorem 8.14 of that paper.

6. Group algebra of the finitary symmetric group

A permutation of the set $\mathbb{N} = \{1, 2, 3, \dots\}$ is called *finitary* if it fixes all but finitely many elements. The finitary symmetric group Σ_∞ is the group of all finitary permutations of $\mathbb{N}$. It can be represented as a union $\Sigma_\infty = \bigcup_{n\geq 1} \Sigma_n$.

A. Zalesskii has shown (see [68] and the references there) that some ring theoretic questions on the group algebras of locally finite groups are closely related to the asymptotic behavior of the branching rules for finite groups. This fundamental observation accounts for the recent noticeable progress in the theory of group algebras of locally finite groups.

We illustrate the ideas for the case of the finitary symmetric group. Given a (two-sided) ideal I in the group algebra $\mathbb{F}\Sigma_\infty$ we can form a family of ideals

$$I_n := I \cap \mathbb{F}\Sigma_n < \mathbb{F}\Sigma_n, \quad n = 1, 2, \dots.$$

This family has the property

$$I_n \cap \mathbb{F}\Sigma_m = I_m, \quad \text{for any } 1 \le m \le n, \tag{5}$$

and I can be reconstructed from it as $I = \bigcup_{n \ge 1} I_n$. On the other hand, given a family of ideals $I_n < \mathbb{F}\Sigma_n$ satisfying the property (5) we may form a union $I = \bigcup_{n \ge 1} I_n$, and then $I \cap \mathbb{F}\Sigma_n = I_n$ for any n. Thus the main problem is how to 'glue' a big ideal from small ones or how to produce and classify the families of ideals satisfying (5).

The right formalism for doing this comes from a sort of 'asymptotic' representation theory of symmetric groups.

Definition 6.1 (A. Zalesskii). Let Φ_n be a set of the isomorphism classes of irreducible $\mathbb{F}\Sigma_n$-modules, $n = 1, 2, \ldots$. The collection $\Phi = \{\Phi_n\}_{n \in \mathbb{N}}$ is called an *inductive system* (for Σ_∞) if for any $m, n \in \mathbb{N}$ with $m < n$ the following two properties hold:

(1) For any $D \in \Phi_n$, all composition factors of the restriction $D \downarrow_{\Sigma_m}$ belong to Φ_m.

(2) For any $E \in \Phi_m$, there exists $D \in \Phi_n$ such that E is a composition factor of $D \downarrow_{\Sigma_m}$.

Theorem 6.2 ([68, 8.1,1.25], see also [7, 2.6]). *There is an order reversing bijection of partially ordered sets between the inductive systems and the semi-primitive ideals of* $\mathbb{F}\Sigma_\infty$.

The ideals of $\mathbb{F}\Sigma_\infty$ are closely related with the theory of PI-rings (see [1, 69, 60, 59, 29, 28]). In [44], A. Kemer has shown that determining the prime ideals in $\mathbb{F}\Sigma_\infty$ would be a crucial step in a classification of the prime varieties of associative algebras.

The ideal structure of $\mathbb{F}\Sigma_\infty$ is very rich. In case where char $\mathbb{F} = 0$ it was described in [59, 28]. Using Theorem 6.2 and the classical branching rule one can easily rederive such a description. On the other hand, the case of positive characteristic seems to be very difficult since a complete modular branching rule is not known. Recently Baranov and the second author [7] have described the *maximal* ideals in $\mathbb{F}\Sigma_\infty$ and $\mathbb{F}A_\infty$ (where A_∞ is the finitary alternating group), provided $p > 2$ (see also [9, 51] for some relevant results).

The following problem has been raised in [69].

Question (A. Zalesskii). Do ideals of $\mathbb{F}\Sigma_\infty$ satisfy the ascending chains condition?

A little more special (but important) question is whether the semiprimitive ideals of $\mathbb{F}\Sigma_\infty$ satisfy A.C.C. By 6.2 this is equivalent to the D.C.C. for inductive systems.

It was observed in [8] that the ideal structure of $\mathbb{F}\Sigma_\infty$ is closely related with the ideal structure of some commutative algebra which appeared in the papers of Georgiev and Mathieu [30, 31] and of Andersen and Paradowski [5]. This observation is based on 5.10(iii) so we mention this as one of the applications of the results described in Section 4.

First we note that if $\mu, \nu \in \Lambda^+(m)$ are p-regular and $(T_m(\mu) \otimes T_m(\nu) : T_m(\lambda)) \neq 0$, then λ is also p-regular. This fact is a folklore (see [8] for one of the proofs). This allows one to define a commutative $\mathbb{Z}$-algebra Q_m as follows. The elements of Q_m are the formal finite $\mathbb{Z}$-linear combinations $\sum a_\lambda T_m(\lambda)$ of indecomposable tilting modules corresponding to the *p-regular* λ, and the multiplication is induced by tensor products. More precisely, we put

$$T_m(\mu)T_m(\nu) = \sum_\lambda a^\lambda_{\mu\nu} T_m(\lambda)$$

if

$$T_m(\mu) \otimes T_m(\nu) \cong \bigoplus_\lambda a^\lambda_{\mu\nu} T_m(\lambda).$$

An ideal I of Q_m is called a *tensor ideal* if $\sum a_\lambda T_m(\lambda) \in I$ and $a_\mu \neq 0$ imply $T_m(\lambda) \in I$. A tensor ideal is called *special* if $T_m(\lambda) \otimes V_m \in I$ implies $T_m(\lambda) \in I$.

Now we define an important invariant of inductive systems.

Definition 6.3. If $\lambda = (\lambda_1 \geq \cdots \geq \lambda_h > 0)$, we call h the *height* of λ and denote it by $h(\lambda)$. If $\Phi = \{\Phi_n\}_{n \in \mathbb{N}}$ is an inductive system we define its height $h(\Phi)$ by setting

$$h(\Phi) = \sup\{h(\lambda) \mid D^\lambda \in \Phi_n \text{ for some } n \in \mathbb{N}\}.$$

The following result shows that all interesting inductive systems have finite height.

Lemma 6.4. [8, 5.2] *Let $\Phi = \{\Phi_n\}_{n \in \mathbb{N}}$ be an inductive system for Σ_∞. Assume that $h(\Phi) = +\infty$. Then $\Phi_n = \mathrm{Irr}\ \Sigma_n$ for all $n \in \mathbb{N}$, where* Irr Σ_n *denotes the set of all isomorphism classes of irreducible $F\Sigma_n$-modules.*

Now we are able to state the main result.

Theorem 6.5. [8] *There is an order reversing bijection between the special tensor ideals of the algebra Q_m and the inductive systems for Σ_∞ of height m.*

Motivated by the question of Zalesskii above, the following is raised in [8]:

Question. Is Q_m noetherian?

We note though that Zalesskii's question is more subtle as, according to 6.5, it has to do with *tensor* ideals, which are the ideals 'respecting' the basis of tilting modules.

References

[1] S. A. Amitsur, The polynomial identities of associative rings, in: Noetherian rings and rings with polynomial identities (A. Goldie and T. Willmore, eds.), Proc. Conf. Univ. Durham, 1979, 1–38.

[2] H. H. Andersen, An inversion formula for the Kazhdan-Lusztig polynomials for affine Weyl groups, Adv. Math. 60 (1986), 125–153.

[3] H. H. Andersen, Filtrations and tilting modules, Ann. Sci. École Norm. Sup. (4) 30 (1997), 353–366.

[4] H. H. Andersen, A sum formula for tilting filtrations, J. Pure Appl. Algebra 152 (2000), 17–40.

[5] H. H. Andersen and J. Paradowski, Fusion categories arising from semisimple Lie algebras, Comm. Math. Phys. 169 (1995), 563–588.

[6] S. Ariki, On the decomposition numbers of the Hecke algebra of $G(m, 1, n)$, J. Math. Kyoto Univ. 36 (1996), 789–808.

[7] A. Baranov and A. Kleshchev, Maximal ideals in modular group algebras of the finitary symmetric and alternating groups, Trans. Amer. Math. Soc. 351 (1999), 595–617.

[8] A. Baranov and A. Kleshchev, Group algebra of the finitary symmetric group and tilting modules, in preparation.

[9] A. Baranov, A. Kleshchev and A. E. Zalesskii, Asymptotic results on modular representations of symmetric groups and almost simple modular group algebras, J. Algebra 219 (1999), 506–530.

[10] J. Brundan, Modular branching rules and the Mullineux map for Hecke algebras of type A, Proc. London Math. Soc. 77 (1998), 551–581.

[11] J. Brundan, Lowering operators for $\mathrm{GL}(n)$ and quantum $\mathrm{GL}(n)$, Proc. Sympos. Pure Math. 63 (1998), 95–114.

[12] J. Brundan, R. Dipper and A. Kleshchev, Quantum linear groups and representations of $\mathrm{GL}_n(q)$ in non-defining characteristics, Mem. Amer. Math. Soc. 149, No 706, American Mathematical Society, Providence, RI, 2001.

[13] J. Brundan and A. Kleshchev, Modular Littlewood-Richardson coefficients, Math. Z. 232 (1999), 287–320.

[14] J. Brundan and A. Kleshchev, On translation functors for general linear and symmetric groups, Proc. Lond. Math. Soc. 80 (2000), 75–106.

[15] J. Brundan and A. Kleshchev, Some remarks on branching rules and tensor products for algebraic groups, J. Algebra 217 (1999), 335–351.

[16] J. Brundan, A. Kleshchev and I. Suprunenko, Semisimple restrictions from $\mathrm{GL}(n)$ to $\mathrm{GL}(n-1)$, J. Reine Angew. Math. 500 (1998), 83–112.

[17] R. W. Carter and G. Lusztig, On the modular representations of the general linear and symmetric groups, Math. Z. 136 (1974), 193–242.

[18] E. Cline, B. Parshall and L. Scott, Abstract Kazhdan-Lusztig theories, Tôhoku Math. J. 45 (1993), 511–534.

[19] E. Cline, B. Parshall, L. Scott and W. van der Kallen, Rational and generic cohomology, Invent. Math. 39 (1977), 143–163.

[20] S. Donkin, Rational representations of algebraic groups: Tensor products and filtrations, Lecture Notes in Math. 1140, Springer-Verlag, Berlin 1985.

[21] S. Donkin, On Schur algebras and related algebras I, J. Algebra 104 (1986), 310–328.

[22] S. Donkin, On tilting modules for algebraic groups, Math. Z. 212 (1993), 39–60.

[23] S. Donkin, The q-Schur algebra, London Math. Soc. Lecture Note Ser. 253, Cambridge University Press, 1998.

[24] S. Donkin and K. Erdmann, Tilting modules, symmetric functions, and the module structure of the free Lie algebra, J. Algebra 203 (1998), 69–90.

[25] S. Doty, Representation theory of reductive normal algebraic monoids, Trans. Amer. Math. Soc. 351 (1999), 2539–2551.

[26] K. Erdmann, Tensor products and dimensions of simple modules for symmetric groups, Manuscripta Math. 88 (1995), 357–386.

[27] O. Foda, B. Leclerc, M. Okado, J.-Y. Thibon and T. Welsh, RSOS models and Jantzen-Seitz representations of Hecke algebras at roots of unity, Lett. Math. Phys. 43 (1998), 31–42.

[28] E. Formanek and D. Lawrence, The group algebra of the infinite symmetric group, Israel J. Math. 23 (1976), 325–331.

[29] E. Formanek and C. Procesi, Mumford's conjecture for the general linear group, Adv. Math. 19 (1976), 292–305.

[30] G. Georgiev and O. Mathieu, Categorie de fusion pour les groupes de Chevalley, C. R. Acad. Sci. Paris Ser. I Math. 315 (1992), 659–662.

[31] G. Georgiev and O. Mathieu, Fusion rings for modular representations of Chevalley groups, in: Mathematical aspects of conformal and topological field theories and quantum groups, South Hadley, MA, 1992 (Paul J. Sally, Jr., et al., eds.), Contemp. Math. 175, Amer. Math. Soc., Providence, RI, 1994, 89–100.

[32] V. Ginzburg, N. Reshetikhin and E. Vasserot, Quantum groups and flag varieties, in: Mathematical aspects of conformal and topological field theories and quantum groups, South Hadley, MA, 1992 (Paul J. Sally, Jr., et al., eds.), Contemp. Math. 175, Amer. Math. Soc., Providence, RI, 1994, 101–130.

[33] F. Goodman and H. Wenzl, Crystal bases of quantum affine algebras and affine Kazhdan-Lusztig polynomials, Internat. Math. Res. Notices 5 (1999), 251–275.

[34] J. A. Green, Polynomial representations of GL_n, Lecture Notes in Math. 830, Springer-Verlag, Berlin 1980.

[35] J. E. Humphreys, Modular representations of classical Lie algebras and semisimple groups, J. Algebra 19 (1971), 51–79.

[36] J. E. Humphreys, Introduction to Lie algebras and representation theory, sixth edition, Springer-Verlag, New York 1996.

[37] G. D. James, The representation theory of the symmetric groups, Lecture Notes in Math. 682, Springer-Verlag, Berlin 1978.

[38] G. D. James, The representation theory of the symmetric groups, Proc. Sympos. Pure Math. 47 (1987), 111–126.

[39] G. D. James and A. Kerber, The representation theory of the symmetric group, Addison-Wesley, London 1981.

[40] J. C. Jantzen, Representations of algebraic groups, Academic Press, Orlando 1987.

[41] J. C. Jantzen and G. M. Seitz, On the representation theory of the symmetric groups, Proc. London Math. Soc. 65 (1992), 475–504.

[42] V. G. Kac and A. K. Raina, Bombay lectures on highest weight representations of infinite-dimensional Lie algebras, Adv. Ser. Math. Phys. 2, World Scientific Publishing Co., Inc., Teaneck, NJ, 1987.

[43] M. Kashiwara, T. Miwa and E. Stern, Decomposition of q-deformed Fock spaces, Selecta Math. 1 (1995), 787–805.

[44] A. Kemer, On the multilinear components of the regular prime varieties, in: Methods in ring theory, Levico Terme, 1997 (Vesselin S. Drensky et al., eds.), Lecture Notes in Pure and Appl. Math. 198, Dekker, New York 1998, 171–183.

[45] A. Kleshchev, On restrictions of irreducible modular representations of semisimple algebraic groups and symmetric groups to some natural subgroups I, Proc. London Math. Soc. 69 (1994), 515–540.

[46] A. Kleshchev, Branching rules for modular representations of symmetric groups I, J. Algebra 178 (1995), 493–511.

[47] A. Kleshchev, Branching rules for modular representations of symmetric groups II, J. Reine Angew. Math. 459 (1995), 163–212.

[48] A. Kleshchev, Branching rules for modular representations of symmetric groups III: some corollaries and a problem of Mullineux, J. London Math. Soc. 54 (1996), 25–38.

[49] A. Kleshchev, Branching rules for modular representations of symmetric groups IV, J. Algebra 201 (1998), 547–572.

[50] A. Kleshchev, On decomposition numbers and branching coefficients for symmetric and special linear groups, Proc. London Math. Soc. 75 (1997), 497–558.

[51] A. Kleshchev, Completely splittable representations of symmetric groups, J. Algebra 181 (1996), 584–592.

[52] P. Landrock, Finite Group Algebras and their Modules, Cambridge University Press, Cambridge 1983.

[53] B. Leclerc and J.-Y. Thibon, Canonical bases of q-deformed Fock spaces, Internat. Math. Res. Notices 9 (1996), 447–456.

[54] A. Lascoux, B. Leclerc and J.-Y. Thibon, Hecke algebras at roots of unity and crystal bases of quantum affine algebras, Comm. Math. Phys. 181 (1996), 205–263.

[55] O. Mathieu, Filtrations of G-modules, Ann. Sci. École Norm. Sup. 23 (1990), 625–644.

[56] O. Mathieu and G. Papadopoulo, A character formula for a family of simple modular representations of GL_n, Comment. Math. Helv. 74 (1999), 280–296.

[57] G. McNinch, Filtrations and positive characteristic Howe duality, Math. Z. 235 (2000), 651–685.

[58] K. Misra and T. Miwa, Crystal base for the basic representation of $U_q(\hat{\mathfrak{sl}}_n)$, Commun. Math. Phys. 134 (1990), 79–88.

[59] Yu. P. Razmyslov, Trace identities of full matrix algebras over a field of characteristic zero, Math. USSR. Izv. 8 (1974), 727–760.

[60] Yu. P. Razmyslov, Identities of algebras and their representations, Nauka, Moscow 1989 (in Russian).

[61] A. Ryba, Fibonacci representations of the symmetric groups, J. Algebra 170 (1994), 678–686.

[62] G. de B. Robinson, Representation theory of the symmetric group, Mathematical Expositions 12, University of Toronto Press, Toronto, 1961.

[63] O. Schiffmann, An equality of bases of the Fock space (appendix to [67]).

[64] W. Soergel, Kazhdan-Lusztig Polynome und eine Kombinatorik für Kipp-Moduln, Represent. Theory 1 (1997), 37–68.

[65] W. Soergel, Charakterformeln für Kipp-Moduln über Kac-Moody-Algebren, Represent. Theory 1 (1997), 115–132.

[66] Wang J-P., Sheaf cohomology on G/B and tensor products of Weyl modules, J. Algebra 77 (1982), 162–185.

[67] M. Varagnolo and E. Vasserot, On the decomposition matrices of the quantized Schur algebra, Duke Math. J. 100 (1999), 267–297.

[68] A. E. Zalesskii, Group rings of simple locally finite groups, in: Finite and locally finite groups, Kluwer Acad. Publ., 1995, 219–246.

[69] A. E. Zalesskii, Modular group rings of the finitary symmetric groups, Israel J. Math. 96 (1996), 609-621.

Department of Mathematics, University of Oregon, Eugene, Oregon 97403, U.S.A.

E-mail: brundan@darkwing.uoregon.edu

E-mail: klesh@math.uoregon.edu

Block theory via stable and Rickard equivalences

Raphaël Rouquier

Contents

1 Introduction 102

2 Symmetric algebras, functors and equivalences 103
- 2.1 Notation and conventions . . . 103
- 2.2 Functors and adjunctions . . . 104
- 2.3 Equivalences . . . 108

3 Some steps in block theory 112
- 3.1 The group algebra . . . 112
- 3.2 Blocks . . . 115

4 The Brauer functor 117
- 4.1 p-permutation modules . . . 117
- 4.2 The Brauer morphism . . . 118
- 4.3 Defect of blocks . . . 118

5 Rickard equivalences 119
- 5.1 An example: A_5 in characteristic 2 . . . 119
- 5.2 Broué's conjecture . . . 121
- 5.3 Splendid stable equivalences and local Rickard equivalences . . . 124

6 Blocks with defect group $\mathbb{Z}/p^a \times \mathbb{Z}/p^b$ 126
- 6.1 Cyclic defect groups . . . 126
- 6.2 Blocks with abelian defect group of rank 2 . . . 127
- 6.3 Blocks with Klein four defect groups . . . 128
- 6.4 Extension to blocks with defect group $\mathbb{Z}/2 \times \mathbb{Z}/2 \times \mathbb{Z}/2$ . . . 130

7 Local constructions 131
- 7.1 Gluing G-sets . . . 131
- 7.2 p-permutation modules . . . 134
- 7.3 Geometric stable and Rickard equivalences . . . 136

Appendix 137
A.1 Splendid equivalences . . . 138
A.2 Nilpotent blocks, p-rank 2 and Puig's finiteness conjecture . . . 139
A.3 Further categories of sheaves . . . 141
A.4 p-extensions . . . 142

References 144

1. Introduction

We present some topics of modular representation theory of finite groups, based on functorial methods, and motivated by Broué's abelian defect group conjecture.

In the first three sections, we review some classical material. In §2, we define various types of equivalence for symmetric algebras induced by tensoring with complexes of bimodules, following some discussion of general adjunction properties of such functors. In §3, we give some basic properties of group algebras: blocks, normal p'-subgroups and the case of TI Sylow p-subgroups. Finally, in §4, we deal more specifically with (direct summands of) permutation modules, where the Brauer functor allows us to transfer local information.

In §5, we discuss Rickard equivalences in block theory. After a detailed example, we consider Broué's abelian defect conjecture and its refinements (splendidness, equivariance with respect to p'-automorphism groups, central extensions by p-groups). In §5.3, we come to a crucial result: a splendid complex induces a stable equivalence if and only if it induces (via the Brauer functor) local Rickard equivalences (from this point, we consider only principal blocks).

The results in §6 and §7 are new. We use the results of §5.3 in §6.2 to construct stable equivalences between a principal block with defect group $\mathbb{Z}/p^a \times \mathbb{Z}/p^b$ and the principal block of the normalizer of a defect group (in the case where $(a, b) \neq (1, 1)$, we need the Z_p^*-theorem which, for p odd, depends on the classification of finite simple groups). We use here a new construction of stable equivalences given by complexes. In §6.3, we go one step further to lift these stable equivalences to Rickard equivalences when, in addition, $p = 2$ (thus solving Broué's conjecture for principal blocks with defect group $\mathbb{Z}/2 \times \mathbb{Z}/2$). In §6.4, we construct stable equivalences for principal blocks with defect group elementary abelian of order 8.

§7 is devoted to the study of a locally determined category of p-permutation modules with additional structure. We explain how this can be used to glue local Rickard equivalences into a stable equivalence: as a consequence, we prove that Broué's abelian defect group conjecture would follow (inductively) from the possibility of lifting stable equivalences to Rickard equivalences. This requires additional structure on the Rickard complexes. The construction of §6.2 appears as a special case. Our belief is that these methods reduce Broué's conjecture to a problem of "representation theory of algebras" where the groups will not be useful anymore, namely the problem of lifting certain stable equivalences to Rickard equivalences.

In the appendix, we explain some aspects of the theory for nonprincipal blocks [Rou3].

This article is based on, and extends, a set of five lectures given at the Symposium on the Modular Representation Theory of Finite Groups, University of Virginia, Charlottesville, in May 1998.

I thank J. Alperin, S. Bouc, M. Broué, N. Kunugi and J. Rickard for their useful discussions. This paper owes a lot to M. Collins for his persistent encouragement and for his suggestions of improvement.

2. Symmetric algebras, functors and equivalences

In this section, we explain what standard concepts of category theory become for module categories of symmetric algebras. §2.2 is largely inspired by Broué's notes [Br3], where more details are to be found.

In §2.3, we define various notions of equivalences and gather various properties.

2.1. Notation and conventions

Let $\mathcal{O}$ be a noetherian local commutative ring (all rings are with identity) with residue field k. All $\mathcal{O}$-algebras considered will be free and finitely generated over $\mathcal{O}$. Let A an $\mathcal{O}$-algebra.

All A-modules considered will be left modules, finitely generated over the underlying coefficient ring $\mathcal{O}$. Complexes of A-modules will always be bounded. We identify the category A-mod of A-modules with the full subcategory of complexes of A-modules concentrated in degree 0.

We denote by A° the algebra opposite to A. It is A as an $\mathcal{O}$-module, but the multiplication of a and b in A° is ba. Note that a left A-module is the same as a right A°-module and if B is a B-algebra, an (A, B)-bimodule is an $(A \otimes_{\mathcal{O}} B^\circ)$-module.

Similarly, if G is a group, we define the group G° opposite to G, with the same set of elements as G but with multiplication of g and h given by hg. The group algebra $\mathcal{O}G^\circ$ of G° is the algebra $(\mathcal{O}G)^\circ$.

We will often write $\otimes$ for $\otimes_{\mathcal{O}}$. For M an A-module, we denote by M^* the A°-module $\operatorname{Hom}_{\mathcal{O}}(M, \mathcal{O})$.

By *the* (A, A)-bimodule A, we mean the regular bimodule given by left and right multiplication.

2.2. Functors and adjunctions

2.2.1. Basic adjunction. Let A and B be two $\mathcal{O}$-algebras and M an (A, B)-bimodule.

We have an isomorphism of $\mathcal{O}$-modules

$$\begin{aligned}\gamma_M(V, U) : \operatorname{Hom}_A(M \otimes_B V, U) &\xrightarrow{\sim} \operatorname{Hom}_B(V, \operatorname{Hom}_A(M, U))\\ f &\mapsto (v \mapsto (m \mapsto f(m \otimes v)))\\ \text{with inverse} \quad (m \otimes v \mapsto g(v)(m)) &\leftarrow\!\shortmid\; g\end{aligned}$$

for any A-module U and B-module V.

Denote by Φ and Ψ respectively the functors

$$\Phi = M \otimes_B - : B\text{-mod} \to A\text{-mod} \text{ and } \Psi = \operatorname{Hom}_A(M, -) : A\text{-mod} \to B\text{-mod}.$$

We have an isomorphism functorial in U and V:

$$\operatorname{Hom}_A(\Phi(V), U) \xrightarrow{\sim} \operatorname{Hom}_B(V, \Psi(U)).$$

We say that the functor Φ is *left adjoint* to the functor Ψ (or Ψ is *right adjoint* to Φ or (Φ, Ψ) is an *adjoint pair*) when there is such an isomorphism.

2.2.2. Projective modules. Let U and V be two A-modules. Consider the $\mathcal{O}$-linear map

$$\begin{aligned}\tau_{U,V} : \operatorname{Hom}_A(U, A) \otimes_A V &\longrightarrow \operatorname{Hom}_A(U, V)\\ f \otimes v &\mapsto (u \mapsto f(u)v)\end{aligned}$$

If U or V is projective, then $\tau_{U,V}$ is an isomorphism: the result is clear when one of the modules is A, thus when it is A^n and finally when it is any direct summand of A^n.

We have a converse to this property: if $\tau_{U,V}$ is an isomorphism for every V, then U is projective. More precisely, we have

Proposition 2.1. *Let S be a simple A-module with a projective cover P_S. Then $\tau_{U,S}$ is non-zero if and only if U has a direct summand isomorphic to P_S.*

Proof. Note that $\tau_{P_S,S} \neq 0$; hence $\tau_{U,S} \neq 0$ if P_S is a direct summand of U.

Let $f : P_S \to S$ be an essential map : this is a surjective morphism whose restriction to a proper submodule of P_S is not surjective anymore. We have a commutative diagram

$$\begin{array}{ccccc}\operatorname{Hom}_A(U, P_S) & \xrightarrow{\operatorname{Hom}_A(U,f)} & \operatorname{Hom}_A(U, S) & & \\ \tau_{U,P_S}\uparrow\simeq & & \uparrow\tau_{U,S} & & \\ \operatorname{Hom}_A(U, A) \otimes_A P_S & \xrightarrow[1\otimes f]{} & \operatorname{Hom}_A(U, A) \otimes_A S & \longrightarrow & 0\,.\end{array}$$

If $\tau_{U,S} \neq 0$, then $\mathrm{Hom}_A(U, f)$ is non-zero. Therefore there is $g : U \to P_S$ whose composite with f is non-zero, and thus surjective. Since f is essential, it follows that g is surjective and splits since P_S is projective. □

When $\mathcal{O}$ is henselian (e.g. complete), then all A-modules have projective covers.

Let M be an (A, B)-bimodule. Let U be an A-module and V a B-module. By §2.2.1, we have an isomorphism of (A, A)-bimodules

$$\mathrm{Hom}_{B^\circ}(M, \mathrm{Hom}_{\mathcal{O}}(V, U)) \xrightarrow{\sim} \mathrm{Hom}_{\mathcal{O}}(M \otimes_B V, U).$$

This induces an $\mathcal{O}$-module isomorphism

$$\mathrm{Hom}_{A \otimes B^\circ}(M, \mathrm{Hom}_{\mathcal{O}}(V, U)) \xrightarrow{\sim} \mathrm{Hom}_A(M \otimes_B V, U).$$

If U and V are free over $\mathcal{O}$ and M is projective (or flat) as a B°-module, then we deduce, for all i,

$$\mathrm{Ext}^i_{A \otimes B^\circ}(M, \mathrm{Hom}_{\mathcal{O}}(V, U)) \xrightarrow{\sim} \mathrm{Ext}^i_A(M \otimes_B V, U).$$

Note that we have similar statements relating $\mathrm{Hom}_{A \otimes B^\circ}$ with Hom_{B°, by considering M as a (B°, A°)-bimodule.

Lemma 2.2. *Assume $\mathcal{O} = k$ is a field and assume the centers of the endomorphism rings of the simple A-modules and the simple B-modules are separable extensions of k. Let M be an (A, B)-bimodule.*

Then M is a projective (A, B)-bimodule if and only if $M \otimes_B V$ is a projective A-module for every B-module V and $U \otimes_A M$ is a projective B°-module for every A°-module U.

Proof. The hypothesis ensures that the largest semi-simple quotients of A and B are products of central simple algebras over separable extensions of k. Now the tensor product over k of two such simple algebras is a semi-simple algebra. It follows that given S a simple A-module and T a simple B°-module, the $(A \otimes B^\circ)$-module $\mathrm{Hom}_k(T^*, S) \simeq S \otimes T$ is semi-simple (note that every simple $(A \otimes B^\circ)$-module occurs as a direct summand of such a module for some S, T). In particular, M is projective if and only if $\mathrm{Ext}^i_{A \otimes B^\circ}(M, S \otimes T) = 0$ for all $i > 0$ and S, T simple.

Assume $M \otimes_B V$ is a projective A-module for every B-module V and $U \otimes_A M$ is a projective B°-module for every A°-module U.

The case $V = B$ shows that M is projective as an A-module. Hence, for $i > 0$, we have

$$\mathrm{Ext}^i_{A \otimes B^\circ}(M, S \otimes T) \simeq \mathrm{Ext}^i_{B^\circ}(S \otimes_A M, T) = 0.$$

It follows that M is projective.

The converse is clear. □

2.2.3. Symmetric algebras. Assume A is a *symmetric algebra*, i.e., is endowed with an $\mathcal{O}$-linear map $t = t_A : A \to \mathcal{O}$ which is a trace ($t(aa') = t(a'a)$) and such that the morphism of (A, A)-bimodules

$$\begin{aligned} \hat{t} : A &\xrightarrow{\sim} \mathrm{Hom}_{\mathcal{O}}(A, \mathcal{O}) \\ a &\mapsto (a' \mapsto t(aa')) \end{aligned}$$

is an isomorphism.

This last isomorphism is equivalent to the requirement that, given an $\mathcal{O}$-basis $\{a_i\}$ of A, there is another basis $\{a'_i\}$ such that $t(a_i a'_j) = \delta_{ij}$.

When $\mathcal{O}$ is a field, the algebra A is in particular self-injective, i.e., the injective modules are the projective modules.

Given an A-module U, we have an isomorphism of right A-modules

$$\begin{aligned} \hat{t}_U : \mathrm{Hom}_A(U, A) &\xrightarrow{\sim} U^* = \mathrm{Hom}_{\mathcal{O}}(U, \mathcal{O}) \\ f &\mapsto tf \\ \text{with inverse} \quad (x \mapsto \sum_i a'_i u(a_i x)) &\leftarrow\!\shortmid u. \end{aligned}$$

Lemma 2.3. *Let M be an $(A \otimes B^\circ)$-module, projective as a B°-module, and V be a B-module. Then we have an isomorphism of A°-modules*

$$\tau_{M,A\otimes V} : \mathrm{Hom}_{A\otimes B^\circ}(M, A \otimes B^\circ) \otimes_{A\otimes B^\circ} (A \otimes V) \xrightarrow{\sim} \mathrm{Hom}_{A\otimes B^\circ}(M, A \otimes V).$$

Proof. We have a commutative diagram

$$\begin{array}{ccc} \mathrm{Hom}_{A\otimes B^\circ}(M, A \otimes B^\circ) \otimes_{A\otimes B^\circ} (A \otimes V) & \xrightarrow{f\otimes(a\otimes x)\mapsto(t_A\otimes 1)(fa)\otimes x} & \mathrm{Hom}_{B^\circ}(M, B^\circ) \otimes_{B^\circ} V \\ {\scriptstyle \tau_{M,A\otimes V}}\downarrow & & \downarrow{\scriptstyle \tau_{\mathrm{Res}_{B^\circ} M,V}} \\ \mathrm{Hom}_{A\otimes B^\circ}(M, A \otimes V) & \xrightarrow[f\mapsto(t_A\otimes 1)f]{} & \mathrm{Hom}_{B^\circ}(M, V) \end{array}$$

where the horizontal maps are isomorphisms. Since $\tau_{\mathrm{Res}_{B^\circ} M,V}$ is an isomorphism, we are done. □

2.2.4. Exact bimodules. Assuming A symmetric and M projective as an A-module, we have constructed isomorphisms of functors

$$\Psi = \mathrm{Hom}_A(M, -) \xrightarrow{\sim} \mathrm{Hom}_A(M, A) \otimes_A - \xrightarrow{\sim} M^* \otimes_A -.$$

In particular, the functor $M \otimes_B -$ is left adjoint to $M^* \otimes_A -$. If in addition M is projective as a right B-module and B is symmetric, then $M \otimes_B -$ is right adjoint to $M^* \otimes_A -$.

We say that M is an *exact* (A, B)-bimodule if it is projective as an A-module and as a right B-module (i.e., if the functors $\mathrm{Hom}_A(M, -)$ and $\mathrm{Hom}_{B^\circ}(M, -)$ are exact).

Proposition 2.4. *If A and B are symmetric $\mathcal{O}$-algebras and M is an exact (A, B)-bimodule, then the functor $M \otimes_B -$ is left and right adjoint to $M^* \otimes_A -$.*

Note that M can be projective as an A-module and as a right B-module without being projective as an $(A \otimes B^\circ)$-module. In the special case where $A = B$ and $M = A$ is the regular bimodule, then M is indeed projective as a right and as a left A-module, but is not projective as an $(A \otimes A^\circ)$-module in general: when $\mathcal{O}$ is a field, A is projective as an $(A \otimes A^\circ)$-module if and only if A is semi-simple separable (i.e., A is a product of matrix algebras over separable field extensions of $\mathcal{O}$), cf. Lemma 2.2.

2.2.5. Units and counits. Since Φ is left adjoint to Ψ, we have an isomorphism $\operatorname{Hom}(\Phi \circ \Psi, I_A) \xrightarrow{\sim} \operatorname{Hom}(\Psi, \Psi)$, where I_A is the identity functor of A-mod. The morphism $\varepsilon : \Phi \circ \Psi \to I_A$ corresponding to the identity morphism $\Psi \to \Psi$ is called the *counit*. Using the isomorphism $\operatorname{Hom}(\Phi, \Phi) \xrightarrow{\sim} \operatorname{Hom}(I_B, \Psi \circ \Phi)$, we obtain the *unit* $\eta : I_B \to \Psi \circ \Phi$.

Note that the functor Φ is an equivalence of categories B-mod $\to$ A-mod if and only if η and ε are isomorphisms, and then Ψ is an inverse to Φ.

In terms of morphisms of bimodules, the counit is the morphism of (A, A)-bimodules

$$\varepsilon_M : M \otimes_B M^* \to A,\ m \otimes x \mapsto \hat{t}^{-1}_{\operatorname{Res}_A M}(x)(m)$$

and the unit is the morphism of (B, B)-bimodules

$$\eta_M : B \to M^* \otimes_A M,\ b \mapsto (\hat{t}_{\operatorname{Res}_A M} \otimes 1)\tau^{-1}_{\operatorname{Res}_A M, \operatorname{Res}_A M}(b \cdot 1_M).$$

2.2.6. Complexes. Let C be a complex of A-modules. We denote by d_C its differential, with degree i part $d_C^i : C^i \to C^{i+1}$.

Let C^* the complex of A°-modules given by

$$(C^*)^i = (C^{-i})^* \text{ and } d^i_{C^*} = (-1)^{i+1}(d_C^{-i-1})^*.$$

Let D be a complex of A° modules. We denote by $C \otimes_A D$ the complex given by

$$(C \otimes_A D)^i = \bigoplus_{r+s=i} C^r \otimes D^s \text{ and } d^i_{C\otimes_A D} = \sum_{r+s=i} d^r_C \otimes 1 + (-1)^r 1 \otimes d^s_D.$$

(Let us recall that all our complexes are bounded).

A complex C of (A, B)-bimodules gives rise to a functor $C \otimes_B -$ from the category $C^b(B)$ of complexes of B-modules to the category $C^b(A)$ of complexes of A-modules.

The results of §2.2.1–2.2.5 generalize to complexes. Given a complex C of (A, B)-bimodules which are projective as A-modules, there is a canonical morphism $\varepsilon_C : C \otimes_B C^* \to A$ and a canonical morphism $\eta_C : B \to C^* \otimes_A C$, which are units and counits of the adjoint pair $(C \otimes_B -, C^* \otimes_A -)$.

2.3. Equivalences

Let A and B be two symmetric $\mathcal{O}$-algebras. We define three types of equivalence. The usual Morita equivalences are a special case of Rickard equivalences. The Rickard equivalences are in turn a special case of the even weaker type of stable equivalences.

2.3.1. Morita. Let M be an exact (A, B)-bimodule.

The following assertions are equivalent.

(i) We have isomorphisms

$$M \otimes_B M^* \simeq A \text{ as } (A, A)\text{-bimodules},$$

$$M^* \otimes_A M \simeq B \text{ as } (B, B)\text{-bimodules}.$$

(ii) The morphisms ε_M and η_{M^*} are isomorphisms of (A, A)-bimodules and η_M, ε_{M^*} are isomorphisms of (B, B)-bimodules

$$\eta_{M^*} : A \xrightarrow{\sim} M \otimes_B M^*, \quad \varepsilon_M : M \otimes_B M^* \xrightarrow{\sim} A,$$

$$\eta_M : B \xrightarrow{\sim} M^* \otimes_A M, \quad \varepsilon_{M^*} : M^* \otimes_A M \xrightarrow{\sim} B.$$

When these conditions are satisfied, we say that M induces a *Morita equivalence* between A and B. This is equivalent to the requirement that $M \otimes_B -$ is an equivalence between A-mod and B-mod.

2.3.2. Rickard. We now take C a complex of exact (A, B)-bimodules.

The following assertions are equivalent.

(i) We have isomorphisms

$$C \otimes_B C^* \simeq A \oplus Z_1 \text{ as complexes of } (A, A)\text{-bimodules}$$

$$C^* \otimes_A C \simeq B \oplus Z_2 \text{ as complexes of } (B, B)\text{-bimodules}$$

where A and B are viewed as complexes concentrated in degree 0 and Z_1 and Z_2 are homotopy equivalent to 0.

(ii) The morphisms η_C, η_{C^*} (resp. ε_C and ε_{C^*}) are split injections (resp. surjections) with cokernel (resp. kernel) homotopy equivalent to 0.

When these conditions are satisfied, we say that C induces a *Rickard equivalence* between A and B or that C is a *Rickard complex*. These conditions are equivalent to the requirement that $C \otimes_B -$ is an equivalence between the homotopy categories of complexes of B-modules and A-modules.

Note that if $C = C_1 \oplus C_2$ with C_2 homotopy equivalent to 0, then C induces a Rickard equivalence if and only if C_1 induces a Rickard equivalence.

2.3.3. Stable. The following assertions are equivalent.

(i) We have isomorphisms

$$C \otimes_B C^* \simeq A \oplus Z_1' \text{ as complexes of } (A, A)\text{-bimodules}$$

$$C^* \otimes_A C \simeq B \oplus Z_2' \text{ as complexes of } (B, B)\text{-bimodules}$$

where Z_1' and Z_2' are homotopy equivalent to complexes of projective bimodules.

(ii) The morphisms η_C, η_{C^*} (resp. ε_C and ε_{C^*}) are split injections (resp. surjections) with cokernel (resp. kernel) homotopy equivalent to a complex of projective bimodules.

When these conditions are satisfied, we say that C induces a *stable equivalence* between A and B.

Actually, we want a slightly more general definition: D induces a stable equivalence whenever $D \oplus A \otimes_{\mathcal{O}} B$ satisfies the equivalent conditions above; then, when $\mathcal{O}$ is a field, $D = 0$ induces a stable equivalence between any two semisimple separable $\mathcal{O}$-algebras.

Note that if $C = C_1 \oplus C_2$ with C_2 homotopy equivalent to a complex of projective bimodules, then C induces a stable equivalence if and only if C_1 induces a stable equivalence.

The situation more commonly considered, after Broué, is the case where $C = M$ is a complex with only one term in degree 0.

In that case, we can restate the equivalences as follows:

(i) We have isomorphisms

$$M \otimes_B M^* \oplus \text{ projective } \simeq A \oplus \text{ projective}$$

$$M^* \otimes_A M \oplus \text{ projective } \simeq B \oplus \text{ projective.}$$

(ii) The morphisms η_M, η_{M^*}, ε_M and ε_{M^*} are split with projective kernels and cokernels.

This implies that $M \otimes_B -$ induces an equivalence between the $\mathcal{O}$-stable categories of B-modules and A-modules. The $\mathcal{O}$-stable category of A-modules is the quotient of A-mod by the full subcategory of $\mathcal{O}$-projective A-modules (direct summands of modules $A \otimes_{\mathcal{O}} U$ for some $\mathcal{O}$-module U). Under separability assumptions (e.g., $\mathcal{O}$ is a field and centers of endomorphism algebras of simple modules are separable extensions of $\mathcal{O}$), the conditions are equivalent to the fact that $M \otimes_B -$ induces an equivalence of stable categories.

Note that if $M = M_1 \oplus M_2$ with M_2 projective, then M induces a stable equivalence if and only if M_1 induces a stable equivalence.

Note that $M \otimes_B V$ is projective or zero if and only if V is projective or zero and that $M \otimes_B V$ is the direct sum of an indecomposable non-projective module with a projective module if V is indecomposable non-projective.

The following result shows that (for $\mathcal{O}$ henselian), when M has no projective direct summand, it sends a simple module to an indecomposable module.

Proposition 2.5. *Assume $\mathcal{O}$ is a field. Let S be a simple A-module with a projective cover P_S and T a simple B-module with a projective cover P_T. Then P_S is a direct summand of $M \otimes_B T$ if and only if $P_S \otimes P_T^*$ is a direct summand of M.*

Proof. We have a commutative diagram

$$\begin{array}{ccc}
\mathrm{Hom}_{A\otimes B^\circ}(M, A\otimes B^\circ)\otimes_{A\otimes B^\circ}\mathrm{Hom}_{\mathcal{O}}(T,S) & \xrightarrow{\tau_{M,\mathrm{Hom}_{\mathcal{O}}(T,S)}} & \mathrm{Hom}_{A\otimes B^\circ}(M,\mathrm{Hom}_{\mathcal{O}}(T,S)) \\
\uparrow{\scriptstyle \simeq} & & \\
\mathrm{Hom}_{A\otimes B^\circ}(M, A\otimes B^\circ)\otimes_{A\otimes B^\circ}\mathrm{Hom}_{\mathcal{O}}(T,A)\otimes_A S & & \\
{\scriptstyle \tau_{M,\mathrm{Hom}_{\mathcal{O}}(T,A)}\otimes 1}\downarrow & & {\scriptstyle \simeq}\downarrow{\scriptstyle \gamma_T(M,S)^{-1}} \\
\mathrm{Hom}_{A\otimes B^\circ}(M,\mathrm{Hom}_{\mathcal{O}}(T,A))\otimes_A S & & \\
{\scriptstyle \gamma_T^{-1}(M,A)\otimes 1}\downarrow{\scriptstyle \simeq} & & \\
\mathrm{Hom}_A(M\otimes_B T, A)\otimes_A S & \xrightarrow[\tau_{M\otimes_B T,S}]{} & \mathrm{Hom}_A(M\otimes_B T, S)
\end{array}$$

where the top vertical map is induced by the product

$$\mathrm{Hom}_{\mathcal{O}}(T,A)\otimes_A S \to \mathrm{Hom}_{\mathcal{O}}(T,S).$$

We apply Lemma 2.3 to M and T^*: the map $\tau_{M,\mathrm{Hom}_{\mathcal{O}}(T,A)}$ is an isomorphism. Consequently, $\tau_{M,\mathrm{Hom}_{\mathcal{O}}(T,S)}$ is non-zero if and only if $\tau_{M\otimes_B T,S}$ is non-zero and the proposition follows from Proposition 2.1. □

2.3.4. Composition. Let A' be a symmetric $\mathcal{O}$-algebra, M an exact (A, B)-bimodule and N an exact (B, A')-bimodule. If M and N induce Morita equivalences, then $M \otimes_B N$ induces a Morita equivalence between A and A'. Rickard equivalences and stable equivalences can be similarly composed.

2.3.5. Comparison. If M induces a Morita equivalence, then it induces a Rickard equivalence. If C induces a Rickard equivalence, then it induces a stable equivalence.

Let M be an exact (A, B)-bimodule inducing a stable equivalence. Assume A and B have no projective direct summands as bimodules. Then M induces a Morita equivalence if and only if $M \otimes_B S$ is simple for every simple B-module S (for then we know that B is a direct summand of $M^* \otimes_A M$ and $M^* \otimes_A M \otimes_B S$ is indecomposable for every simple B-module S, so $M^* \otimes_A M \simeq B$. We have $M \otimes_B M^* \simeq A \oplus Z$, whence $M \otimes_B M^* \otimes_A Z = 0$; thus $Z = 0$).

Let C be a complex of exact (A, B)-bimodules. Assume that all terms of C are projective but C^r and that C^r induces a stable equivalence. Then C induces a stable equivalence.

Let $\Omega_{A\otimes A^\circ}$ be the kernel of the multiplication map $A \otimes A \to A$. This is an (A, A)-bimodule inducing a self-stable equivalence of A.

For U an A-module, we denote by $\Omega_A U$ (or ΩU) an A-module without $\mathcal{O}$-projective direct summand such that $\Omega_{A\otimes A^\circ} \otimes_A U = \Omega U \oplus \mathcal{O}$-projective. Let $f : P_U \to U$ be a surjective map with P_U an $\mathcal{O}$-projective A-module. If f splits as a morphism of $\mathcal{O}$-modules, then $\ker f \oplus \mathcal{O}$-projective $\simeq \Omega U \oplus \mathcal{O}$-projective.

We define inductively $\Omega^n U$ as $\Omega(\Omega^{n-1}U)$ for n positive. Similarly, using $\Omega^{-1}_{A\otimes A^\circ} = \Omega^*_{A\otimes A^\circ}$, we define $\Omega^{-1}U$ and $\Omega^{-n}U$ for n positive. Finally, $\Omega^0 U = U$.

For M an exact (A, B)-bimodule, the (A, B)-bimodule $(A \otimes A^\circ) \otimes_A M$ is projective. Hence

$$\Omega_{A\otimes A^\circ} \otimes_A M \oplus \text{ projective} \simeq \Omega_{A\otimes B^\circ} M \oplus \text{ projective}.$$

Let M be an exact (A, B)-bimodule. Then

$$\Omega^n_{A\otimes A^\circ} A \otimes_A M \oplus \text{ projective} \simeq \Omega^n_{A\otimes B^\circ} M \oplus \text{ projective} .$$

So, if M induces a stable equivalence, then $\Omega^n_{A\otimes B^\circ} M$ also induces a stable equivalence.

The next proposition explains how to construct a Rickard equivalence from a Morita equivalence by truncating a projective resolution of the bimodule.

Proposition 2.6. *Let M be an exact (A, B)-bimodule. Let C be an complex of exact (A, B)-bimodules with homology only in degree 0, isomorphic to M, with zero terms outside $\{0, \ldots, r-1, r\}$ and with projective terms in degrees $0, \ldots, r-1$ where r is an integer.*

If M induces a Morita equivalence, then C induces a Rickard equivalence.

Proof. Assume first r is non-positive. We have

$$C^r \oplus \text{ projective} \sim \Omega^{-r}_{A\otimes B^\circ} M \oplus \text{ projective};$$

therefore C^r induces a stable equivalence. Since C^i is projective for $i \neq r$, it follows that C induces a stable equivalence. In particular, the kernel of ε_C is homotopy equivalent to a complex of projective modules Z.

The homology of C is projective over B; thus the homology of $C \otimes_B C^*$ is isomorphic to $H^0(C) \otimes_B H^0(C)^*$ (in degree 0). Since $H^0(C) \simeq M$ induces a Morita equivalence, it follows that $C \otimes_B C^*$ has homology only in degree 0, isomorphic to A. More precisely, the kernel of ε_C has zero homology.

The complex Z is a (bounded) complex of projective modules with zero homology, whence it is homotopy equivalent to 0.

Similarly, one shows that the kernel of ε_{C^*} is homotopy equivalent to 0.

The case where r is positive follows from the negative case by replacing A, B, M, C and r by B, A, M^*, C^* and $-r$. □

2.3.6. Extension of scalars. Let $\mathcal{O}'$ be a commutative $\mathcal{O}$-algebra. Let $A' = \mathcal{O}' \otimes_{\mathcal{O}} A$, $B' = \mathcal{O}' \otimes_{\mathcal{O}} B$: these are symmetric $\mathcal{O}'$-algebras.

An isomorphism $A \xrightarrow{\sim} B$ gives rise to an isomorphism $A' \xrightarrow{\sim} B'$ by extending scalars from $\mathcal{O}$ to $\mathcal{O}'$. More generally, given an (A, B)-bimodule M inducing a Morita equivalence between A and B, the (A', B')-bimodule $A' \otimes_A M \otimes_B B'$ induces a Morita equivalence between A' and B'. We have similar statements for Rickard and stable equivalence.

3. Some steps in block theory

3.1. The group algebra

Let us start gathering some properties that do not involve blocks. We take special care to provide explicit isomorphisms when studying the TI case in §3.1.5. This way, we avoid use of the Krull–Schmidt Theorem and we can work over a non-complete ring $\mathcal{O}$.

3.1.1. Symmetric algebra structure. We have an $\mathcal{O}$-linear trace on the group algebra

$$t : \mathcal{O}G \to \mathcal{O},\ g \mapsto \delta_{1g} \text{ for } g \in G.$$

Since $t(g'g^{-1}) = \delta_{gg'}$, the form is symmetrizing. The basis dual to $\{g\}_{g\in G}$ is $\{g^{-1}\}_{g\in G}$.

3.1.2. Let H be a subgroup of G and $M = \mathcal{O}G$ the exact $(\mathcal{O}H, \mathcal{O}G)$-bimodule where the actions are given by multiplication. The functor $\operatorname{Res}_H^G = M \otimes_{\mathcal{O}G} -$ is the restriction functor from $\mathcal{O}G$-mod to $\mathcal{O}H$-mod. It is an exact functor.

We have an isomorphism $\hat{t} : \mathcal{O}G \xrightarrow{\sim} M^*$, where $\mathcal{O}G$ is the $(\mathcal{O}G, \mathcal{O}H)$-bimodule with actions given by multiplication. The corresponding functor $\operatorname{Ind}_H^G = M^* \otimes_{\mathcal{O}H} -$ is the induction functor from $\mathcal{O}H$-mod to $\mathcal{O}G$-mod. It is also an exact functor and Ind_H^G is left and right adjoint to Res_H^G.

3.1.3. The counit ε_{M^*} is the surjective morphism given by multiplication

$$\varepsilon_{M^*} : \mathcal{O}G \otimes_{\mathcal{O}H} \mathcal{O}G \to \mathcal{O}G,\quad g \otimes g' \mapsto gg'.$$

Assume $[G : H]$ is invertible in $\mathcal{O}$. Then

$$\mathcal{O}G \to \mathcal{O}G \otimes_{\mathcal{O}H} \mathcal{O}G,\ g \mapsto \frac{1}{[G:H]} g \sum_{x\in G/H} x \otimes x^{-1}$$

is a splitting to the surjection, i.e., the morphism of functors $\varepsilon : \operatorname{Ind}_H^G \operatorname{Res}_H^G \to 1_G$ is a split surjection.

Now let U and V be two $\mathcal{O}G$-modules. Then U is a direct summand of $\mathrm{Ind}_H^G \mathrm{Res}_H^G U$. So

$$\mathrm{Ext}^1_{\mathcal{O}G}(U, V) \leq \mathrm{Ext}^1_{\mathcal{O}G}(\mathrm{Ind}_H^G \mathrm{Res}_H^G U, V) \simeq \mathrm{Ext}^1_{\mathcal{O}H}(\mathrm{Res}_H^G U, \mathrm{Res}_H^G V),$$

where the isomorphism comes from the fact that Ind_H^G is an exact functor which is a left adjoint to the exact functor Res_H^G.

As a special case, let us take for H a Sylow p-subgroup of G and for $\mathcal{O}$ a field k of characteristic p.

If H is trivial, we deduce that all Ext^1-groups are zero in kG-mod. So we obtain Maschke's theorem.

Proposition 3.1. *If k is a field and $|G|$ is invertible in k, then kG is semi-simple.*

More generally the "homological complexity" of kG-mod is measured by a Sylow p-subgroup P:

- kG has finite representation type (i.e., there are only finitely many isomorphism classes of indecomposable modules) if and only if kP has finite representation type; this is known to happen exactly when P is cyclic.

- kG is tame (i.e., indecomposable modules are in some sense classifiable) if and only if kP is tame; that happens exactly when $p = 2$ and P is a dihedral, semidihedral or generalized quaternion group.

Note that in all other cases kG is wild. So instead of pursuing the unreasonable task of describing kG-mod completely, we will try to compare it with module categories of smaller groups.

3.1.4. Mackey's formula. Composition of an induction functor followed by a restriction functor is described by Mackey's formula as a sum of compositions of a restriction functor followed by an induction functor.

Let H and H' be two subgroups of G. Then $\mathcal{O}G = \bigoplus_{g\in H'\backslash G/H} \mathcal{O}H'gH$ is a decomposition as $(\mathcal{O}H', \mathcal{O}H)$-bimodules.

Let $K = H' \cap gHg^{-1}$. Then we have an isomorphism of $(\mathcal{O}H', \mathcal{O}H)$-bimodules

$$\mathcal{O}H' \otimes_{\mathcal{O}K} (\mathcal{O}H)^g \xrightarrow{\sim} \mathcal{O}H'gH, \; x \otimes y \mapsto xgy$$

where $(\mathcal{O}H)^g = \mathcal{O}H$ as a right $\mathcal{O}H$-module and the action of $a \in \mathcal{O}K$ is given by left multiplication by $g^{-1}ag$.

So we have constructed an isomorphism of $(\mathcal{O}H', \mathcal{O}H)$-bimodules

$$\mathcal{O}G \xrightarrow{\sim} \bigoplus_{g\in H'\backslash G/H} \mathcal{O}H' \otimes_{\mathcal{O}(H'\cap gHg^{-1})} (\mathcal{O}H)^g.$$

In terms of functors, this is the usual Mackey formula

$$\mathrm{Res}^G_{H'} \circ \mathrm{Ind}^G_H \xrightarrow{\sim} \bigoplus_{g\in H'\backslash G/H} \mathrm{Ind}^{H'}_{H'\cap gHg^{-1}} \circ \mathrm{Res}^{gHg^{-1}}_{H'\cap gHg^{-1}} \circ g^*$$

where $g^* : \mathcal{O}H\text{-mod} \to \mathcal{O}(gHg^{-1})\text{-mod}$ is restriction via the isomorphism

$$gHg^{-1} \to H, \quad x \mapsto g^{-1}xg.$$

3.1.5. TI Sylow *p*-subgroups. We will see here our first comparison result.

Let us assume that G has trivial intersection (TI) Sylow p-subgroups: given two distinct Sylow p-subgroups P and Q, then $P \cap Q = \{1\}$.

Let P be a Sylow p-subgroup of G and $H = N_G(P)$. We denote by $\mathbb{Z}_p$ the localization of $\mathbb{Z}$ at the prime p. Let N be the $(\mathbb{Z}_pG, \mathbb{Z}_pH)$-bimodule $\mathbb{Z}_pG$. Then $N^* = \mathbb{Z}_pG$ (the identification is made via $\hat{t}$).

We have a split exact sequence of $(\mathbb{Z}_pH, \mathbb{Z}_pH)$-bimodules (cf. §3.1.4)

$$0 \to \mathbb{Z}_pH \to \mathbb{Z}_pG \to \bigoplus_g \mathbb{Z}_pH \otimes_{\mathbb{Z}_pK} (\mathbb{Z}_pH)^g \to 0,$$

where $g \in H \setminus G/H$, $g \notin H$ and $K = H \cap gHg^{-1}$ and where the first map is the inclusion.

For $g \in G, g \notin H$, we have $P \cap gPg^{-1} = \{1\}$; thus K is a p'-group. It follows that $\mathbb{Z}_pH$ is a projective $(\mathbb{Z}_pH, \mathbb{Z}_pK)$-bimodule and $(\mathbb{Z}_pH)^g$ a projective $(\mathbb{Z}_pK, \mathbb{Z}_pH)$-bimodule, whence $\mathbb{Z}_pH \otimes_{\mathbb{Z}_pK} (\mathbb{Z}_pH)^g$ is a projective $(\mathbb{Z}_pH, \mathbb{Z}_pH)$-bimodule.

Since the counit

$$\eta_N : \mathbb{Z}_pH \to N^* \otimes_{\mathbb{Z}_pG} N = \mathbb{Z}_pG$$

is the inclusion, we have a split exact sequence of $(\mathbb{Z}_pH, \mathbb{Z}_pH)$-bimodules

$$0 \to \mathbb{Z}_pH \xrightarrow{\eta_N} N^* \otimes_{\mathbb{Z}_pG} N \to Z \to 0,$$

where Z is projective.

Since $[G : H]$ is invertible in $\mathbb{Z}_p$, the surjection $\varepsilon_N : N \otimes_{\mathbb{Z}_pH} N^* \to \mathbb{Z}_pG$ splits. Let Z' be its kernel.

The composition

$$N^* \xrightarrow{\eta_N\otimes 1} N^* \otimes_{\mathbb{Z}_pG} N \otimes_{\mathbb{Z}_pH} N^* \xrightarrow{1\otimes\varepsilon_N} N^*$$

is the identity. So

$$N^* \otimes_{\mathbb{Z}_pG} Z' = \ker(1 \otimes \varepsilon_N) \simeq \mathrm{coker}(\eta_N \otimes 1) = Z \otimes_{\mathbb{Z}_pH} N^*,$$

whence

$$N \otimes_{\mathbb{Z}_pH} N^* \otimes_{\mathbb{Z}_pG} Z' \simeq N \otimes_{\mathbb{Z}_pH} Z \otimes_{\mathbb{Z}_pH} N^*.$$

As $\mathbb{Z}_pG$ is a direct summand of $N \otimes_{\mathbb{Z}_pH} N^*$, it follows that Z' is a direct summand of $N \otimes_{\mathbb{Z}_pH} Z \otimes_{\mathbb{Z}_pH} N^*$. That last module is projective since Z is projective: this shows the projectivity of Z'.

We have obtained the following isomorphisms

$$N \otimes_{\mathbb{Z}_pH} N^* \simeq \mathbb{Z}_pG \oplus \text{ projective (as } (\mathbb{Z}_pG, \mathbb{Z}_pG)\text{-bimodules)}$$

$$N^* \otimes_{\mathbb{Z}_pG} N \simeq \mathbb{Z}_pH \oplus \text{ projective (as } (\mathbb{Z}_pH, \mathbb{Z}_pH)\text{-bimodules).}$$

So we have

Proposition 3.2. *The bimodule $\mathbb{Z}_pG$ induces a stable equivalence between $\mathbb{Z}_pG$ and $\mathbb{Z}_pH$.*

3.2. Blocks

The representation theory of $\mathcal{O}G$ reduces naturally to the study of the representations of the blocks of $\mathcal{O}G$. Some blocks can have a much simpler structure than others. Furthermore, most interesting equivalences arise between blocks, not between the whole group algebras.

3.2.1. A *block idempotent* b of $\mathcal{O}G$ is a primitive idempotent of the center $Z(\mathcal{O}G)$ of $\mathcal{O}G$: $b^2 = b \neq 0$ and there do not exist idempotents b_1 and b_2 of $Z(\mathcal{O}G)$ with $b_1b_2 = 0$ and $b = b_1 + b_2$. Let $\mathcal{B}$ be the set of block idempotents of $\mathcal{O}G$. Then we have

$$Z(\mathcal{O}G) = \bigoplus_{b \in \mathcal{B}} bZ(\mathcal{O}G).$$

This is the unique decomposition of $Z(\mathcal{O}G)$ as a direct sum of local rings. Note that $1 = \sum_b b$.

We now have the block decomposition of the group algebra

$$\mathcal{O}G = \bigoplus_b b\mathcal{O}G.$$

This is the unique decomposition of $\mathcal{O}G$ as a direct sum of indecomposable $\mathcal{O}$-algebras or, equivalently, the unique decomposition of $\mathcal{O}G$ as a direct sum of indecomposable $(\mathcal{O}G, \mathcal{O}G)$-bimodules. The (non-unitary) subalgebras $b\mathcal{O}G$ of $\mathcal{O}G$ are the *blocks* of $\mathcal{O}G$.

We now have a decomposition $\mathcal{O}G\text{-mod} = \bigoplus_b b\mathcal{O}G\text{-mod}$: every $\mathcal{O}G$-module M splits as $M = \bigoplus_b bM$ where $bM = b\mathcal{O}G \otimes_{\mathcal{O}G} M$. In particular, a non-zero indecomposable module belongs to a unique block. The *principal block* of $\mathcal{O}G$ is the block containing the trivial $\mathcal{O}G$-module.

Assume $|G|$ is invertible in $\mathcal{O}$. Then $e_1 = \frac{1}{|G|}\sum_{g\in G} g$ is the principal block idempotent and $e_1\mathcal{O}G \simeq \mathcal{O}$. If in addition $\mathcal{O}$ is a field, then the blocks are simple algebras.

3.2.2. Normal p'-subgroups. Let $H = O_{p'}(G)$, the largest normal subgroup of G whose order is prime to p and $e = \frac{1}{|H|}\sum_{h\in H} h$. Then e is an idempotent of $Z(\mathbb{Z}_p G)$.

Let $\bar{G} = G/H$. We have an isomorphism

$$e\mathbb{Z}_p G \xrightarrow{\sim} \mathbb{Z}_p\bar{G}, \quad eg \mapsto gH.$$

We have compared $\mathbb{Z}_p\bar{G}$ with a direct summand $e\mathbb{Z}_p G$ (i.e., a sum of blocks) of $\mathbb{Z}_p G$. This is compatible with blocks.

Hypothesis 1. For the remainder of the article, we assume that the residue field k of $\mathcal{O}$ has characteristic p.

The isomorphism $e\mathcal{O}G \xrightarrow{\sim} \mathcal{O}\bar{G}$ induces a bijection between the set of block idempotents of $\mathcal{O}\bar{G}$ and the set of those block idempotents b of $\mathcal{O}G$ such that $be = b$. We then have an isomorphism between the corresponding blocks of $\mathcal{O}G$ and $\mathcal{O}\bar{G}$. For example, we obtain an isomorphism between the principal blocks of $\mathcal{O}G$ and $\mathcal{O}\bar{G}$.

Note that the discussion above remains unchanged if we take for H any normal subgroup of G with order prime to p.

As a special case, assume G is p-nilpotent, i.e., $G = H \rtimes P$ where P is a Sylow p-subgroup. Then we have an isomorphism between $\mathcal{O}P$ and the principal block of $\mathcal{O}G$.

3.2.3. Blockwise version of the TI equivalence. We go back to the assumption of §3.1.5 that G has TI Sylow p-subgroups and p divides the order of G. Let f be a block idempotent of $\mathcal{O}H$ (recall that $H = N_G(P)$ where P is a Sylow p-subgroup of G). There is a unique block idempotent e of $\mathcal{O}G$ such that $eNf = e\mathcal{O}Gf$ is not a projective $(\mathcal{O}G, \mathcal{O}H)$-bimodule.

Then

Proposition 3.3. *The bimodule $e\mathcal{O}Gf$ induces a stable equivalence between $e\mathcal{O}G$ and $f\mathcal{O}H$.*

Note that the blocks of $\mathcal{O}G$ which do not correspond to blocks of $\mathcal{O}H$ are stably equivalent to 0.

In general, $e\mathcal{O}G$ and $f\mathcal{O}H$ are not Morita equivalent (i.e., you cannot get rid of the projective "remainder"), although they might be in some exceptional cases. Let us give two such cases.

Assume $p = 3$, $G = \mathfrak{S}_4$ and e is the principal block idempotent of $\mathbb{Z}_p G$. Then $P \simeq \mathbb{Z}/3$ and $H \simeq \mathfrak{S}_3$. By §3.2.2, we have an isomorphism $e\mathbb{Z}_p G \simeq \mathbb{Z}_p H$.

Assume now $p = 3$, $G = \mathfrak{S}_5$ and e is the principal block idempotent. We have $P \simeq \mathbb{Z}/3$, $H \simeq \mathfrak{S}_3 \times \mathbb{Z}/2$ and $f\mathbb{Z}_p H \simeq \mathbb{Z}_p\mathfrak{S}_3$. One can check that $e\mathbb{Z}_p G$ is Morita equivalent to $f\mathbb{Z}_pH$, but the algebras are not isomorphic (they have different $\mathbb{Z}_p$-ranks!); there is a direct summand M of $e\mathbb{Z}_pGf$ inducing such a Morita equivalence.

3.2.4. The TI case suggests that isomorphisms or Morita equivalences are too narrow concepts in order to compare blocks. On the other hand, it is difficult to deduce much numerical information from the existence of a stable equivalence although it is expected that the number of non-projective simple modules will be invariant when $\mathcal{O}$ is a field (Auslander's conjecture).

4. The Brauer functor

The Brauer functor is a fundamental tool to pass from global to local data.

4.1. p-permutation modules

4.1.1. Let Q be a p-subgroup of G. We denote by Br_Q the *Brauer functor* Br_Q : $\mathcal{O}G$-mod $\to kN_G(Q)$-mod, defined as follows.

For U an $\mathcal{O}G$-module, define

$$\mathrm{Br}_Q(U) = U^Q / \Big(\big(\sum_{P<Q} \mathrm{Tr}_P^Q U^P \big) + \mathfrak{p}U^Q \Big),$$

where the trace map $\mathrm{Tr}_P^Q : U^P \to U^Q$ between fixed point sets is given by $v \mapsto \sum_{g\in Q/P} gU$, and where $\mathfrak{p}$ is the maximal ideal of $\mathcal{O}$.

We will also consider the extension of Br_Q to the category of complexes of $\mathcal{O}G$-modules.

4.1.2. Let Ω be a G-set and $\mathcal{O}\Omega$ the corresponding permutation $\mathcal{O}G$-module. Then the inclusion $\mathcal{O}(\Omega^Q) \hookrightarrow (\mathcal{O}\Omega)^Q$ induces an isomorphism $k(\Omega^Q) \xrightarrow{\sim} \mathrm{Br}_Q(\mathcal{O}\Omega)$.

Let H be a subgroup of G. Then $(G/H)^Q \neq \emptyset$ if and only if Q is conjugate to a subgroup of H. So $\mathrm{Br}_Q(\mathrm{Ind}_H^G \mathcal{O}) \neq 0$ if and only if Q is conjugate to a subgroup of H.

Let V be a $\mathcal{O}N_G(Q)/Q$-module. We have (by adjunction) a morphism

$$\mathrm{Res}_{N_G(Q)}^{N_G(Q)/Q} V \to (\mathrm{Ind}_{N_G(Q)}^G \mathrm{Res}_{N_G(Q)}^{N_G(Q)/Q} V)^Q$$

and hence, by composition, a morphism

$$\mathrm{Res}_{N_G(Q)}^{N_G(Q)/Q} V \to \mathrm{Br}_Q \mathrm{Ind}_{N_G(Q)}^G \mathrm{Res}_{N_G(Q)}^{N_G(Q)/Q} V.$$

This gives a morphism of endofunctors of $\mathcal{O}N_G(Q)/Q$-mod

$$1_{\mathcal{O}N_G(Q)/Q\text{-mod}} \to \mathrm{Br}_Q \mathrm{Ind}_{N_G(Q)}^G \mathrm{Res}_{N_G(Q)}^{N_G(Q)/Q}.$$

Furthermore, this is an isomorphism when applied to projective $\mathcal{O}N_G(Q)/Q$-modules.

4.1.3. The Brauer functor is of particular interest when applied to *p-permutation modules* (direct summands of permutation modules).

If U is an indecomposable p-permutation $\mathcal{O}G$-module, then there is a minimal subgroup P of G such that the surjection $\mathcal{O}G \otimes_{\mathcal{O}P} \mathcal{O}G \otimes_{\mathcal{O}G} U \to U$ splits. This is a p-subgroup of G, called a *vertex* of U. It is unique up to conjugation. It is also characterized (up to conjugation) as the minimal subgroup of G such that U is a direct summand of a module induced from P or as the maximal subgroup of G such that $\mathrm{Br}_P(U) \neq 0$.

4.2. The Brauer morphism

For H a group, we define a subgroup ΔH of $H \times H^\circ$ by $\Delta H = \{(x, x^{-1}) | \, x \in H\}$. Let Q be a p-subgroup of G. The surjection

$$\mathrm{br}_Q : (\mathcal{O}G)^{\Delta Q} \to \mathrm{Br}_{\Delta Q}(\mathcal{O}G) = kC_G(Q)$$

is the *Brauer morphism*. This is a morphism of $((\mathcal{O}G)^{\Delta Q}, (\mathcal{O}G)^{\Delta Q})$-bimodules, and hence a morphism of algebras. It restricts to a (not necessarily surjective) morphism

$$\mathrm{br}_Q : Z(\mathcal{O}G) \to Z(kC_G(Q)).$$

Let $z \in Z(\mathcal{O}G)$. Then multiplication by z defines an endomorphism of any $\mathcal{O}G$-module U and the corresponding endomorphism of $\mathrm{Br}_Q(U)$ is multiplication by $\mathrm{br}_Q(z)$.

For example, if b is an idempotent of $Z(\mathcal{O}G)$, then we can consider the $b\mathcal{O}G$-module $bU = b\mathcal{O}G \otimes_{\mathcal{O}G} U$. We have

$$\mathrm{Br}_Q(bU) = \mathrm{br}_Q(b) \cdot \mathrm{Br}_Q(U).$$

If b is the principal block idempotent of $\mathcal{O}G$, then $\mathrm{br}_Q(b)$ is the principal block idempotent of $kC_G(Q)$.

4.3. Defect of blocks

It is now time to turn to defect groups of blocks!

Let e be a block idempotent of $\mathcal{O}G$. A *defect group* of $e\mathcal{O}G$ is a subgroup D of G such that ΔD is a vertex of the $\mathcal{O}(G \times G^\circ)$-module $e\mathcal{O}G$ — i.e., this is a subgroup of G minimal with respect to the property that the multiplication map

$$e\mathcal{O}G \otimes_{\mathcal{O}D} \mathcal{O}G \to e\mathcal{O}G$$

splits. This is also a subgroup of G maximal with respect to the property that $\mathrm{br}_D(e) \neq 0$.

If $e\mathcal{O}G$ is the principal block, then D is a Sylow p-subgroup.

We can now refine the discussion of §3.1.3: the complexity of ekG is accounted for by D (ekG is semi-simple (and then simple) if and only if $D = 1$, has finite representation type if and only if D is cyclic, etc.).

The following conditions for a block are equivalent:

(i) $D = 1$;

(ii) $e\mathcal{O}G$ is a projective $(\mathcal{O}G, \mathcal{O}G)$-bimodule;

(iii) $e\mathcal{O}G$ is stably equivalent to 0.

When $\mathcal{O} = k$, this is furthermore equivalent to the fact that ekG is a simple algebra.

Defining the (numerical) defect of $e\mathcal{O}G$ to be $\log_p |D|$, we see that the blocks fulfilling those conditions are the blocks with defect 0.

5. Rickard equivalences

From here on, we will consider the usual setting for modular representation theory.

Hypothesis 2. We assume $\mathcal{O}$ is a discrete valuation ring containing all $|I|$-th roots of unity, for all the finite groups I to be considered.

5.1. An example: A_5 in characteristic 2

Let G be the alternating group A_5, $p = 2$ and D be a Sylow 2-subgroup of G ($D \simeq \mathbb{Z}/2 \times \mathbb{Z}/2$). Let $H = N_G(D)$: we have $H \simeq A_4$. Let E be a cyclic subgroup of order 3 of H. Then $H = D \rtimes E$. The algebra $\mathcal{O}H$ is indecomposable.

Let e be the principal block idempotent of $\mathcal{O}G$. Then $(1-e)\mathcal{O}G$ is a block of defect zero — it is actually a 5-dimensional matrix algebra over $\mathcal{O}$. The Sylow 2-subgroups of G are TI; thus we know from §3.2.3 that the bimodule $M = e\mathcal{O}G$ induces a stable equivalence between $A = e\mathcal{O}G$ and $B = \mathcal{O}H$.

The non-trivial simple B-modules lift to B-modules free over $\mathcal{O}$, whereas the non-trivial simple A-modules V_1 and V_2 do not lift to $\mathcal{O}$-free A-modules; in particular, A and B are not Morita equivalent (the algebras $k \otimes A$ and $k \otimes B$ are not Morita equivalent either: they have distinct Cartan matrices).

The module $M^* \otimes_A V_i = \operatorname{Res}_H^G V_i$ is an indecomposable two-dimensional B-module: let S_i be its unique simple submodule. Then S_1 and S_2 are the non-trivial simple B-modules.

Lemma 5.1. *A projective cover of M is*

$$P_M = P_{k_A} \otimes P_{k_B}^* \oplus P_{V_1} \otimes P_{S_1}^* \oplus P_{V_2} \otimes P_{S_2}^*,$$

where we denote by P_L a projective cover of the module L.

Proof. Let V be a simple A-module and S a simple B-module. We have an isomorphism of (B, B)-bimodules (§2.2.1)

$$\mathrm{Hom}_k(V^* \otimes_A M, S^*) \simeq \mathrm{Hom}_A(M, \mathrm{Hom}_k(V^*, S^*)),$$

whence

$$\mathrm{Hom}_{B^\circ}(V^* \otimes_A M, S^*) \simeq \mathrm{Hom}_{A\otimes B^\circ}(M, \mathrm{Hom}_k(V^*, S^*)) \simeq \mathrm{Hom}_{A\otimes B^\circ}(M, V \otimes S^*).$$

Finally,

$$\mathrm{Hom}_{B^\circ}(V^* \otimes_A M, S^*) \simeq \mathrm{Hom}_B(S, M^* \otimes_A V). \qquad \square$$

Let $f : P_M \to M$ be a surjection and let δ be its restriction to $R = P_{V_1} \otimes P^*_{S_1} \oplus P_{V_2} \otimes P^*_{S_2}$.

Let C be the complex

$$C = 0 \to R \xrightarrow{\delta} M \to 0,$$

where M is in degree 0.

As shown by Rickard, we have

Proposition 5.2. *The complex C induces a Rickard equivalence between A and B.*

Proof. Let us consider the double complex

$$\begin{array}{ccccccc}
 & & 0 & & 0 & & \\
 & & \downarrow & & \downarrow & & \\
0 & \longrightarrow & R \otimes_B M^* & \xrightarrow{\delta\otimes 1} & M \otimes_B M^* & \longrightarrow & 0 \\
 & & \downarrow{\scriptstyle 1\otimes\delta^*} & & \downarrow{\scriptstyle 1\otimes\delta^*} & & \\
0 & \longrightarrow & R \otimes_B R^* & \xrightarrow[\delta\otimes 1]{} & M \otimes_B R^* & \longrightarrow & 0 \\
 & & \downarrow & & \downarrow & & \\
 & & 0 & & 0\,. & &
\end{array}$$

We have

$$R \otimes_B M^* \simeq M \otimes_B R^* \simeq \bigoplus_{i,j} P_{V_i} \otimes P^*_{V_j},$$

$$R \otimes_B R^* \simeq \bigoplus_{i} P_{V_i} \otimes P^*_{V_i} \oplus \bigoplus_{i,j} P_{V_i} \otimes P^*_{V_j}$$

$$\text{and } M \otimes_B M^* \simeq A \oplus \bigoplus_{i\neq j} P_{V_i} \otimes P^*_{V_j}.$$

We have a split surjection $f \otimes 1 : P_M \otimes_B R^* \to M \otimes_B R^*$. Since $(P_M/R) \otimes_B R^*$ and $M \otimes_B R^*$ have no common non-zero direct summand, it follows that the map $\delta \otimes 1 : R \otimes_B R^* \to M \otimes_B R^*$ is still a split surjection.

Similarly, $1 \otimes \delta^* : R \otimes_B M^* \to R \otimes_B R^*$ is a split injection.

Let us consider now the complex $C \otimes_B C^*$, i.e., the total complex associated to the double complex above,

$$C \otimes_B C^* = 0 \to R \otimes_B M^* \overset{\delta\otimes 1 - 1\otimes\delta^*}{\longrightarrow} M \otimes_B M^* \oplus R \otimes_B R^* \overset{1\otimes\delta^* + \delta\otimes 1}{\longrightarrow} M \otimes_B R^* \to 0.$$

This complex is homotopy equivalent to its 0-th homology and

$$H^0(C \otimes_B C^*) \oplus R \otimes_B M^* \oplus M \otimes_B R^* \simeq R \otimes_B R^* \oplus M \otimes_B M^*.$$

It follows that $H^0(C \otimes_B C^*) \simeq A$; thus $C \otimes_B C^*$ is homotopy equivalent to A.

A similar proof shows that $C^* \otimes_A C$ is homotopy equivalent to B. □

This means that we have been able to get rid of the projective "remainder" by suitably modifying M into C. In order to achieve this, we had to move from modules to complexes of modules — more precisely, to the homotopy category of complexes of modules.

5.2. Broué's conjecture

We present here the abelian defect conjecture of Broué and its expected compatibilities with p'-outer automorphism groups and central extensions by p-groups.

5.2.1. Let us now fix our objects of study.

Hypothesis 3. From now on, we assume Hypothesis 2 and we denote by G a finite group, by e a block idempotent of $\mathcal{O}G$ and by D a defect group of $e\mathcal{O}G$. We put $H = N_G(D)$ and we denote by f the block idempotent of $\mathcal{O}H$ corresponding to e (it is the unique block idempotent with the property that $e\mathcal{O}G$ is a direct summand of $\mathrm{Ind}_{H\times H^\circ}^{G\times G^\circ} f\mathcal{O}H$). We put $A = e\mathcal{O}G$ and $B = f\mathcal{O}H$.

Following Rickard, we say that a complex C of $(e\mathcal{O}G, f\mathcal{O}H)$-bimodules is *splendid* if its components are p-permutation modules whose indecomposable summands have vertices contained in ΔD (note that the components are then exact bimodules). The relevance of this definition will appear in §5.3.

We can now state

Conjecture 5.3 (Broué). *Assume D is abelian. Then $e\mathcal{O}G$ and $f\mathcal{O}H$ are splendidly Rickard equivalent.*

Some remarks.

- See the Appendix for comments on the notion of splendidness.
- It is unclear whether there should be a more natural equivalence.

- Not every equivalence is splendid.
- The form of the conjecture given here is a refinement due to Rickard.
- The conjecture is known to fail when D is not abelian, even if the Sylow p-subgroups of G are TI, as in the case $G = \mathrm{Sz}(8)$ and $p = 2$. It remains an open problem to find an extension of the conjecture to blocks with non-abelian defect groups.

When e and f are principal block idempotents, then it is conjectured that there is a splendid Rickard complex C with $C \otimes_{f\mathcal{O}H} \mathcal{O} \simeq \mathcal{O}$. Such an equivalence is called a *normalized equivalence*. For example, the construction of §5.1 gives a positive answer for $G = A_5$, $p = 2$ and e the principal block idempotent.

5.2.2. Let us try to give the current status of the conjecture.
The conjecture holds for

- D cyclic [Ri1, Li1, Rou2] and $D \simeq \mathbb{Z}/2 \times \mathbb{Z}/2$ ([Ri4, Li2, Li3, Rou3] and §6.3 for principal blocks)
- G p-solvable [Da1, Pu1, HaLi]
- G a connected reductive algebraic group over $\mathbf{F}_q$ and p divides $q-1$ but p does not divide the order of the Weyl group [Pu2]
- G a symmetric group and $D \simeq \mathbb{Z}/p \times \mathbb{Z}/p$ [Ch1]
- several more cases where G a symmetric group, e.g., when $D \simeq (\mathbb{Z}/p)^r$ with $r \leq 5$ [Ri2] and [ChKe]
- $G = \mathrm{GL}_n(q)$, $p \nmid q$ and D has p-rank 2 [HiMi, Tu] and [BoRou],

for the principal blocks of

- $G = {}^2G_2(q)$ and $p = 2$ [Ok1]
- $G = \mathrm{Sp}_4(q)$, $q \equiv 2, 5 \pmod 9$ and $p = 3$ [Ok1]
- $G = \mathrm{PSU}_3(q^2)$, $q \equiv 2, 5 \pmod 9$ and $p = 3$ [KoKu1]
- $G = \mathrm{PSL}_3(q)$, $q \equiv 4, 7 \pmod 9$ and $p = 3$ [Ku]
- $G = A_7, A_8, M_{11}, M_{22}, M_{23}, \mathrm{PSL}_3(4)$ and HS, $p = 3$ [Ok1]
- $G = J_2$ and $G = \mathrm{Sp}_4(4)$ and $p = 5$ [Holl]
- $G = J_1$ and $p = 2$ [GoOk]
- any group G with $D \simeq \mathbb{Z}/3 \times \mathbb{Z}/3$ [KoKu2]
- $G = \mathrm{PSL}_2(p^n)$ ([Ch2] for $n = 2$, [Rou1] for $p^n = 8$ and [Ok2] in general)

- $G = \mathrm{GL}_4(q)$ and $G = \mathrm{GL}_5(q)$, $q \equiv 2, 5 \pmod 9$ and $p = 3$ [KoMi]
- $G = \mathrm{SU}_3(q^2)$, $p > 3$ and $p|q+1$ [KuWa],

and for the non-principal blocks of

- $G = ON$ and $p = 3$ [KoKuWa]
- $G = HS$ and $p = 3$ [Holm, KoKuWa]
- $G = 2.J_2$ and $p = 5$ [Holl]
- $G = \mathrm{SL}_2(p^2)$ [Holl].

5.2.3. We now consider automorphisms.

Hypothesis 3′. Hypothesis 3 holds and we let X be a finite group containing G as a normal subgroup and $Y = N_X(D)$. We assume that e is X-invariant. Then $X/G = Y/H$, and we assume that this group F is a p'-group. We put $\Delta = \{(g,h) \in X \times Y^\circ \mid (gG, hH^\circ) \in \Delta F\}$.

Then it is conjectured that there is a complex C of $\mathcal{O}\Delta$-modules whose restriction to $e\mathcal{O}G \otimes (f\mathcal{O}H)^\circ$ is a splendid Rickard complex. By Marcus [Ma] (or [Rou2, Lemma 2.8]), the complex $\mathrm{Ind}_\Delta^{X\times Y^\circ} C$ is then a splendid Rickard complex.

Remark 5.4. If F is not a p'-group, the same proof shows only that $\mathrm{Ind}_\Delta^{X\times Y^\circ} C$ will induce an equivalence between the derived categories of $e\mathcal{O}X$ and $f\mathcal{O}Y$.

Let us state some simple facts related to the extension problem.

Let M be an indecomposable direct summand of the $(e\mathcal{O}G \otimes (f\mathcal{O}H)^\circ)$-module $e\mathcal{O}Gf$ with vertex ΔD and M' with $e\mathcal{O}Gf = M \oplus M'$. Then the indecomposable summands of M' have vertices strictly contained in ΔD. The action of Δ on $\mathcal{O}X$ restricts to an action on $e\mathcal{O}Gf$ extending the natural action of $G \times H^\circ$. It follows that M and M' extend uniquely to $\mathcal{O}\Delta$-modules $\tilde{M}$ and $\tilde{M}'$ with $e\mathcal{O}Gf \simeq \tilde{M} \oplus \tilde{M}'$.

Let M be an $\mathcal{O}\Delta$-module. If $f : P_M \to M$ is a projective cover of M, then $\mathrm{Res}^\Delta_{G\times H^\circ} f$ is a projective cover of $\mathrm{Res}^\Delta_{G\times H^\circ} M$ (if $\mathrm{Res}^\Delta_{G\times H^\circ} \ker f$ has a projective direct summand, then so does $\ker f$).

5.2.4. Finally, we come to central extensions.

Hypothesis 3″. Hypothesis 3′ holds and we consider $\tilde{X}$ a finite group with a normal p-subgroup P such that $\tilde{X}/P = X$. Let $\tilde{G}$ be the inverse image of G in $\tilde{X}$. We assume P is central in $\tilde{G}$. The block idempotent e of $\mathcal{O}G$ lifts to a block idempotent $\tilde{e}$ of $\mathcal{O}\tilde{G}$ (Hensel's lemma in $Z(\mathcal{O}\tilde{G})$). Let $\tilde{Y}$ (resp. $\tilde{H}$) be the inverse image of Y (resp. H) in $\tilde{X}$. Let $\tilde{f}$ be the block idempotent of $\mathcal{O}\tilde{H}$ lifting f. Let $\tilde{\Delta}$ be the inverse image of Δ in $\tilde{X} \times \tilde{Y}^\circ$. Note that ΔP is normal in $\tilde{\Delta}$.

We have a commutative diagram:

$$\begin{array}{ccccccccc} & & 1 & & 1 & & & & \\ & & \downarrow & & \downarrow & & & & \\ & & P & = & P & & & & \\ & & \downarrow & & \downarrow & & & & \\ 1 & \longrightarrow & \tilde{G} & \longrightarrow & \tilde{X} & \longrightarrow & F & \longrightarrow & 1 \\ & & \downarrow & & \downarrow & & \| & & \\ 1 & \longrightarrow & G & \longrightarrow & X & \longrightarrow & F & \longrightarrow & 1 \\ & & \downarrow & & \downarrow & & & & \\ & & 1 & & 1\,. & & & & \end{array}$$

It is conjectured that there exists a complex $\tilde{C}$ of $\mathcal{O}(\tilde{\Delta}/\Delta P)$-modules that are projective for $\mathcal{O}\tilde{G}$ and $\mathcal{O}\tilde{H}^{\circ}$ such that $\mathrm{Res}_{G\times H^{\circ}}(\tilde{C}\otimes_{P\times P^{\circ}}\mathcal{O})$ is a splendid Rickard complex.

By [Rou2, Lemma 2.11], the complex $\mathrm{Res}^{\tilde{\Delta}/\Delta P}_{\tilde{G}\times\tilde{H}^{\circ}}\tilde{C}$ is then a splendid Rickard complex for $(\tilde{e}\mathcal{O}\tilde{G},\ \tilde{f}\mathcal{O}\tilde{H})$ and $\mathrm{Ind}^{\tilde{X}\times\tilde{Y}^{\circ}}_{\tilde{\Delta}}\mathrm{Res}^{\tilde{\Delta}/\Delta P}_{\tilde{\Delta}}\tilde{C}$ a splendid Rickard complex for $(\tilde{e}\mathcal{O}\tilde{X},\ \tilde{f}\mathcal{O}\tilde{Y})$.

Remark 5.5. This refined conjecture holds when D is cyclic or $D\simeq(\mathbb{Z}/2)^2$.

5.3. Splendid stable equivalences and local Rickard equivalences

The following result, which is a variation on a classical theme, is a cornerstone to our approach. It is the generalization from the case of local Morita equivalences to the case of local Rickard equivalences of [Br2, Theorem 6.3]. The first implication is due to Rickard and was the motivation for the introduction of the special class of spendid complexes. Given a global splendid stable equivalence, we obtain local Rickard equivalences. The second half shows that in order to check that a global splendid complex induces a stable equivalence, it suffices to check that the associated local complexes induce Rickard equivalences. This follows quickly from a result of Bouc.

From now on (except in the Appendix), we will consider only principal blocks. For the general case as well as for more details, see the Appendix and [Rou3].

Hypothesis 4. Hypothesis 3 holds and we assume furthermore than e is the principal block idempotent of $\mathcal{O}G$.

Now, D is a Sylow p-subgroup of G.

We take a subgroup K of G containing D and controlling the fusion of p-subgroups in G (i.e., for $P \leq D$ and $g \in G$ such that $gPg^{-1} \leq D$, then there exists $h \in K$ and $z \in C_G(P)$ such that $g = hz$). By Burnside's lemma, when D is abelian, we can take $K = H = N_G(D)$.

Let us denote by b the principal block idempotent of $\mathcal{O}K$. For $Q \leq D$, we denote also by e_Q (resp. b_Q) the principal block idempotent of $kC_G(Q)$ (resp. $kC_K(Q)$).

Theorem 5.6. *Let C be a splendid complex of $(e\mathcal{O}G, b\mathcal{O}K)$-bimodules. The following assertions are equivalent.*

(i) *C induces a stable equivalence between $e\mathcal{O}G$ and $b\mathcal{O}K$.*

(ii) *For every non-trivial subgroup Q of D, the complex $\mathrm{Br}_{\Delta Q}(C)$ induces a Rickard equivalence between $e_QkC_G(Q)$ and $b_QkC_K(Q)$.*

(ii′) *For every subgroup Q of order p in D, the complex $\mathrm{Br}_{\Delta Q}(C)$ induces a Rickard equivalence between $e_QkC_G(Q)$ and $b_QkC_K(Q)$.*

Proof. Let us recall the results of Rickard [Ri4, proof of Theorem 4.1].

- The components of the complex $C \otimes_{\mathcal{O}K} C^*$ are relatively ΔD-projective.
- For $Q \leq D$, we have an isomorphism of complexes of $(kC_G(Q), kC_K(Q))$-bimodules
$$\mathrm{Br}_{\Delta Q}(C \otimes_{\mathcal{O}K} C^*) \simeq C_Q \otimes_{kC_K(Q)} C_Q^*$$
where $C_Q = \mathrm{Br}_{\Delta Q}(C)$ is a splendid complex of $(e_QkC_G(Q), f_QkC_H(Q))$-bimodules. More precisely, let X_Q be the cone of the adjunction morphism $C_Q \otimes_{kC_K(Q)} C_Q^* \to e_QkC_G(Q)$. Then $X_Q \simeq \mathrm{Br}_{\Delta Q}(X)$, where X is the cone of the adjunction morphism $C \otimes_{\mathcal{O}K} C^* \to e\mathcal{O}G$.

By [Bou, Proposition 7.9] (cf. [Rou3] for the extension from k to $\mathcal{O}$), a complex Z is homotopy equivalent to a complex of projective modules if and only if, for every non-trivial subgroup Q of D, the complex $\mathrm{Br}_{\Delta Q}(Z)$ is homotopy equivalent to 0 (using that $\mathrm{Br}_P(Z) = 0$ if P is not contained in ΔD up to conjugacy).

We have a similar statement concerning $C^* \otimes_{\mathcal{O}G} C$ and the equivalence between (i) and (ii) follows.

The implication (ii′) $\Rightarrow$ (ii) follows by induction from (ii) $\Rightarrow$ (i) and from the isomorphism
$$\mathrm{Br}_Q(\mathrm{Br}_P(V)) \simeq \mathrm{Br}_Q(V)$$
when $P \lhd Q$. □

6. Blocks with defect group $\mathbb{Z}/p^a \times \mathbb{Z}/p^b$

In this section we assume that Hypothesis 4 holds. We recall that $A = e\mathcal{O}G$ and $B = f\mathcal{O}H$.

6.1. Cyclic defect groups

In this section, we recall the construction of splendid Rickard complexes for principal blocks with cyclic defect groups. Let us assume in §6.1 that D cyclic.

6.1.1. Let $\pi : P_{e\mathcal{O}Gf} \to e\mathcal{O}Gf$ be a projective cover of $e\mathcal{O}Gf$.

In [Rou1, Theorem 4.1], we have constructed a direct summand N of $P_{e\mathcal{O}Gf}$ with the following property. Let ϕ be the restriction of π to N and $C = 0 \to N \xrightarrow{\phi} e\mathcal{O}Gf \to 0$, where $e\mathcal{O}Gf$ is in degree 0. Then C induces a normalized splendid Rickard equivalence between $e\mathcal{O}G$ and $f\mathcal{O}H$.

Assume now Hypothesis 3′ of §5.2.3. We have [Ma, §5.5]:

Lemma 6.1. *The complex C extends to a complex of $\mathcal{O}\Delta$-modules.*

Proof. Let us decompose the $\mathcal{O}\Delta$-module $e\mathcal{O}Gf$ as $e\mathcal{O}Gf = M \oplus M'$ with M' projective and M without projective direct summand. By §5.2.3, the module $\operatorname{Res}^{\Delta}_{G\times H^\circ} M$ has no projective direct summand. Let P_M be a projective cover of M. Then $\operatorname{Res}^{\Delta}_{G\times H^\circ} P_M$ is a projective cover of $\operatorname{Res}^{\Delta}_{G\times H^\circ} M$. The direct summand N of $P_{e\mathcal{O}Gf} \simeq \operatorname{Res}^{\Delta}_{G\times H^\circ}(P_M \oplus M')$ arises as $P_0 \oplus \operatorname{Res}^{\Delta}_{G\times H^\circ} M'$, where P_0 is a direct summand of $\operatorname{Res}^{\Delta}_{G\times H^\circ} P_M$. The construction of P_M, using the Brauer tree, shows it is invariant under Δ, whence there is a direct summand N_0 of P_M with $\operatorname{Res}^{\Delta}_{G\times H^\circ} N_0 \simeq P_0$. □

6.1.2. We assume finally Hypothesis 3″ of §5.2.4.

There is a projective $\mathcal{O}(\tilde{\Delta}/\Delta P)$-module $\tilde{N}$ such that $\mathcal{O} \otimes_{\mathcal{O}(P\times P^\circ)} \tilde{N} = N$. The composition of $\phi : N \to e\mathcal{O}Gf$ with the canonical map $\tilde{N} \to N$ factors through the canonical map $\tilde{e}\mathcal{O}\tilde{G}\tilde{f} \to e\mathcal{O}Gf$ as $\tilde{\phi} : \tilde{N} \to \tilde{e}\mathcal{O}\tilde{G}\tilde{f}$. This last map lifts ϕ, i.e., $\phi = 1 \otimes \tilde{\phi}$. By restriction, we obtain a morphism $\tilde{\phi} : \tilde{N} \to \tilde{e}\mathcal{O}\tilde{G}\tilde{f}$ of $\mathcal{O}\tilde{\Delta}$-modules.

We now define the complex of $\mathcal{O}\tilde{\Delta}$-modules

$$\tilde{C} = 0 \to \tilde{N} \xrightarrow{\tilde{\phi}} \tilde{e}\mathcal{O}\tilde{G}\tilde{f} \to 0$$

where $\tilde{e}\tilde{G}\tilde{f}$ is in degree 0. Then $\mathcal{O} \otimes_{\mathcal{O}(P\times P^\circ)} \tilde{C} \simeq C$.

So by §5.2.4, we have

Theorem 6.2. *The complex $\operatorname{Res}^{\tilde{\Delta}}_{\tilde{G}\times\tilde{H}^\circ} \tilde{C}$ induces a normalized splendid Rickard equivalence between $\tilde{e}\mathcal{O}\tilde{G}$ and $\tilde{e}\mathcal{O}\tilde{H}$.*

6.2. Blocks with abelian defect group of rank 2

6.2.1. In this subsection, assume that D is elementary abelian of order p^2.

Let P be a subgroup of order p of D. We have $N_G(P)/C_G(P) = N_H(P)/C_H(P)$; this is a p'-group.

We are in the setting of §6.1.2. We add an index $?_P$ to an object $?$ from this section to avoid confusion. We have $G_P = C_G(P)/P$, $H_P = C_H(P)/P$, $X_P = N_G(P)/P$, $Y_P = N_H(P)/P$, $\tilde{G}_P = C_G(P)$, $\tilde{H}_P = C_H(P)$, $\tilde{X}_P = N_G(P)$ and $\tilde{Y}_P = N_H(P)$. We nevertheless denote by e_P and f_P the principal block idempotents of $\mathcal{O}C_G(P)$ and $\mathcal{O}C_H(P)$. We have $\tilde{\Delta}_P = N_{G\times H^\circ}(\Delta P)$. We have a projective $\mathcal{O}(\tilde{\Delta}_P/\Delta P)$-module $\tilde{N}_P$ and a map $\tilde{\phi}_P : \tilde{N}_P \to e_P\mathcal{O}C_G(P)f_P$ with the property that the complex $\tilde{C}_P = 0 \to \tilde{N}_P \xrightarrow{\tilde{\phi}_P} e_P\mathcal{O}C_G(P)f_P \to 0$ induces a Rickard equivalence between $e_P\mathcal{O}C_G(P)$ and $f_P\mathcal{O}C_H(P)$.

Let $V_P = \operatorname{Ind}_{\tilde{\Delta}_P}^{G\times H^\circ} \operatorname{Res}_{\tilde{\Delta}_P}^{\tilde{\Delta}_P/\Delta P} \tilde{N}_P$, where we denote by $\operatorname{Res}_{\tilde{\Delta}_P}^{\tilde{\Delta}_P/\Delta P}$ the restriction through the canonical map $\mathcal{O}\tilde{\Delta}_P \to \mathcal{O}(\tilde{\Delta}_P/\Delta P)$, also called inflation. The morphism

$$e_P\mathcal{O}C_G(P)f_P \to \operatorname{Res}_{\tilde{\Delta}_P}^{G\times H^\circ} e\mathcal{O}Gf$$

(coming from the inclusion $e_P\mathcal{O}C_G(P)f_P \to \mathcal{O}G$) induces by adjunction a morphism

$$\alpha_P : \operatorname{Ind}_{\tilde{\Delta}_P}^{G\times H^\circ} e_P\mathcal{O}C_G(P)f_P \to e\mathcal{O}Gf$$

and $\operatorname{Br}_{\Delta P}(\alpha_P)$ is an isomorphism. Let $\psi_P = \alpha_P \operatorname{Ind}_{\tilde{\Delta}_P}^{G\times H^\circ}(\tilde{\phi}_P) : V_P \to e\mathcal{O}Gf$.

Let

$$C = 0 \to \bigoplus_P V_P \xrightarrow{\sum_P \psi_P} e\mathcal{O}Gf \to 0$$

where P runs over the subgroups of order p of D up to H-conjugacy (the term $e\mathcal{O}Gf$ is in degree 0).

Theorem 6.3. *The complex C induces a normalized splendid stable equivalence between $e\mathcal{O}G$ and $f\mathcal{O}H$.*

Proof. The complex C is splendid since V_P is a sum of p-permutation modules with vertex ΔP and $e\mathcal{O}Gf$ is a p-permutation module induced from ΔD. Let P be a subgroup of order p of D. For Q a subgroup of order p of D, we have $\operatorname{Br}_{\Delta P}(V_Q) = 0$ unless ΔQ is $(G\times H^\circ)$-conjugate to ΔP, i.e., Q is H-conjugate to P. Now we have $\operatorname{Br}_{\Delta P}(\psi_P) = \tilde{\phi}_P$ (cf. §4.1.2).

It follows that $\operatorname{Br}_{\Delta P}(C) \simeq k \otimes \tilde{C}_P$ induces a Rickard equivalence between $e_PkC_G(P)$ and $f_PkC_H(P)$.

Since C is splendid, the theorem follows now from Theorem 5.6, (ii′) ⇒ (i). □

Remark 6.4. A similar construction works for nonprincipal blocks (cf. the Appendix and [Rou3]).

6.2.2. In this subsection, assume that D is abelian and has p-rank 2.

Let Q be a subgroup of order p of D. We have $D = (D \cap Z(C_H(Q))) \times [D, C_H(Q)]$. Let $P = D \cap Z(C_H(Q))$. Then $Q \leq P \leq D$, D/P is cyclic and $C_G(P)$ controls fusion of p-subgroups in $C_G(Q)$. This implies $C_G(Q) = O_{p'}C_G(Q) \cdot C_G(P)$, by the Z_p^*-theorem (given a finite group G and a p-subgroup P such that $C_G(P)$ controls fusion of p-subgroups in G, then $G = O_{p'}G \cdot C_G(P)$). This implies also $N_G(Q) = O_{p'}C_G(Q) \cdot N_G(P)$.

When $P = D$ (the only possible case when $p = 2$), the results above are easy since $C_G(Q)$ is p-nilpotent by Burnside's Theorem. Let e_Q and f_Q be the principal block idempotents of $\mathcal{O}C_G(Q)$ and $\mathcal{O}C_H(Q)$. The construction of §3.2.2 provides an extension of $\mathrm{Res}^{\tilde{\Delta}_P/\Delta P}_{\tilde{\Delta}_P/\Delta Q} \tilde{N}_P$ to an $(e_Q \otimes f_Q)\mathcal{O}(\tilde{\Delta}_Q/\Delta Q)$-module $\tilde{N}_Q$ and $\tilde{\phi}_P$ lifts to a morphism $\tilde{\phi}_Q : \tilde{N}_Q \to e_Q\mathcal{O}C_G(Q)f_Q$. Now, we continue as in §6.2.1 and construct V_Q, α_Q and ψ_Q.

We put

$$C = 0 \to \bigoplus_Q V_Q \xrightarrow{\sum_Q \psi_Q} e\mathcal{O}Gf \to 0$$

where Q runs over the subgroups of order p in D up to H-conjugacy (the term $e\mathcal{O}Gf$ is in degree 0).

The same proof as in Theorem 6.3 leads to the following result.

Theorem 6.5. *The complex C induces a normalized splendid stable equivalence between $e\mathcal{O}G$ and $f\mathcal{O}H$.*

Remark 6.6. When $N_G(D)/C_G(D)$ acts freely on $D - \{1\}$, then this result is due to Puig [Pu3, Corollary 6.7]. This is always the case when $p = 2$.

6.3. Blocks with Klein four defect groups

6.3.1. In this part we will make more explicit the constructions of §6.2 for the case $D = \mathbb{Z}/2 \times \mathbb{Z}/2$, where some simplifications occur. Then we will show how to construct a Rickard equivalence from the stable equivalence. The reason why the method does not apply for any other D of rank 2 is that there are too many indecomposable kD-modules (the type is wild).

Throughout §6.3, we assume D is elementary abelian of order 4.

6.3.2. Stable equivalence. Let P be a subgroup of order 2 of D. The complex C_P of §6.2.1 (i.e., the complex C of §6.1.1 constructed for the group $C_G(P)/P$) has homology only in degree 0: this homology is a direct summand of $\bar{e}_P\mathcal{O}(C_G(P)/P)\bar{f}_P$ and it induces a Morita equivalence between $\bar{e}_P\mathcal{O}(C_G(P)/P)$ and $\bar{f}_P\mathcal{O}(C_H(P)/P)$ (here, $\bar{e}_P$ and $\bar{f}_P$ are the principal block idempotents of $\mathcal{O}(C_G(P)/P)$ and $\mathcal{O}(C_H(P)/P)$).

The complex $\tilde{C}_P$ inducing a Rickard equivalence between $e_P\mathcal{O}C_G(P)$ and $f_P\mathcal{O}C_H(P)$ then has homology only in degree 0, i.e., there is a direct summand of $e_P\mathcal{O}C_G(P)f_P$ inducing a Morita equivalence between $e_P\mathcal{O}C_G(P)$ and $f_P\mathcal{O}C_H(P)$.

Finally, the complex C constructed in §6.2.1 has homology only in degree 0: this is a bimodule N isomorphic to a direct summand of $e\mathcal{O}Gf$. Let M be a direct summand of N such that $N = M \oplus$ projective and M has no projective direct summand. Then M induces a stable equivalence between $A = e\mathcal{O}G$ and $B = f\mathcal{O}H$.

Let us state this as

Proposition 6.7. *There is a direct summand of $e\mathcal{O}Gf$ inducing a normalized splendid stable equivalence between $e\mathcal{O}G$ and $f\mathcal{O}H$.*

6.3.3. The following result now solves Conjecture 5.3 for $D \simeq \mathbb{Z}/2 \times \mathbb{Z}/2$.

Theorem 6.8. *There is a normalized splendid Rickard equivalence between $e\mathcal{O}G$ and $f\mathcal{O}H$ induced by a complex C such that C^{-1} is projective, $C^0 = M$ and $C^i = 0$ for $i \neq 0, -1$.*

Remark 6.9. For non-principal blocks, see the Appendix and [Rou3]. The existence of a Rickard equivalence was established by Linckelmann in [Li2, Corollary 1.5], but no equivalence was constructed.

The rest of this section is devoted to the proof of this theorem.

The group $E = H/C_G(D)$ is a p'-subgroup of the automorphism group of D, whence it is the trivial group or a cyclic group of order 3.

There is a normalized Morita equivalence between B and $B' = \mathcal{O}D \rtimes E$ (cf. for example [Rou2, Proposition 2.15]). Let W be a (B, B')-bimodule inducing such an equivalence. Let $L = M \otimes_B W$: this is an indecomposable (A, B')-bimodule inducing a stable equivalence.

Note that $L \otimes_{B'} \mathcal{O} \simeq \mathcal{O}$.

6.3.4. The nilpotent case. Assume $E = 1$. Then L induces a Morita equivalence between A and B' (cf. §2.3.5), whence $C = M$ induces a Morita equivalence and hence a Rickard equivalence between A and B.

6.3.5. The A_4-case. Let us now consider the case $|E| = 3$. Then $D \rtimes E \simeq A_4$.

Let S be a non-trivial simple B'-module. We have

$$\mathrm{Hom}_A(L \otimes_{B'} S, k) \simeq \mathrm{Hom}_{B'}(S, L^* \otimes_A k) = \mathrm{Hom}_{B'}(S, k) = 0.$$

Let V_1 be a simple quotient of $L \otimes_{B'} S$ (this is not the trivial module).

We have an isomorphism $\mathrm{Hom}_{B'}(L^* \otimes_A V_1, k) \simeq \mathrm{Hom}_A(V_1, k) = 0$, and similarly $\mathrm{Hom}_{B'}(k, L^* \otimes_A V_1) = 0$. It follows that k is not a composition factor of $L^* \otimes_A V_1$ (this module has no projective direct summand and hence has Loewy length at most 2). Consequently, $L^* \otimes_A V_1$ has a unique simple quotient S_1. Let S_2 be a simple

B'-module not isomorphic to k or S_1. Then $L^* \otimes_A V_1 = S_1$ or $L^* \otimes_A V_1$ is an extension of S_1 by S_2.

Furthermore, we have $\mathrm{Hom}_A(V_1, L \otimes_{B'} S_2) \simeq \mathrm{Hom}_{B'}(L^* \otimes_A V_1, S_2) = 0$ and $\mathrm{Hom}_A(k, L \otimes_{B'} S_2) = 0$. So there is a simple submodule V_2 of $L \otimes_{B'} S_2$ that is not isomorphic to k or V_1.

If $L^* \otimes_A V_1 = S_1$, then we have $L^* \otimes_A V_2 = S_2$, so $L \otimes_{B'} -$ send simple modules to simple modules, whence L induces a Morita equivalence between A and B' (cf. 2.3.5). So in this case a solution to Theorem 6.8 is provided by $C = M$.

Assume $L^* \otimes_A V_1$ is an extension of S_1 by S_2. Then $L^* \otimes_A V_2$ is an extension of S_2 by S_1. Now we are in a situation similar to 5.1: a projective cover of L is

$$P_L = P_{k_A} \otimes P^*_{k_B} \oplus P_{V_1} \otimes P^*_{S_2} \oplus P_{V_2} \otimes P^*_{S_1}.$$

Let δ be the restriction of a surjective map $P_L \to L$ to $R = P_{V_1} \otimes P^*_{S_1} \oplus P_{V_2} \otimes P^*_{S_2}$ and C' be the complex

$$C' = 0 \to R \xrightarrow{\delta} L \to 0$$

with L in degree 0.

The same proof as that of Proposition 5.2 shows that C' induces a Rickard equivalence between A and B'. So

$$C = C' \otimes_{B'} W^* = 0 \to R \otimes_{B'} W^* \to M \to 0$$

provides a solution to Theorem 6.8. □

6.3.6. As in §6.1.1 and 6.1.2, one checks that the construction can be done compatibly with p'-outer automorphism groups and central extensions by p-groups, as conjectured in §5.2.3 and 5.2.4.

6.4. Extensions to blocks with defect group $\mathbb{Z}/2 \times \mathbb{Z}/2 \times \mathbb{Z}/2$

In this subsection, we assume that Hypothesis 4 holds with $D \simeq \mathbb{Z}/2 \times \mathbb{Z}/2 \times \mathbb{Z}/2$.

Let M be the unique indecomposable direct summand of $e\mathcal{O}Gf$ with vertex ΔD. A construction similar to that in §6.2.1 (we now use §6.3 instead of §6.1) provides, for every subgroup P of order 2 of D, a relatively ΔP-projective p-permutation $(e\mathcal{O}G \otimes (f\mathcal{O}H)^\circ)$-module V_P and a map $\psi_P : V_P \to M$. Consider now the complex

$$C = 0 \to \bigoplus_P V_P \xrightarrow{\sum_P \psi_P} M \to 0$$

where P runs over the subgroups of order 2 of D up to H-conjugacy (the term M is in degree 0).

A proof analogous to that of Theorem 6.3 shows

Theorem 6.10. *The complex C induces a normalized splendid stable equivalence between $e\mathcal{O}G$ and $f\mathcal{O}H$.*

Note that we have a block of wild type (unlike the case where D has order 4) and we do not know how to lift this to a Rickard equivalence (without using a case by case proof based on the classification of finite simple groups with 2-Sylow subgroups elementary abelian of order at most 8).

7. Local constructions

In this section, we develop a formalism for gluing Rickard complexes and apply it here only to principal blocks. In §5.3, we constructed local Rickard equivalences from a global stable equivalence, using the Brauer functor. Here, we try to provide some converse, namely the construction of a global stable equivalence from a suitable family of local Rickard equivalences. This generalizes the construction of §6.2.

A more detailed study is being conducted in [Rou3] (cf. also the Appendix), where various categories of "sheaves" over the poset of p-subgroups will be considered, giving rise for example to a local construction of the stable category of all modules or of p-permutation modules.

Our approach here consists of constructing locally a "subcategory" of the category of p-permutation modules.

7.1. Gluing G-sets

Let G be a finite group and k a field of characteristic $p > 0$.

Let $\mathcal{T} = \mathcal{T}_G$ be the category of p-subgroups of G, with maps the inclusions. There is an action of G by conjugation on $\mathcal{T}$.

Let $\mathcal{F}$ be a G-stable full subcategory of $\mathcal{T}$.

7.1.1. We define the category $\mathcal{E}(\mathcal{F}) = \mathcal{E}_G(\mathcal{F})$ as the full subcategory of the category $\mathcal{E}$ of G-sets of objects whose point-stabilizers are in $\mathcal{F}$.

Given X, Y in $\mathcal{E}(\mathcal{F})$, then a morphism $f : X \to Y$ is an isomorphism if and only if, for every P in $\mathcal{F}$, the morphism $f^P : X^P \to Y^P$ is an isomorphism.

7.1.2. Sheaves. Assume now that $\mathcal{F}$ is closed above (i.e., if P is in $\mathcal{F}$ and Q is a p-subgroup containing P, then Q is in $\mathcal{F}$).

Let us denote by $\mathcal{L}(\mathcal{F})$ the category of G-equivariant presheaves of (finite) sets over $\mathcal{F}$, i.e., the category of G-equivariant contravariant functors $\mathcal{F} \to$ sets.

Its objects are families $\Omega = (\Omega_P, \phi(Q, R), [g, S])_{P,Q,R,S,g}$ where P, Q, R, S run over the objects of $\mathcal{F}$ with $Q \subseteq R$ and g over G. Here, Ω_P is a set, $\phi(Q, R)$ is a map from Ω_R to Ω_Q and $[g, S]$ is an isomorphism $\Omega_S \xrightarrow{\sim} \Omega_{{}^gS}$. Furthermore, the following conditions should be fulfilled:

(i) $\phi(R, R) = 1_{\Omega_R}$;

(ii) $\phi(Q, R)\phi(R, S) = \phi(Q, S)$;

(iii) $[g, {}^hP][h, P] = [gh, P]$;

(iv) we have a commutative diagram

$$\begin{array}{ccc} \Omega_Q & \xrightarrow{\phi(P,Q)} & \Omega_P \\ \big\downarrow{\scriptstyle [g,Q]} & & \big\downarrow{\scriptstyle [g,P]} \\ \Omega_{{}^gQ} & \xrightarrow[\phi({}^gP,{}^gQ)]{} & \Omega_{{}^gP}\,. \end{array}$$

Note that the maps $[g, P]$ for $g \in N_G(P)$ give a structure of a $N_G(P)$-set to Ω_P: we have a functor $?_P : \mathcal{L}(\mathcal{F}) \to N_G(P)$sets.

We say that Ω is a *sheaf* if, for H a subgroup of G and P an object of $\mathcal{F}$ normal in H, then

$$\phi(P, H) \text{ is an isomorphism } \Omega_H \xrightarrow{\sim} (\Omega_P)^H \text{ if } H \text{ is a } p\text{-group}$$

$$\text{and } (\Omega_P)^H = 0 \text{ otherwise.}$$

We denote by $\mathcal{S}(\mathcal{F})$ the full subcategory of $\mathcal{L}(\mathcal{F})$ consisting of sheaves.

For a sheaf Ω, the group P acts trivially on Ω_P and the maps $\Phi(P, Q)$ are inclusions, since every map in $\mathcal{F}$ is a composition of normal inclusions.

7.1.3. Some functors. Let $\mathcal{G}$ be a full subcategory of $\mathcal{F}$ closed above.

We denote by $\mathrm{Res}_{\mathcal{G}}^{\mathcal{F}}$ the restriction functor $\mathcal{L}(\mathcal{F}) \to \mathcal{L}(\mathcal{G})$.

We need to construct also a left adjoint $\mathrm{Ind}_{\mathcal{G}}^{\mathcal{F}}$ from $\mathcal{L}(\mathcal{G})$ to $\mathcal{L}(\mathcal{F})$.

For $\Omega \in \mathcal{L}(\mathcal{G})$, and $P \in \mathcal{F}$, the set $\Omega'_P = (\mathrm{Ind}_{\mathcal{G}}^{\mathcal{F}} \Omega)_P$ is the direct limit of the restriction of Ω to the full subcategory of $\mathcal{G}$ of objects containing P: this is the quotient of $\coprod_{Q \in \mathcal{G}, P \subseteq Q} \Omega_Q$ by the (coarsest) equivalence relation that identifies x and $\phi(R, Q)(x)$ for $x \in \Omega_Q$ and $R \hookrightarrow Q$ in $\mathcal{G}$. We denote by $\lambda_{P,Q}$ the canonical map $\Omega_Q \to \Omega'_P$ for $P \subseteq Q$.

The equivalence relation admits the following easier description when $\Omega \in \mathcal{S}(\mathcal{G})$: we have $\lambda_{P,Q}(a) = \lambda_{P,Q'}(a')$ if and only if there exists $R \in \mathcal{G}$ and $b \in \Omega_R$ with $Q \subseteq R$ and $Q' \subseteq R$ such that $a = \phi(Q, R)(b)$ and $a' = \phi(Q', R)(b)$.

To prove this claim, it is enough to consider the case where there is S normal in Q and in Q' with $\phi(S, Q)(a) = \phi(S, Q')(a')$. Let us denote by c this element of $\Omega_S^Q \cap \Omega_S^{Q'} = \Omega_S^R$ (we put $R = \langle Q, Q' \rangle$). Since Ω is a sheaf, this shows that R is a p-group and that there is $b \in \Omega_R$ such that $c = \phi(S, R)(b)$. Now $\phi(S, Q)\phi(Q, R)(b) = \phi(S, Q)(a)$. Since $\phi(S, Q)$ is an inclusion, we obtain $a = \phi(Q, R)$. Similarly, $a' = \phi(Q, R')(b)$.

There is a morphism of functors $\mathrm{Ind}_{\mathcal{G}}^{\mathcal{F}} \mathrm{Res}_{\mathcal{G}}^{\mathcal{F}} \to 1$ making the following diagram commutative

$$\begin{array}{ccc} (\mathrm{Ind}_{\mathcal{G}}^{\mathcal{F}} \mathrm{Res}_{\mathcal{G}}^{\mathcal{F}} \Omega)_P & \longrightarrow & \Omega_P \\ \uparrow & \nearrow_{\{\phi(P,Q)\}_Q} & \\ \coprod_{Q\in\mathcal{G}, P\subseteq Q} \Omega_Q & & \end{array}$$

and a morphism of functors $1 \to \mathrm{Res}_{\mathcal{G}}^{\mathcal{F}} \mathrm{Ind}_{\mathcal{G}}^{\mathcal{F}}$ given by

$$\Omega_Q \xrightarrow{\lambda_{Q,Q}} (\mathrm{Res}_{\mathcal{G}}^{\mathcal{F}} \mathrm{Ind}_{\mathcal{G}}^{\mathcal{F}} \Omega)_Q.$$

They make $\mathrm{Ind}_{\mathcal{G}}^{\mathcal{F}}$ a left adjoint of $\mathrm{Res}_{\mathcal{G}}^{\mathcal{F}}$.

The functor $\mathrm{Res}_{\mathcal{G}}^{\mathcal{F}}$ clearly restricts to a functor $\mathcal{S}(\mathcal{F}) \to \mathcal{S}(\mathcal{G})$.

Lemma 7.1. *The functor* $\mathrm{Ind}_{\mathcal{G}}^{\mathcal{F}}$ *restricts to a functor from* $\mathcal{S}(\mathcal{G})$ *to* $\mathcal{S}(\mathcal{F})$.

Proof. Let $\Omega \in \mathcal{S}(\mathcal{G})$ and $\Omega' = \mathrm{Ind}_{\mathcal{G}}^{\mathcal{F}} \Omega$. Let $P \in \mathcal{F}$ and $x \in \Omega'_P$. Let $Q \in \mathcal{G}$ be maximal such that $x = \lambda_{P,Q}(a)$ for some element $a \in \Omega_Q$. Let $g \in N_G(P)$ such that $g(x) = x$. Then $\lambda_{P,{}^gQ}([g, Q](a)) = \lambda_{P,Q}(a)$. It follows that there is R in $\mathcal{G}$ containing Q and Q^g and $b \in \Omega_R$ such that $\phi(Q, R)(b) = a$ and $\phi({}^gQ, R)(b) = [g, Q](a)$.

The maximality of Q shows that $R = Q = Q^g$, whence $g \in N_G(Q)$ and $a \in (\Omega_Q)^{\langle g, Q\rangle}$. So $\langle g, Q\rangle$ is a p-group and a is in the image of $\phi(Q, \langle g, Q\rangle)$. The maximality of Q shows that $g \in Q$.

Since Ω'_P is a direct limit over a transitive system of injections, the map $\lambda_{P,Q}$: $\Omega_Q \to \Omega'_P$ is injective. It follows that Ω' is a sheaf. □

Remark 7.2.

- The adjunction between Res and Ind restricts also to sheaves.
- The adjunction morphism $1 \to \mathrm{Res}_{\mathcal{G}}^{\mathcal{F}} \mathrm{Ind}_{\mathcal{G}}^{\mathcal{F}}$ between functors from $\mathcal{S}(\mathcal{G})$ to itself is an isomorphism.
- For Ω in $\mathcal{S}(\mathcal{G})$ and $P \in \mathcal{F}$, the point-stabilizers of the $N_G(P)$-set $(\mathrm{Ind}_{\mathcal{G}}^{\mathcal{F}} \Omega)_P$ are in $\mathcal{G}$.

Let $X \in \mathcal{E}$. Let $\Omega_P = X^P$ and, for $P \subseteq Q$, let $\phi(P, Q)$ be the inclusion $X^Q \hookrightarrow X^P$. We define $[g, P] : X^P \to X^{{}^gP}$ by $x \mapsto gx$. Then $(\Omega_P, \phi(Q, R), [g, S])$ is an element of $\mathcal{S}(\mathcal{T})$. This gives a functor $\mathrm{Br} : \mathcal{E} \to \mathcal{S}(\mathcal{T})$ which is canonically inverse to $?_1 : \mathcal{S}(\mathcal{T}) \to \mathcal{E}$.

Let $\mathrm{Br}(\mathcal{F})$ be the restriction of $\mathrm{Res}_{\mathcal{F}}^{\mathcal{T}} \mathrm{Br}$ to $\mathcal{E}(\mathcal{F})$.

Theorem 7.3. *The functor* $\mathrm{Br}(\mathcal{F})$ *is an equivalence of categories* $\mathcal{E}(\mathcal{F}) \xrightarrow{\sim} \mathcal{S}(\mathcal{F})$ *with inverse* $?_1 \cdot \mathrm{Ind}_{\mathcal{F}}^{\mathcal{T}}$.

Proof. To simplify notation, we put $A = \mathrm{Br}(\mathcal{F})$ and $B = ?_1 \cdot \mathrm{Ind}_{\mathcal{F}}^{\mathcal{T}}$. We know already that B is left adjoint to A and that the adjunction morphism $1 \to AB$ is an isomorphism.

Now the counit of adjunction $BA \to 1$ becomes an isomorphism after composing with A since the composite $A \to ABA \to A$ is the identity and the first map is already known to be an isomorphism. This means that, given X in $\mathcal{E}(\mathcal{F})$, the counit $BA(X) \to X$ becomes an isomorphism after taking fixed points by a subgroup in $\mathcal{F}$. Since $BA(X)$ and X are in $\mathcal{E}(\mathcal{F})$, it follows that the counit is an isomorphism (cf. §7.1.1). □

Of special interest is the case where $\mathcal{F}$ consists of the non-trivial p-subgroups of G: Theorem 7.3 says that the category $\mathcal{E}(\mathcal{F})$ of G-sets whose stabilizers are non-trivial p-groups is "locally determined".

7.1.4. Let $\mathcal{F}$ be a G-stable full subcategory of $\mathcal{T}$ and $\bar{\mathcal{F}}$ be the closure of $\mathcal{F}$, i.e., the full subcategory of $\mathcal{T}$ with objects the p-subgroups that contain some object of $\mathcal{F}$.

We define $\mathcal{S}(\mathcal{F})$ as the full subcategory of $\mathcal{S}(\bar{\mathcal{F}})$ with objects the Ω such that $\Omega_P = \emptyset$ for $P \notin \mathcal{F}$. Then $\mathrm{Br}(\bar{\mathcal{F}})$ restricts to an equivalence $\mathrm{Br}(\mathcal{F}) : \mathcal{E}(\mathcal{F}) \to \mathcal{S}(\mathcal{F})$.

7.2. *p*-permutation modules

7.2.1. Let $\tilde{\mathcal{E}}(\mathcal{F})$ be the Karoubian envelope of $\mathcal{E}(\mathcal{F})$: this is the category obtained from $\mathcal{E}(\mathcal{F})$ by k-linearizing and then adding images of idempotents. Its objects are pairs (X, e) where X is an object of $\mathcal{E}(\mathcal{F})$ and e is an idempotent of the k-algebra of the monoid $\mathrm{End}(X)$.

The space $\mathrm{Hom}((X, e), (X', e'))$ is the subspace $e'(k^{\mathrm{Hom}(X,X')})e$ of $k^{\mathrm{Hom}(X,X')}$. Similarly, we have a category $\tilde{\mathcal{S}}(\mathcal{F})$ obtained from $\mathcal{S}(\mathcal{F})$ by k-linearizing and then adding images of idempotents as above.

The functor $\mathrm{Br}(\mathcal{F})$ gives rise to a functor $\tilde{\mathcal{E}}(\mathcal{F}) \to \tilde{\mathcal{S}}(\mathcal{F})$. From Theorem 7.3 we can deduce

Corollary 7.4. *The functor* $\mathrm{Br}(\mathcal{F})$ *is an equivalence* $\tilde{\mathcal{E}}(\mathcal{F}) \xrightarrow{\sim} \tilde{\mathcal{S}}(\mathcal{F})$.

We have a faithful functor $\rho : \tilde{\mathcal{E}}(\mathcal{T}) \to kG\text{-perm}$, $(\Omega, e) \mapsto k\Omega e$. If Ω is free, then we have an isomorphism

$$\mathrm{Hom}_{\tilde{\mathcal{E}}(\mathcal{T})}((\Omega, e), (\Omega', e')) \xrightarrow{\sim} \mathrm{Hom}_{kG\text{-perm}}(k\Omega e, k\Omega' e').$$

The category $\tilde{\mathcal{E}}(\mathcal{T})$ consists of certain p-permutation modules with additional structure and the maps between them are those which can be "constructed" from maps between G-sets. A complex of objects of $\tilde{\mathcal{E}}(\mathcal{T})$ will be called a *geometrical* complex for kG.

7.2.2. Given $X, Y \in \tilde{\mathcal{E}}$ such that the G-set underlying X is free, we have an isomorphism

$$\operatorname{Hom}_{\tilde{\mathcal{E}}}(X, Y) \xrightarrow{\sim} \operatorname{Hom}_{kG}(\rho(X), \rho(Y)).$$

Consequently, every projective kG-module arises as $\rho(X)$ for some $X \in \tilde{\mathcal{E}}$ with an underlying G-set free.

7.2.3. Let $\mathcal{F}$ be a G-stable full subcategory of $\mathcal{T}_G$ closed above. Let Q be a maximal p-subgroup of G outside $\mathcal{F}$. Let $\mathcal{F}'$ be the full subcategory of $\mathcal{T}_G$ with objects those of $\mathcal{F}$ together with the conjugates of Q. Let $\mathcal{H}$ be the full subcategory of $\mathcal{T}_{N_G(Q)}$ with objects those p-subgroups of $N_G(Q)$ containing Q.

Let us construct a category $\mathcal{C}$. Its objects are families $(\Omega, V, \{\phi_R\}_{Q<R})$ where $\Omega \in \mathcal{S}(\mathcal{F})$, $V \in \mathcal{E}_{N_G(Q)}(\mathcal{H})$ and $\phi_R : \Omega_R \to V$ is a map of sets satisfying

(i) given S in $\mathcal{F}$ containing R, we have $\phi_S = \phi_R \phi(R, S)$,

(ii) given $h \in N_G(Q)$, we have a commutative diagram

$$\begin{array}{ccc} \Omega_R & \xrightarrow{\phi_R} & V \\ {\scriptstyle [h,R]}\downarrow & & \downarrow{\scriptstyle h} \\ \Omega_{{}^hR} & \xrightarrow[\phi_{{}^hR}]{} & V, \end{array}$$

and

(iii) given R a p-subgroup with $Q \lhd R$ and $Q \neq R$, we have an isomorphism $\phi_R : \Omega_R \xrightarrow{\sim} V^R$.

A morphism $(\Omega, V, \{\phi_R\}_{Q\leq R}) \to (\Omega', V', \{\phi'_R\}_{Q\leq R})$ is a pair (Λ, f) consisting of a morphism $\Lambda : \Omega \to \Omega'$ and a map $f : V \to V'$ such that, for all R containing Q, the following diagram is commutative

$$\begin{array}{ccc} \Omega_R & \xrightarrow{\phi_R} & V \\ {\scriptstyle \Lambda_R}\downarrow & & \downarrow{\scriptstyle f} \\ \Omega'_R & \xrightarrow[\phi'_R]{} & V'. \end{array}$$

Lemma 7.5. *The functor $\mathcal{S}(\mathcal{F}') \to \mathcal{C}$ given by $\Omega \mapsto (\operatorname{Res}^{\mathcal{F}'}_{\mathcal{F}} \Omega, \Omega_Q, \{\phi(Q, R)\})$ is an equivalence.*

Let us give a useful application of this lemma. Let $X \in \tilde{\mathcal{S}}(\mathcal{F}')$ and $Y \in \tilde{\mathcal{E}}_{N_G(Q)}(\mathcal{H})$ with $\operatorname{Res}^{\mathcal{H}}_{\mathcal{H}-\{Q\}} \operatorname{Br} X_Q \xrightarrow{\sim} \operatorname{Res}^{\mathcal{H}}_{\mathcal{H}-\{Q\}} \operatorname{Br} Y$. Then there is $X' \in \tilde{\mathcal{S}}(\mathcal{F}')$ with $\operatorname{Res}^{\mathcal{F}'}_{\mathcal{F}} X' \xrightarrow{\sim} \operatorname{Res}^{\mathcal{F}'}_{\mathcal{F}} X$ and $X'_Q \xrightarrow{\sim} Y$.

7.3. Geometric stable and Rickard equivalences

7.3.1. We assume Hypotheses 4 (the blocks are principal) and 3′ of §5.2.3 (cf. the Appendix and [Rou3] for nonprincipal blocks). Assume furthermore that D is abelian.

Denote by $\mathcal{F}$ the full subcategory of p-subgroups of Δ contained in ΔD up to conjugacy.

We may refine Broué's conjecture 5.3 to one for the pair (X, G):

Conjecture 7.6. *There is a complex C in $\tilde{\mathcal{E}}_\Delta(\mathcal{F})$ such that* $\mathrm{Res}^{\Delta}_{G\times H^\circ}\,\rho(C)$ *is a Rickard complex for $A \otimes B^\circ$.*

Note that if C is such a complex, then $\mathrm{Res}^{\Delta}_{G\times H^\circ}\,\rho(C)$ is a splendid Rickard complex, i.e., a Rickard complex of p-permutation $k(G \times H^\circ)$-modules with vertices contained in ΔD. So Conjecture 7.6 indeed implies Broué's conjecture.

7.3.2. Let us assume Hypothesis 3″. We denote by $\tilde{D}$ the inverse image of D in $\tilde{X}$. Let $\tilde{\mathcal{F}}$ be the full subcategory of p-subgroups of $\tilde{\Delta}$ containing ΔP and contained in $\Delta\tilde{D}$ up to conjugacy.

One can ask the following question about $(\tilde{X}, P, G)$:

Question 7.7. Let C be a complex in $\tilde{\mathcal{E}}_{\tilde{\Delta}}(\tilde{\mathcal{F}})$ such that $\mathrm{Res}^{\Delta}_{G\times H^\circ}\,k\Delta\otimes_{k\tilde{\Delta}}\rho(C)$ induces a stable equivalence between A and B.

Is there a complex C' in $\tilde{\mathcal{E}}_{\tilde{\Delta}}(\tilde{\mathcal{F}})$ such that

(i) $\mathrm{Res}^{\Delta}_{G\times H^\circ}\,k\Delta\otimes_{k\tilde{\Delta}}\rho(C')$ induces a Rickard equivalence between A and B, and

(ii) $\mathrm{Res}^{\tilde{\mathcal{F}}}_{\tilde{\mathcal{F}}-\{\Delta P\}}\,\mathrm{Br}\,C' \simeq \mathrm{Res}^{\tilde{\mathcal{F}}}_{\tilde{\mathcal{F}}-\{\Delta P\}}\,\mathrm{Br}\,C$?

This question has a positive answer when $|D| = p$ (more generally, for D cyclic) or when D is a Klein four group.

Let us justify this claim, for a particular choice of C (when this will be used in the application of Theorem 7.8, it will be possible to ensure that C is of that type). We assume C is concentrated in degree 0 and $M = \rho(C^0)$ is a direct summand of $k\tilde{X}$, as a $k\tilde{\Delta}$-module. Then it follows from §6.1 and §6.3 that there is a projective $k(\tilde{\Delta}/\Delta P)$-module R' and a morphism

$$f : R = \mathrm{Res}^{\tilde{\Delta}/\Delta P}_{\tilde{\Delta}}\,R' \to M$$

such that the complex $0 \to R \xrightarrow{f} M \to 0$ induces a Rickard equivalence between the principal blocks of $\tilde{G}$ and $\tilde{H}$. By §7.2.2, there is an object V in $\tilde{\mathcal{E}}_{\tilde{\Delta}/\Delta P}$ and a morphism $\alpha : V \to C^0$ with $\rho(\alpha) = f$. Now $C' = 0 \to V \xrightarrow{\alpha} C^0 \to 0$ answers Question 7.7.

Note that when D is not abelian, there are stable equivalences that do not lift to Rickard equivalences (that happens for the principal block of $G = \mathrm{Sz}(8)$ for $p = 2$).

7.3.3. We can now state our main result.

Theorem 7.8. *Assume Question* 7.7 *has a positive answer for* $(N_X(Q), Q, C_G(Q)/Q)$ *for all non-trivial subgroups* Q *of* D. *Then there is a complex* C *in* $\tilde{\mathcal{E}}_\Delta(\mathcal{F})$ *such that* $\mathrm{Res}^{\Delta}_{G\times H^\circ} \rho(C)$ *induces a stable equivalence between* $e\mathcal{O}G$ *and* $f\mathcal{O}H$.

We recover the existence of stable equivalences for $D \simeq \mathbb{Z}/p \times \mathbb{Z}/p$ (§6.2) and for D elementary abelian of order 8 (§6.4).

This result readily implies

Corollary 7.9. *Assume Question* 7.7 *has a positive answer for*

$$(N_X(Q), Q, C_G(Q)/Q)$$

for all subgroups Q *of* D. *Then Conjecture* 7.6 *holds for* (X, G).

7.3.4. *Proof of Theorem* 7.8. Let $\mathcal{G}$ be a full subcategory of $\mathcal{F} - \{1\}$ closed under conjugation and such that, given Q in $\mathcal{G}$ and R in $\mathcal{F}$ with $Q \le R$, then R is in $\mathcal{G}$. We will construct by induction on $|\mathcal{G}|$ a complex $X_{\mathcal{G}}$ in $\tilde{\mathcal{S}}_\Delta(\mathcal{G})$ such that

$$k(\bar{C}_G(Q) \times \bar{C}_H(Q)^\circ) \otimes_{k(C_G(Q)\times C_H(Q)^\circ)} \mathrm{Res}^{N_\Delta(\Delta Q)}_{C_G(Q)\times C_H(Q)^\circ} \rho\left((X_{\mathcal{G}})_{\Delta Q}\right)$$

is a Rickard complex for the principal blocks of $\bar{C}_G(Q) = C_G(Q)/Q$ and $\bar{C}_H(Q) = C_H(Q)/Q$, for all $Q \le D$ with $\Delta Q \in \mathcal{G}$.

Assume we have a complex $X_{\mathcal{G}}$ satisfying the induction hypothesis and let $Q \ne 1$ be a maximal subgroup of D with $\Delta Q \notin \mathcal{G}$. Let $\mathcal{G}' = \mathcal{G} \cup \{(\Delta Q)^x\}_{x\in\Delta}$.

Let $Y = \mathrm{Ind}^{\mathcal{G}'}_{\mathcal{G}} X_{\mathcal{G}}$. Let $Z = Y_{\Delta Q}$, an object of $\tilde{\mathcal{E}}_{N_\Delta(\Delta Q)}(\mathcal{H})$, where $\mathcal{H}$ is the full subcategory of $\mathcal{T}_{N_\Delta(\Delta Q)}$ of p-subgroups containing ΔQ and contained in ΔD up to conjugacy.

For every p-subgroup R of D strictly containing Q, we have $\mathrm{Br}_{\Delta R}\, \rho(Z) \simeq \rho\left((X_{\mathcal{G}})_{\Delta R}\right)$, whence $\mathrm{Br}_{\Delta R/\Delta Q}\, C$ induces a Rickard equivalence between the principal blocks of $\bar{C}_G(R)$ and $\bar{C}_H(R)$, where

$$C = \mathrm{Res}^{N_\Delta(\Delta Q)/(Q\times Q^\circ)}_{\bar{C}_G(Q)\times \bar{C}_H(Q)^\circ}\, kN_\Delta(\Delta Q)/(Q \times Q^\circ) \otimes_{kN_\Delta(\Delta Q)} \rho(Z).$$

By Theorem 5.6, it follows that C induces a stable equivalence between the principal blocks of $\bar{C}_G(Q)$ and $\bar{C}_H(Q)$.

Since we are assuming a positive answer to Question 7.7 by hypothesis for $(N_X(Q), Q, C_G(Q)/Q)$, it follows that there exists a complex Z' in $\tilde{\mathcal{E}}_{N_\Delta(\Delta Q)}(\mathcal{H})$ and an isomorphism

$$\mathrm{Res}^{\mathcal{H}}_{\mathcal{H}-\{\Delta_Q\}} \mathrm{Br}\, Z' \xrightarrow{\sim} \mathrm{Res}^{\mathcal{H}}_{\mathcal{H}-\{\Delta_Q\}} \mathrm{Br}\, Z$$

such that $\mathrm{Res}^{N_\Delta(\Delta Q)/(Q\times Q^\circ)}_{\bar{C}_G(Q)\times\bar{C}_H(Q)^\circ}\, kN_\Delta(\Delta Q)/(Q\times Q^\circ) \otimes_{kN_\Delta(\Delta Q)} \rho(Z')$ induces a Rickard equivalence between the principal blocks of $\bar{C}_G(Q)$ and $\bar{C}_H(Q)$.

By §7.2.3, there is a complex Y' in $\tilde{\mathcal{S}}_\Delta(\mathcal{G}')$ with $Y'_{\Delta Q} \xrightarrow{\sim} Z'$ and $Y'_R \xrightarrow{\sim} (X_{\mathcal{G}})_R$ for $R \in \mathcal{G}$. Now $X_{\mathcal{G}'} = Y'$ satisfies the required properties. □

Appendix

Our aim here is to present some of the main points of the work in progress of [Rou3]. It completes the results described previously, mainly by extending most of these to nonprincipal blocks.

We will be working in the Appendix with Hypothesis 3 unless otherwise stated.

A.1. Splendid equivalences

A.1.1. Previous constructions. As explained by Broué in [Br1], a Rickard equivalence between A and B induces a perfect isometry between the blocks — a character correspondence with signs satisfying certain arithmetical properties. Now, Broué introduced also in [Br1] the class of isotypies, which consist of a compatible system (for the generalized decomposition maps) of perfect isometries for local subgroups. In [Ri4], Rickard proved that, by adding the assumption of splendidness to a Rickard equivalence, one then gets such an isotypie. Actually, Rickard proves more. Namely, he shows how to get Rickard equivalences for local subgroups (this is Theorem 5.6, (i) $\Rightarrow$ (ii)). Unfortunately, Rickard's construction was done only for principal blocks. Later, Harris [Ha] showed that, if H is a subgroup of G, then Rickard's definition of splendidness still provides Rickard equivalences for local subgroups (i.e., Theorem 5.6,(i) $\Rightarrow$ (ii) holds for nonprincipal blocks and without the assumption that K controls fusion of p-subgroups in G). As for a general definition of splendidness, this has been given by Linckelmann in [Li5].

In order to define splendid equivalences (or even isotypies), one needs an equivalence between the Brauer categories of the two blocks (we will recall later what they are). One needs also an identification of the defect groups of the blocks. Then Linckelmann's definition of splendidness is relative to a choice of source idempotents for the blocks, which makes it difficult to check on examples.

For principal blocks, the situation is simpler because of Brauer's third main theorem and because the assumption on the Brauer categories means that the groups have the same fusion of p-subgroups. More generally, Harris pointed out that it is still enough to assume some relative projectivity on the terms of the complex, provided one of the groups controls the fusion of p-subgroups in the other group — but this is not a very satisfactory assumption, being non-symmetric and involving more than just information about the Brauer categories of the blocks. There, one trick (due to Rickard) is needed, namely, that a complex of exact (A, B)-bimodules C induces a Rickard equivalence if and only if one of the two isomorphisms in §2.3.2, (i) holds.

A.1.2. A new approach. What is not natural in the previous approaches is the identification *a priori* of the defect groups of the blocks, as pointed out by Broué.

Let D be a defect group of A and D' a defect group of B. Let C be an indecomposable complex of p-permutation (A, B)-bimodules inducing a stable equivalence.

One shows that there is an isomorphism $\phi : D \xrightarrow{\sim} D'$ such that the terms of C are projective relatively to $\Delta_\phi(D) = \{(x, \phi(x)) | x \in D\}$. Now let (D, e_D) be a maximal e-subpair. There is an equivalence $F : \mathrm{Br}(e) \xrightarrow{\sim} \mathrm{Br}(f)$ with the following property: Given a subpair $(Q, b_Q) \leq (D, b_D)$, the complex $e_Q \operatorname{Br}_{\Delta_\phi(Q)}(C)$ induces a Rickard equivalence between $e_Q kC_G(Q)$ and $f_{Q'} kC_H(Q')$, where $(Q', f_{Q'}) = F(Q, e_Q)$. This is the generalization of Theorem 5.6, (i) $\Rightarrow$ (ii).

We recall that the Brauer category $\mathrm{Br}(e)$ of $e\mathcal{O}G$ has for objects the e-subpairs and $\mathrm{Hom}((Q, e_Q), (R, e_R))$ is the set of homorphisms $Q \to R$ which are the composition of conjugation $Q \to Q^g$, $x \mapsto g^{-1}xg$ followed by an inclusion $Q^g \subseteq R$, for some $g \in G$ such that $g^{-1} e_Q g = e_R$.

One can say much more about the vertices of the indecomposable terms in C. One constructs a theory of vertex-subpairs for indecomposable modules. Then one shows that the vertex-subpairs of the terms have the form $(\Delta_\phi(Q), e_Q \otimes f_{\phi(Q)})$, where $(Q, e_Q) \leq (D, e_D)$ and $(\phi(Q), f_{\phi(Q)}) = F(Q, e_Q)$. This last property finally tells us that the complex is splendid in the sense of Linckelmann.

Now, what should be a splendid complex? It depends on the data! As long as there has been no chosen isomorphism $\phi : D \xrightarrow{\sim} D'$ between defect groups of the two blocks, this should be any indecomposable complex of p-permutation modules (this makes it easy to check that the known examples of Rickard complexes are splendid). Once such an isomorphism is chosen, we should ask that the terms are projective relative to $\Delta_\phi D$. If furthermore maximal subpairs (D, e_D) and $(D', f_{D'})$ are chosen, one should put the more precise assumption on the vertex-subpairs.

All of this is compatible with the previous definitions of Rickard, Harris and Linckelmann. This should also be seen as a special (but more explicit) case of the theory of basic equivalences of Puig [Pu4].

A.2. Nilpotent blocks, p-rank 2 and Puig's finiteness conjecture

Let P be an abelian p-group. We know that given an indecomposable bimodule M inducing a stable equivalence between ekG and kP, there is an integer n such that $\Omega^n M$ induces a Rickard equivalence (it is crucial, here, to assume P is abelian). This follows from Dade's classification of endotrivial modules for abelian p-groups [Da2] (we use the fact that $M^* \otimes_{ekG} V$ is endotrivial if V is simple). A difficulty arises when n is negative: starting from a geometrical complex giving the stable equivalence, we cannot, in general, get a geometrical complex giving a Rickard equivalence. Note that when P is cyclic, we can always take $n = 0$ or $n = 1$, so we assume now P is not cyclic.

Now take for M an indecomposable p-permutation module. Assume the block ekG is self-dual. Then "the" simple ekG-module V is self-dual; hence $M^* \otimes_{ekG} V$ is also self-dual. Since $M^* \otimes_{ekG} V \simeq \Omega^{-n} k$, it follows that $\Omega^{2n} k \simeq k$. As P is not cyclic, this forces $n = 0$, so that M already induces a Morita equivalence.

Let us explain how the problem should be solved in general. We work over a semi-local ring $\mathcal{O} = \mathbb{Z}_p[\zeta]$ where ζ is a suitable root of unity. The complex conjugation induces an automorphism of $\mathcal{O}$ and the semi-linear anti-automorphism of $\mathcal{O}G$ given by $ag \mapsto \bar{a}g^{-1}$ for $a \in \mathcal{O}$ and $g \in G$ stabilizes all the blocks of $\mathcal{O}G$. Now, the extension of the theory to the base ring $\mathcal{O}$ should provide the conclusion that M is always a Morita equivalence! Note that similar considerations should solve some cases of Puig's finiteness conjecture. Examples are blocks with defect group $\mathbb{Z}/2 \times \mathbb{Z}/2$ and (non-abelian defect) nilpotent blocks, where the problem is to show that the endopermutation module involved has finite order in the Dade group.

This would give a proof of Conjecture 7.6 for nilpotent blocks with abelian defect and would show that there are stable equivalences for blocks with abelian defect groups and nilpotent local structure: the existence of such stable equivalences is due to Puig [Pu3], who shows how to glue endopermutation modules — this was a important source of inspiration for this work on local constructions.

When the defect groups are abelian with p-rank 2, we obtain a geometrical stable equivalence between A and B (here, the problem of "negative n" does not arise, since the local blocks are p-central extensions of blocks with cyclic defect). We then get a splendid Rickard equivalence for blocks with defect group $\mathbb{Z}/2 \times \mathbb{Z}/2$ (without using complex conjugation, it is not clear that this comes from a geometrical complex). Let us explain this last result more precisely.

Assume $D \simeq \mathbb{Z}/2 \times \mathbb{Z}/2$. Let M be the unique indecomposable direct summand of $e\mathcal{O}Gf$ with vertex ΔD. Let $E = N_G(D, e_D)/C_G(D)$, where (D, e_D) is a maximal e-subpair.

If $|E| = 1$, then there is an integer n such that $\Omega^n M$ induces a Morita equivalence between $e\mathcal{O}G$ and $f\mathcal{O}H$. This can be realized using a Rickard complex of p-permutation modules, by truncating a projective resolution (or a relatively injective resolution) of M (note that by the discussion above, we should have $n = 0$).

If $|E| = 3$, then there is an integer n and a direct summand R of a projective cover of $\Omega^n M$ such that the complex $0 \to R \to \Omega^n M \to 0$ induces a Rickard equivalence. As above, this can be realized by a Rickard complex of p-permutation modules (although, again, we should already have $n = 0$).

In order to be able to construct stable equivalences for blocks with defect group $\mathbb{Z}/2 \times \mathbb{Z}/2 \times \mathbb{Z}/2$ by gluing local Rickard equivalences coming from the construction above, we would need $n = 0$.

For nilpotent blocks with non-abelian defect groups, we have a Morita equivalence between A and kD, as shown by Puig, but this cannot be realized, in general, by a complex of p-permutation modules. A common generalization of this case and of the abelian defect case is the case where the hyperfocal subgroup of the defect group (as defined by Puig in [Pu5]) is abelian: it is tempting to ask whether there is still a Rickard equivalence between A and B. Such an equivalence would not be splendid in general, but should be basic in the sense of Puig [Pu4].

A.3. Further categories of sheaves

A.3.1. Stable category of p-permutation modules. In order to construct global complexes of p-permutation modules by gluing local complexes, we can look for a local description of the category of p-permutation modules. The projective modules will not be (directly) reflected locally, so we should rather look for a local description of the stable category of p-permutation kG-modules, defined as the quotient category of the category of p-permutation modules by the full subcategory of projective modules.

This can be achieved as follows: a p-permutation module gives rise, via the Brauer functor, to a family of p-permutation modules for $kN_G(Q)/Q$, where Q runs over the non-trivial p-subgroups of G, together with certain isomorphisms. Conversely, the data of a compatible family of $kN_G(Q)/Q$-modules comes from a p-permutation kG-module, unique up to a unique isomorphism in the stable category. To make all of this precise, one needs first to check various natural properties of the Brauer functor and then to define a category of "sheaves" of p-permutation modules over the p-subgroups complex, where the transitions maps account for isomorphisms $\mathrm{Br}_Q \mathrm{Br}_P V \xrightarrow{\sim} \mathrm{Br}_Q V$ when $P \lhd Q$.

We can use this construction to glue a compatible family of complexes of p-permutation modules for various $kN_G(Q)/Q$ ($Q \neq 1$). What we get is not quite a complex of p-permutation modules, but only a graded p-permutation kG-module with an endomorphism d of degree 1 such that d^2 is a projective map — it is not clear how to get a genuine complex (i.e., with $d^2 = 0$) from such a complex.

This construction has nevertheless some interesting and useful consequences. It permits to get direct sum decompositions of certain full subcategories of the stable category of p-permutation modules defined by conditions on vertex-subpairs.}

A.3.2. Complexes of geometrical origin. The idea we pursue here, following a suggestion of Alperin, is to study a category of complexes that behaves like the complexes of chains of a finite simplicial complex acted upon by G — we want nevertheless to replace the assumption that the terms are permutation modules by the fact that they are p-permutation modules. We introduce a category where the objects are p-permutation modules M with additional structure, namely, compatible splittings of the canonical morphisms $M^P \to \mathrm{Br}_P M$ for every p-subgroup M.

The induction and restriction functors can be extended to this category, and the induction is left adjoint to the restriction. Something new happens nevertheless, namely the Brauer functor has now a left adjoint! One can consider (fairly complicated) categories of presheaves of objects of such categories, over the p-subgroup complex.

Nevertheless, one can deal with an easier subcategory. This category can be given the structure of an exact category, by deciding that the exact sequences are those sequences that are split exact when we only keep the p-permutation modules underlying the objects. It turns out that the full subcategory of projective objects for this structure of exact category is related to the category $\tilde{\mathcal{E}}$ of §7.2.

A.4. *p*-extensions

A.4.1. Inductive approach to Broué's conjecture. In the inductive approach to Broué's conjecture described in §7.3 the problem is to give a positive answer to Question 7.7. Let us recall that inductive approach.

First start with defect 0, then proceed by induction on the order of the defect group:

- Construct liftings of Rickard equivalences through p-central extensions (in order to go from a Rickard equivalence between blocks of $C_G(P)/P$ and $C_H(P)/P$, which exists by the induction hypothesis, to one between blocks of $C_G(P)$ and $C_H(P)$ for $P \neq 1$).
- Use outer automorphism equivariance to extend the Rickard equivalence (in order to obtain a Rickard equivalence between blocks of $N_G(P)$ and $N_H(P)$).
- Gluing: construct a global stable equivalence from the compatible system of local Rickard equivalences.
- Lift the stable equivalence to a derived equivalence.

The introduction of geometrical complexes in §7 was needed in order to achieve the gluing step.

We want to explain two facts here. First, the lifting problem through central p-extensions can be handled *a priori*. Then, in order to lift the stable equivalence to a Rickard equivalence, one need not worry about geometrical complexes, as long as the lifting is of a particular type. So, in order to solve Conjecture 7.6, it is enough to give a positive answer to a weaker form of Question 7.7. Note that this works as well for non-principal blocks. It is not necessary to worry about p-central extensions. More precisely, a positive answer to the question can be deduced from the case where the central p-subgroup P of Hypothesis $3''$ is trivial and one can forget about the geometrical complexes and even work directly with stable equivalences induced by bimodules.

A.4.2. Let us review first the general problem of p-extensions. We assume Hypothesis $3''$ but, to simplify, we assume $X = G$, i.e., we forget about automorphisms.

Consider a Rickard complex C of $(e\mathcal{O}G, f\mathcal{O}H)$-bimodules. Under what condition does there exist a Rickard complex $\tilde{C}$ of $(\tilde{e}\mathcal{O}\tilde{G}, \tilde{f}\mathcal{O}\tilde{H})$-bimodules with ΔP acting trivially, such that $\mathcal{O}(G\times H^\circ)\otimes_{\mathcal{O}(\tilde{G}\times\tilde{H}^\circ)}\tilde{C} \simeq C$? As pointed out in §5.2.4, it is enough to construct a complex $\tilde{C}$ of $(\tilde{e}\mathcal{O}\tilde{G}, \tilde{f}\mathcal{O}\tilde{H})$-bimodules with ΔP acting trivially, and such that $\mathcal{O}(G \times H^\circ) \otimes_{\mathcal{O}(\tilde{G}\times\tilde{H}^\circ)} \tilde{C} \simeq C$.

The problem of lifting from $\mathcal{O}G$ to $\mathcal{O}\tilde{G}$ is similar to the one of lifting from kG to $\mathcal{O}G$: in both cases, we obtain an algebra as a quotient of the other algebra by an ideal I generated by central elements contained in the radical (the set of $x-1$ for $x \in P$ in the first case, the radical of $\mathcal{O}$ in the second case) and the algebra is complete for the toplogy defined by that ideal. We can then apply lifting methods, similar to that of Rickard [Ri3]. The only problem is to lift the individual terms of the complex to modules that are free as left $\mathcal{O}P$-modules and are acted on trivially by ΔP. Once this is done, we lift the differential d of the complex C to get a graded endomorphism $\tilde{d}$. The square $\tilde{d}^2$ is not zero, but it so modulo I. Thanks to the vanishing of the module of homomorphisms from C to $C[2]$ in the homotopy category of complexes of $(e\mathcal{O}G, f\mathcal{O}H)$-bimodules, we can change $\tilde{d}$ to get a new lifting $\tilde{d}_1$ of d such that $\tilde{d}_1^2$ is zero modulo I^2. We go on and, since $I^n = 0$ for n large enough, we eventually get a genuine differential and the complex $\tilde{C}$ is constructed.

Note that this strategy works even if P is a normal but non-central p-subgroup.

When the complex C is splendid and the inverse image of D in $\tilde{G}$ is abelian, one shows that it is always possible to lift the individual terms of the complex. Here, we consider the case where $H = N_G(D)$ but this applies as well to the case where we consider any other finite group H and any block idempotent f of $\mathcal{O}H$. Then we only require C to be a complex of p-permutation modules and we use the results of §A.1.2.

A.4.3. Let us come to a more concrete problem. Suppose we are given a complex C of $(e\mathcal{O}G, f\mathcal{O}H)$-bimodules inducing a stable equivalence and a complex $\tilde{C}$ of $(\tilde{e}\mathcal{O}\tilde{G}, \tilde{f}\mathcal{O}\tilde{H})$-bimodules with ΔP acting trivially, such that

$$\mathcal{O}(G \times H^\circ) \otimes_{\mathcal{O}(\tilde{G} \times \tilde{H}^\circ)} \tilde{C} \simeq C.$$

Consider a (bounded) complex of exact $(e\mathcal{O}G, f\mathcal{O}H)$-bimodules all of whose non-zero terms are projective, except for the one of smallest degree d, $M = C^d$. Then M induces a stable equivalence between $e\mathcal{O}G$ and $f\mathcal{O}H$.

Suppose there is a complex Z of projective $(e\mathcal{O}G, f\mathcal{O}H)$-bimodules and a morphism $\psi : Z \to M$ (where M is seen as a complex concentrated in degree d) whose cone is a Rickard complex. This means we have been able to lift the stable equivalence induced by M to a Rickard equivalence in a particular way. Then one gets a complex Z' of projective $(e\mathcal{O}G, f\mathcal{O}H)$-bimodules and a morphism $\psi' : Z' \to C$ whose cone is a Rickard complex. Now, there is a complex $\tilde{Z}'$ of exact $(\tilde{e}\mathcal{O}\tilde{G}, \tilde{f}\mathcal{O}\tilde{H})$-bimodules with ΔP acting trivially and a map $\tilde{\psi}' : \tilde{Z}' \to \tilde{C}$ such that $1 \otimes \tilde{\psi}' = \psi'$. The cone $\tilde{C}'$ of $\tilde{\psi}'$ is now a Rickard complex of $(\tilde{e}\mathcal{O}\tilde{G}, \tilde{f}\mathcal{O}\tilde{H})$-bimodules with ΔP acting trivially.

Finally, if $\tilde{C}$ comes from a geometrical complex as in Question 7.7, then we will be able to find another geometrical complex giving rise to $\tilde{C}'$. This means that it is enough to lift, in a particular way, certain stable equivalences between $e\mathcal{O}G$ and $f\mathcal{O}H$ to Rickard equivalences in order to solve Question 7.7 — in general, all of this should be done in a way compatible with the action of X/G.

References

[BoRou] C. Bonnafé and R. Rouquier, Catégories dérivées et variétés de Deligne-Lusztig, preprint, 2001.

[Bou] S. Bouc, Résolutions de foncteurs de Mackey, in: Groups, representations: cohomology, group actions and topology, Proc. Sympos. Pure Math. 63, Amer. Math. Soc., Providence, RI, 1998, 31–83.

[Br1] M. Broué, Isométries parfaites, types de blocs, catégories dérivées, Astérisque 181–182 (1990), 61–92.

[Br2] M. Broué, Equivalences of blocks of group algebras, in: Finite dimensional algebras and related topics, Kluwer Academic Press, 1994, 1–26.

[Br3] M. Broué, On representations of symmetric algebras: An introduction, preprint ETH Zürich, 1990.

[Ch1] J. Chuang, The derived category of some blocks of symmetric groups and a conjecture of Broué, J. Algebra 217 (1999), 114–155.

[Ch2] J. Chuang, Derived equivalence in $SL(2, p^2)$, Trans. Amer. Math. Soc. 353 (2001), 2897–2913.

[ChKe] J. Chuang and R. Kessar, Symmetric groups, wreath products, Morita equivalences, and Broué's abelian defect group conjecture, Bull. London Math. Soc., to appear.

[Da1] E. C. Dade, A correspondence of characters, Proc. Symp. Pure Math. 37, Amer. Math. Soc., Providence, RI, 1980, 401–403.

[Da2] E. C. Dade, Endo-permutation modules over p-groups II, Annals of Math. 108 (1978), 317-346.

[GoOk] H. Gollan and T. Okuyama, Derived equivalences for the smallest Janko group, preprint.

[Ha] M. E. Harris, Splendid derived equivalences for blocks of finite groups, J. London Math. Soc. 60 (1999), 71–82.

[HaLi] M.E. Harris and M. Linckelmann, Splendid derived equivalences for blocks of finite p-solvable groups, J. London Math. Soc. 62 (2000), 85–96

[HiMi] A. Hida and H. Miyachi, Some blocks of finite general linear groups in non-defining characteristic, preprint.

[Holl] M. Holloway, Derived equivalences for group algebras, PhD thesis, Bristol, 2001.

[Holm] T. Holm, Derived categories, derived equivalences and representation theory, in: Proceedings of the Summer School on Representation Theory of Algebras, Finite and Reductive Groups (Cluj-Napoca 1997), Babes-Bolyai University, Cluj, 1998, 33–66.

[KoKu1] S. Koshitani and N. Kunugi, The principal 3-blocks of the 3-dimensional projective special unitary groups in non-describing characteristic, preprint, 1997.

[KoKu2] S. Koshitani and N. Kunugi, Broué's conjecture holds for principal 3-blocks with elementary abelian defect group of order 9, preprint, 2000.

[KoKuWa] S. Koshitani, N. Kunugi and K. Waki, Broué's conjecture for non-principal 3-blocks of finite groups, preprint, 2001.

[KoMi] S. Koshitani and H. Miyachi, The principal 3-blocks of four- and five-dimensional projective special linear groups in non-defining characteristic, J. Algebra 226 (2000), 788–806.

[Ku] N. Kunugi, Morita equivalent 3-blocks of the 3-dimensional projective special linear groups, Proc. London Math. Soc. 80 (2000), 575–589.

[KuWa] N. Kunugi and K. Waki, Derived equivalences for the 3-dimensional special unitary groups in non-defining characteristic, J. Algebra 240 (2001), 251–267.

[Li1] M. Linckelmann, Derived equivalences for cyclic blocks over a p-adic ring, Math. Z. 207 (1991), 293–304.

[Li2] M. Linckelmann, A derived equivalence for blocks with dihedral defect groups, J. Algebra 164 (1994), 244–255.

[Li3] M. Linckelmann, The source algebras of blocks with a Klein four defect group, J. Algebra 167 (1994), 821–854.

[Li4] M. Linckelmann, Stable equivalences of Morita type for self-injective algebras and p-groups, Math. Z. 223 (1996), 87–100.

[Li5] M. Linckelmann, On derived equivalences and structure of blocks of finite groups, Turkish J. Math. 22 (1998), 93–107.

[Ma] A. Marcus, On equivalences between blocks of group algebras: reduction to simple components, J. Algebra 184 (1996), 372–396.

[Ok1] T. Okuyama, Some examples of derived equivalent blocks of finite groups, preprint, 1998.

[Ok2] T. Okuyama, Derived equivalence in $\mathrm{SL}_2(q)$, preprint, 2000.

[Pu1] L. Puig, Local block theory in p-solvable groups, Proc. Sympos. Pure Math. 37, Amer. Math. Soc., Providence, RI, 1980, 385–388.

[Pu2] L. Puig, Algèbres de source de certains blocs des groupes de Chevalley, Astérisque 181–182 (1990), 221–236.

[Pu3] L. Puig, Une correspondance de modules pour les blocs à groupes de défaut abéliens, Geom. Dedicata 37 (1991), 9–43.

[Pu4] L. Puig, On the local structure of Morita and Rickard equivalences between Brauer blocks, Progr. Math. 178, Birkhäuser, 1999.

[Pu5] L. Puig, The hyperfocal subalgebra of a block, Invent. Math. 141 (2000), 365–397.

[Ri1] J. Rickard, Derived categories and stable equivalence, J. Pure Appl. Algebra 61 (1989), 303–317.

[Ri2] J. Rickard, MSRI talk, Berkeley, 6 November 1990.

[Ri3] J. Rickard, Lifting theorems for tilting complexes, J. Algebra 142 (1991), 383–393.

[Ri4] J. Rickard, Splendid equivalences: derived categories and permutation modules, Proc. London Math. Soc. 72 (1996), 331–358.

[Rou1] R. Rouquier, From stable equivalences to Rickard equivalences for blocks with cyclic defect, in: Groups '93 Galway/St. Andrews (Campbell, C. M. et al., eds.) Vol. 2, Lond. Math. Soc. Lect. Note Ser. 212, Cambridge University Press, 1995, 512–523.

[Rou2] R. Rouquier, The derived category of blocks with cyclic defect groups, in: Derived equivalences for group rings, Lecture Notes in Math. 1685, Springer-Verlag, Berlin 1998, 199–220.

[Rou3] R. Rouquier, Local constructions in block theory, in preparation.

[Tu] W. Turner, Equivalent blocks of finite general linear groups in non-describing characteritic, preprint.

UFR de Mathématiques and UMR 7586, Université Paris 7, 2 Place Jussieu, 75251 Paris, France

E-mail: rouquier@math.jussieu.fr

Alperin's weight conjecture in terms of linear source modules and trivial source modules

*Robert Boltje**

Abstract. We extend Knörr's and Robinson's reformulation of the blockwise version of Alperin's weight conjecture by showing that the alternating sum they introduced does not change its value if the terms which count simple modules are replaced by terms which count indecomposable linear source or trivial source modules. A more general result is established for quantities attached to finite groups admitting a defect group theory such that an analogue of Brauer's first main theorem or the Green correspondence holds.

1. Introduction and statement of the main result

Since Alperin first published the weight conjecture in [A87] there have been many equivalent reformulations (e.g. [KR89], [TW90], [T92], [T93], [Bc96]) and also refinements ([D92], [D94]). The aim of this note is yet another equivalent reformulation of Alperin's weight conjecture. Our starting point is the reformulation given by Knörr and Robinson in [KR89] which predicts the vanishing of a certain Lefschetz-type alternating sum whose summands count simple modules of certain subgroups. Here we show that the Knörr–Robinson alternating sum does not change its value if we count indecomposable modules of certain types, like for example trivial source modules or linear source modules. This will be a consequence of a more general result (Corollary 2.4) about alternating sums over the simplices of the Brown complex of invariants of their stabilizers and blocks and p-subgroups of those satisfying certain axiomatic properties (see (2.1) (i), (ii)). If these invariants are numbers of indecomposable lattices, then the axioms are satisfied due to the Green correspondence. In a similar way, a defect group theory of multiplicities of lower defect groups was already used in [S93].

We recall the reformulation of Alperin's weight conjecture given by Knörr and Robinson. The following notation will be needed.

1.1. Throughout this article let G be a finite group, let p be a prime, and let $\mathcal{S}_p(G)$ (resp. $\mathcal{S}_p^*(G)$) denote the partially ordered set of non-trivial (resp. all) p-subgroups

*Research supported by a Heisenberg Fellowship from the DFG

of G, ordered by inclusion. Note that the conjugation action of G on $\mathcal{S}_p^*(G)$ and $\mathcal{S}_p(G)$ respects this partial order. A *chain* of *length* $n \in \mathbb{N}_0$ in $\mathcal{S}_p(G)$ is a strictly increasing sequence $P_0 < \cdots < P_n$ of $n+1$ elements of $\mathcal{S}_p(G)$. Besides these non-trivial chains there is the empty chain $\emptyset$ which has length -1 by definition. Again, G acts by conjugation on the set $\Delta(\mathcal{S}_p(G))$ of all proper chains, and also on $\tilde{\Delta}(\mathcal{S}_p(G)) := \Delta(\mathcal{S}_p(G)) \cup \{\emptyset\}$ leaving the empty chain fixed. For $\sigma \in \tilde{\Delta}(\mathcal{S}_p(G))$ let $|\sigma|$ denote its length and G_σ its stabilizer in G. Thus, $|\emptyset| = -1$ and $G_\emptyset = G$.

1.2. Furthermore, let $\mathcal{O}$ denote a complete discrete valuation ring of characteristic 0 with algebraically closed residue field of prime characteristic p. If σ is a chain from $\tilde{\Delta}(\mathcal{S}_p(G))$ and if b is a block of $\mathcal{O}G_\sigma$, it is shown in [KR89, Lemma 3.2] that the induced block b^G is always defined. For a given sum B of blocks of $\mathcal{O}G$ and a subgroup H of G, we define B_H as the sum of those blocks b of $\mathcal{O}H$ for which b^G is defined and is a summand of B. Of course, $B_G = B$.

1.3. For a sum B of blocks of $\mathcal{O}G$ let $l(B)$ denote the number of isomorphism classes of simple B-modules which coincides with the number $\mathrm{pr}(B)$ of isomorphism classes of indecomposable projective B-modules. Now the reformulation of Alperin's weight conjecture given by Knörr and Robinson (see [KR89, Theorem 3.8]) can be stated as

$$\sum_{\sigma \in \tilde{\Delta}(\mathcal{S}_p(G))/G} (-1)^{|\sigma|} l(B_{G_\sigma}) = 0\,, \tag{1.3.a}$$

whenever B is a block of $\mathcal{O}G$ of positive defect. It is easy to see that in the case where B is a block of defect zero, the alternating sum in Equation (1.3.a) equals -1, which is the contribution of the empty chain. Each chain $\sigma \in \Delta(\mathcal{S}_p(G))$ in this case gives a zero contribution, since $B_{G_\sigma} = 0$ (the defect group of each block summand of B_{G_σ} again has to be a block of defect zero, and such blocks do not exist since $O_p(G_\sigma) \neq 1$).

1.4. We establish the notation for our main result. For each $H \leqslant G$ let $\mathcal{A}(\mathcal{O}H)$ be a finite subset of the set of isomorphism classes $[M]$ of indecomposable $\mathcal{O}H$-lattices M. For a sum B of blocks of $\mathcal{O}H$ and $P \in \mathcal{S}_p^*(H)$ we set

$$\mathcal{A}(B) := \{[M] \subset \mathcal{A}(\mathcal{O}H) \mid 1_B \cdot M = M\}\,,$$
$$\mathcal{A}(B, P) := \{[M] \in \mathcal{A}(B) \mid P \text{ is a vertex of } M\}\,,$$

and

$$a(B) := |\mathcal{A}(B)|\,, \quad a(B, P) := |\mathcal{A}(B, P)|\,.$$

Moreover, if $\mathcal{P}(B)$ denotes the set of isomorphism classes of indecomposable projective B-modules, then $|\mathcal{P}(B)| = \mathrm{pr}(B) = l(B)$.

1.5 Theorem. *Assume the notation of Subsection* 1.4 *and assume that the sets* $\mathcal{A}(\mathcal{O}H)$, $H \leqslant G$, *are closed under* G*-conjugation and under taking Green corre-*

spondents in both directions with respect to normalizers of vertices. Then

$$\sum_{\sigma\in\tilde{\Delta}(\mathcal{S}_p(G))/G} (-1)^{|\sigma|} a(B_{G_\sigma}) = \sum_{\sigma\in\tilde{\Delta}(\mathcal{S}_p(G))/G} (-1)^{|\sigma|} a(B_{G_\sigma}, 1)\,.$$

In particular, if $\mathcal{P}(\mathcal{O}H) \subseteq \mathcal{A}(\mathcal{O}H)$ for each $H \leqslant G$, one has

$$\sum_{\sigma\in\tilde{\Delta}(\mathcal{S}_p(G))/G} (-1)^{|\sigma|} a(B_{G_\sigma}) = \sum_{\sigma\in\tilde{\Delta}(\mathcal{S}_p(G))/G} (-1)^{|\sigma|} \mathrm{pr}(B_{G_\sigma})\,.$$

1.6. An $\mathcal{O}G$-lattice M is called a *linear source* (resp. *trivial source*) $\mathcal{O}G$-module if each of its indecomposable direct summands has a source of $\mathcal{O}$-rank 1 (resp. a trivial source). Equivalently, after restriction to a Sylow p-subgroup P of G, the module M is a monomial (resp. permutation) $\mathcal{O}P$-lattice. Another equivalent condition is that M is isomorphic to a direct summand of a monomial (resp. permutation) $\mathcal{O}G$-lattice. See [Br85] and [Bo98] for proofs of these statements. Note that projective $\mathcal{O}G$-modules are trivial source modules and *a fortiori* linear source modules, and that the classes of indecomposable linear source modules and trivial source modules are stable under taking Green correspondents. For $H \leqslant G$ and for a sum B of blocks of $\mathcal{O}H$ let $\mathrm{lin}(B)$ (resp. $\mathrm{triv}(B)$) denote the number of isomorphism classes of indecomposable linear source (resp. trivial source) $\mathcal{O}H$-modules M with $1_B \cdot M = M$. Now the following corollaries are immediate from Theorem 1.5 and from Subsection 1.3.

1.7 Corollary. *Let B be a block of $\mathcal{O}G$. Then*

$$\begin{aligned}\sum_{\sigma\in\tilde{\Delta}(\mathcal{S}_p(G))/G} (-1)^{|\sigma|} \mathrm{lin}(B_{G_\sigma}) &= \sum_{\sigma\in\tilde{\Delta}(\mathcal{S}_p(G))/G} (-1)^{|\sigma|} \mathrm{triv}(B_{G_\sigma}) \\ &= \sum_{\sigma\in\tilde{\Delta}(\mathcal{S}_p(G))/G} (-1)^{|\sigma|} l(B_{G_\sigma})\,.\end{aligned}$$

1.8 Corollary. *Alperin's weight conjecture is equivalent to the vanishing of either of the alternating sums*

$$\sum_{\sigma\in\tilde{\Delta}(\mathcal{S}_p(G))/G} (-1)^{|\sigma|} \mathrm{lin}(B_{G_\sigma}) \quad \text{and} \quad \sum_{\sigma\in\tilde{\Delta}(\mathcal{S}_p(G))/G} (-1)^{|\sigma|} \mathrm{triv}(B_{G_\sigma})$$

for each finite group G and each block B of $\mathcal{O}G$ of positive defect.

1.9 Remark. It is not difficult to count linear source or trivial source modules (see [BK00, Remark 3.10]). If we fix a p-subgroup P of G and an $\mathcal{O}P$-lattice of $\mathcal{O}$-rank 1, say $\mathcal{O}_\varphi$, for a homomorphism $\varphi\colon P \to \mathcal{O}^\times$, then the set of isomorphism classes of indecomposable linear source $\mathcal{O}G$-modules with vertex P and source $\mathcal{O}_\varphi$ is in natural bijective correspondence to the set of isomorphism classes of indecomposable projective $FN_G(P,\varphi)/P$-modules via a generalized Brauer construction, where F

denotes the residue field of $\mathcal{O}$ and $N_G(P, \varphi)$ denotes the stabilizer of φ in $N_G(P)$. The last set of course has as cardinality the number of p'-conjugacy classes of $N_G(P, \varphi)/P$.

2. Proof of Theorem 1.5

Theorem 1.5 will be an immediate consequence of Corollary 2.4 below.

2.1. We assume the notation from Subsection 1.4 and the hypothesis of Theorem 1.5. For a fixed sum B of blocks of $\mathcal{O}G$ we then obtain integer valued functions $a\colon (H, P) \mapsto a(H, P) := a(B_H, P)$ on the set of pairs (H, P) with $H \leqslant G$ and $P \in \mathcal{S}_p^*(H)$, and $a\colon H \mapsto a(H) := a(B_H)$ on the set of subgroups of G, satisfying the following conditions:

(i) $a(gHg^{-1}, gPg^{-1}) = a(H, P)$,

(ii) $a(H, P) = a(N_H(P), P)$,

(iii) $a(H) = \sum_{P\in\mathcal{S}_p^*(H)/H} a(H, P)$,

for each $g \in G$, $H \leqslant G$, and $P \in \mathcal{S}_p^*(H)$. In fact, (i) and (iii) are obvious, and (ii) follows from the fact that if M is an indecomposable $\mathcal{O}H$-lattice with vertex P and its Green correspondent N lies in the block b of $\mathcal{O}N_H(P)$, then M lies in the block b^H of $\mathcal{O}H$; cf. [F82, Theorem III.7.8], together with the fact that $(b^H)^G$ is defined if and only if b^G is defined and in that case $(b^H)^G = b^G$, cf. [F82, Lemma III.92].

2.2. Very often, invariants of finite groups allow a kind of "defect group" or "vertex" theory and therefore can be refined to invariants of a group together with a p-subgroup. It also happens frequently that such invariants can be further refined to blockwise versions. If this is the case, we assume a block of $\mathcal{O}G$ to be fixed and obtain the following axiomatic setup in which the block notationally does not appear.

We extend the concepts of 2.1 to functions taking values in an arbitrary Abelian group A. Assume that, for each $H \leqslant G$ and $P \in \mathcal{S}_p^*(H)$, we are given elements $a(B, P) \in A$ satisfying the conditions (i) and (ii) in Subsection 2.1. Furthermore, let $a(H) \in A$ be defined by Equation (iii) in Subsection 2.1. For $P \in \mathcal{S}_p^*(G)$ and $H \leqslant G$, we set

$$a(H, [P]_G) := \sum_{\substack{P'\in\mathcal{S}_p^*(H)/H \\ P'=_G P}} a(H, P'), \qquad (2.2.a)$$

where $=_G$ stands for being conjugate under G. Of course, the above sum is zero if no G-conjugate of P is contained in H.

We will first show that the following alternating sum can be rewritten as a sum over $\mathcal{S}_p^*(G)/G$:

$$\sum_{\sigma\in\tilde{\Delta}(\mathcal{S}_p(G))/G} (-1)^{|\sigma|} a(G_\sigma) = \sum_{P\in\mathcal{S}_p^*(G)/G}\ \sum_{\sigma\in\tilde{\Delta}(\mathcal{S}_p(G))/G} (-1)^{|\sigma|} a(G_\sigma, [P]_G)\,. \quad (2.2.b)$$

In fact, if we consider the G-set

$$X := \{(\sigma, P) \mid \sigma \in \tilde{\Delta}(\mathcal{S}_p(G)),\ P \in \mathcal{S}_p^*(G)\}$$

with the componentwise G-action and if we set $a(H, P) := 0$ whenever $H \leqslant G$ and $P \in \mathcal{S}_p^*(G)$ with $P \nleqslant H$ then we obtain:

$$\begin{aligned}
\sum_{\sigma\in\tilde{\Delta}(\mathcal{S}_p(G))/G} (-1)^{|\sigma|} a(G_\sigma) &= \sum_{\sigma\in\tilde{\Delta}(\mathcal{S}_p(G))/G}\ \sum_{P\in\mathcal{S}_p^*(G_\sigma)/G_\sigma} (-1)^{|\sigma|} a(G_\sigma, P)\\
&= \sum_{(\sigma,P)\in X/G} (-1)^{|\sigma|} a(G_\sigma, P)\\
&= \sum_{P\in\mathcal{S}_p^*(G)/G}\ \sum_{\sigma\in\tilde{\Delta}(\mathcal{S}_p(G))/N_G(P)} (-1)^{|\sigma|} a(G_\sigma, P)\\
&= \sum_{P\in\mathcal{S}_p^*(G)/G}\ \sum_{\sigma\in\tilde{\Delta}(\mathcal{S}_p(G))/G} (-1)^{|\sigma|} \sum_{g\in N_G(P)\backslash G/G_\sigma} a(gG_\sigma g^{-1}, P)
\end{aligned}$$

with

$$\begin{aligned}
\sum_{g\in N_G(P)\backslash G/G_\sigma} a(gG_\sigma g^{-1}, P) &= \sum_{\substack{g\in N_G(P)\backslash G/G_\sigma\\ P\leqslant gG_\sigma g^{-1}}} a(gG_\sigma g^{-1}, P)\\
&= \sum_{\substack{s\in G_\sigma\backslash G/N_G(P)\\ sPs^{-1}\leqslant G_\sigma}} a(G_\sigma, sPs^{-1})\\
&= \sum_{\substack{P'\in\mathcal{S}_p^*(G_\sigma)/G_\sigma\\ P'=_G P}} a(G_\sigma, P')\\
&= a(G_\sigma, [P]_G)\,,
\end{aligned}$$

for fixed $\sigma \in \tilde{\Delta}(\mathcal{S}_p(G))$ and $P \in \mathcal{S}_p^*(G)$, and with $s = g^{-1}$.

For any $H \leqslant G$ we denote by $\tilde{\Delta}(\mathcal{S}_p(G))^H$ the set of fixed points of $\tilde{\Delta}(\mathcal{S}_p(G))$ under the conjugation action of H. Note that $\sigma = (P_0 < \cdots < P_n) \in \tilde{\Delta}(\mathcal{S}_p(G))^H$ if and only if $H \leqslant G_\sigma = \bigcap_{i=0}^n N_G(P_i)$.

2.3 Theorem. *Let $P \in \mathcal{S}_p^*(G)$. Assuming the notation and hypothesis of Subsection 2.2, one has*

$$\sum_{\sigma \in \tilde{\Delta}(\mathcal{S}_p(G))/G} (-1)^{|\sigma|} a(G_\sigma, [P]_G) = \sum_{\sigma \in \tilde{\Delta}(\mathcal{S}_p(G))^P/N_G(P)} (-1)^{|\sigma|} a(G_\sigma, P)$$

and, if $P \neq 1$, these alternating sums vanish.

Proof. We start by proving the equality. By the definition in (2.2.a) we have

$$\begin{aligned}\sum_{\sigma \in \tilde{\Delta}(\mathcal{S}_p(G))/G} (-1)^{|\sigma|} a(G_\sigma, [P]_G) &= \sum_{\sigma \in \tilde{\Delta}(\mathcal{S}_p(G))/G} \sum_{\substack{P' \in \mathcal{S}_p^*(G_\sigma)/G_\sigma \\ P' =_G P}} (-1)^{|\sigma|} a(G_\sigma, P') \\ &= \sum_{(\sigma, P') \in Y_P/G} (-1)^{|\sigma|} a(G_\sigma, P')\end{aligned}$$

for the G-subset

$$Y_P := \{(\sigma, P') \in \tilde{\Delta}(\mathcal{S}_p(G)) \times \mathcal{S}_p^*(G) \mid P' =_G P,\ P' \leqslant G_\sigma\}$$

of X. Since G acts transitively on the second component of Y_p, and since $P' \leqslant G_\sigma$ if and only if $\sigma \in \tilde{\Delta}(\mathcal{S}_p(G))^{P'}$, the last sum equals

$$\sum_{\sigma \in \tilde{\Delta}(\mathcal{S}_p(G))^P/N_G(P)} (-1)^{|\sigma|} a(G_\sigma, P),$$

and the equation is proved.

Next, we show that

$$\sum_{\sigma \in \tilde{\Delta}(\mathcal{S}_p(G))^P/N_G(P)} (-1)^{|\sigma|} a(G_\sigma, P) \tag{2.3.a}$$

vanishes if $P \neq 1$. Let $H := N_G(P)$. First note that, by Equation (ii) in Subsection 2.1, the alternating sum (2.3.a) equals

$$\sum_{\sigma \in \tilde{\Delta}(\mathcal{S}_p(G))^P/H} (-1)^{|\sigma|} a(H_\sigma, P). \tag{2.3.b}$$

The H-set $\tilde{\Delta}(\mathcal{S}_p(G))^P$ is partitioned into two H-sets Δ_1 and Δ_2, with $\sigma \in \tilde{\Delta}(\mathcal{S}_p(G))^P$ being contained in Δ_1 if and only if for the maximal element Q in $\sigma \cup \{1\}$ with $P \not\leqslant Q$, also $QP \in \sigma$. If we define $f\colon \tilde{\Delta}(\mathcal{S}_p(G))^P \to \tilde{\Delta}(\mathcal{S}_p(G))^P$ by removing QP from σ if $QP \in \sigma$, and by inserting QP if $QP \notin \sigma$, then it is easy to see that f is an H-equivariant involution ($f \circ f = \mathrm{id}$) which induces H-equivariant bijections between Δ_1 and Δ_2. Moreover, since $(-1)^{|f(\sigma)|} = -(-1)^{|\sigma|}$ and $H_\sigma = H_{f(\sigma)}$, each pair of summands indexed by σ and $f(\sigma)$ adds up to zero and the alternating sum (2.3.b) vanishes. □

2.4 Corollary. *Assuming the hypothesis and notation of Subsection 2.2 one has*

$$\sum_{\sigma\in\tilde{\Delta}(\mathcal{S}_p(G))/G} (-1)^{|\sigma|}a(G_\sigma) = \sum_{\sigma\in\tilde{\Delta}(\mathcal{S}_p(G))/G} (-1)^{|\sigma|}a(G_\sigma, 1)\,.$$

Proof. By Equation (2.2.b) and by Theorem 2.3 we have

$$\begin{aligned}\sum_{\sigma\in\tilde{\Delta}(\mathcal{S}_p(G))/G} (-1)^{|\sigma|}a(G_\sigma) &= \sum_{P\in\mathcal{S}_p^*(G)/G}\ \sum_{\sigma\in\tilde{\Delta}(\mathcal{S}_p(G))/G} (-1)^{|\sigma|}a(G_\sigma, [P]_G)\\ &= \sum_{\sigma\in\tilde{\Delta}(\mathcal{S}_p(G))/G} (-1)^{|\sigma|}a(G_\sigma, 1)\,.\end{aligned}$$

□

2.5 Remark. There is a more direct proof of Corollary 2.4 which rewrites the left hand side as a summation over X/G as in Subsection 2.2 and then shows that the elements (σ, P) in X/G with $P \neq 1$ can be paired in a way that they cancel each other. But the statement in Theorem 2.3 is stronger, and of interest in its own right, since it states that the cancellation even takes place for fixed $P \in \mathcal{S}_p(G)/G$.

2.6 Examples. [S93] gives more examples of invariants $a(H, P)$ which satisfy the axioms 2.1(i) and (ii), namely blockwise multiplicities of lower defect groups (by [F82, Lemma V.10.4]), numbers of blocks (by Brauer's first main theorem), and blockwise numbers of isomorphism classes of indecomposable lattices with cyclic vertices (by the Green correspondence). Summing up the multiplicities of lower defect groups over all blocks, one also sees that the number of conjugacy classes with given defect group satisfies the axioms 2.1(i) and (ii).

We examine one of these examples further. We fix a sum B of blocks of $\mathcal{O}G$ and denote by $n(B_H, P)$ the number of block summands of B_H with defect group P, for a subgroup H of G and $P \in \mathcal{S}_p^*(H)$, where B_H is defined as in Subsection 1.2. Then Corollary 2.4 applies to $a(H, P) := n(B_H, P)$. Moreover, we have $n(B_{G_\sigma}, 1) = 0$ for all non-empty chains σ, since $O_p(G_\sigma) \neq 1$. Therefore, for any block B of $\mathcal{O}G$, Corollary 2.4 implies (cf. [S93, Example 4.2])

$$\sum_{\sigma\in\tilde{\Delta}(\mathcal{S}_p(G))/G} (-1)^{|\sigma|}n(B_{G_\sigma}) = \begin{cases}-1 & \text{if } B \text{ has defect } 0,\\ 0 & \text{if } B \text{ has positive defect,}\end{cases}$$

with

$$n(B_{G_\sigma}) := \sum_{P\in\mathcal{S}_p^*(G_\sigma)/G_\sigma} n(B_{G_\sigma}, P)\,.$$

Comparing this with Equation (1.3.a), we see that Alperin's weight conjecture is equivalent to the equation

$$\sum_{\sigma \in \tilde{\Delta}(\mathcal{S}_p(G))/G} (-1)^{|\sigma|} \big(l(B_{G_\sigma}) - n(B_{G_\sigma}) \big) = 0 \qquad (2.6.a)$$

for *all* blocks B of $\mathcal{O}G$, or also to the equations with $l(B_{G_\sigma})$ replaced by $\mathrm{triv}(B_{G_\sigma})$ or $\mathrm{lin}(B_{G_\sigma})$.

Acknowledgements. It is a pleasure for me to thank the Department of Mathematics of the University of Chicago for their hospitality during the time this note was written. Moreover, I am grateful to B. Külshammer for bringing the article [S93] to my attention and for other valuable comments on an earlier version of this note.

References

[A87] J. L. Alperin, Weights for finite groups, Proc. Sympos. Pure Math. 47 (1987), 369–379.

[Bo98] R. Boltje, Linear source modules and trivial source modules, Proc. Sympos. Pure Math. 63 (1998), 7–30.

[BK00] R. Boltje, B. Külshammer, A generalized Brauer construction and linear source modules, Trans. Amer. Math. Soc. 352 (2000), 3411–3428.

[Bc96] S. Bouc, Construction de foncteurs entre catégories de G-ensembles, J. Algebra 183 (1996), 737–825.

[Br85] M. Broué, On Scott modules and p-permutation modules: An approach through the Brauer morphism, Proc. Amer. Math. Soc. 93 (1985), 401–408.

[D92] E. C. Dade, Counting characters in blocks. I, Invent. Math. 109 (1992), 187–210.

[D94] E. C. Dade, Counting characters in blocks. II, J. Reine Angew. Math. 448 (1994), 97–190.

[F82] W. Feit, The representation theory of finite groups, North Holland, Amsterdam 1982.

[KR89] R. Knörr, G. Robinson, Some remarks on a conjecture of Alperin, J. London Math. Soc. (2) 39 (1989), 48–60.

[S93] P. Symonds, An alternating sum formula for multiplicities of lower defect groups, Bull. London Math. Soc. 25 (2) (1993), 121–124.

[T92] J. Thévenaz, Locally determined functions and Alperin's conjecture, J. London Math. Soc. (2) 45 (1992), 446–468.

[T93] J. Thévenaz, Equivariant K-theory and Alperin's conjecture, J. Pure Appl. Algebra 85 (1993), 185–202.

[TW90] J. Thévenaz, P. Webb, A Mackey functor version of a conjecture of Alperin, Astérisque 181-182 (1990), 263–272.

Department of Mathematics, University of California, Santa Cruz, CA 95064, U.S.A.

E-mail: `boltje@math.ucsc.edu`

Alvis–Curtis duality as an equivalence of derived categories

Marc Cabanes and Jeremy Rickard

Abstract. Let G be the group of rational points of a connected reductive group defined over a finite field of characteristic p, and let $\mathcal{O}$ be any commutative ring in which p is invertible. We prove a conjecture of Broué, that the duality operation of Alvis, Curtis, Kawanaka and Lusztig on the characters of G is induced by a self-equivalence of the derived category of $\mathcal{O}G$-modules. We also prove that this equivalence of derived categories is compatible with Harish-Chandra induction and truncation.

1. Introduction

Let G be a finite reductive group of characteristic p.

Alvis–Curtis duality, introduced independently by Alvis, Curtis, Kawanaka, and Lusztig around 1980, is an isometric involution

$$D_G : K_0(\mathbb{C}G) \longrightarrow K_0(\mathbb{C}G)$$

of the group of generalized characters of G (see for instance [7], Chapter 8).

In Section 2 (2) of the paper [2] in which he introduced the notion of perfect isometries and his conjectures on derived equivalence of blocks, Broué showed that in all characteristics other than p, D_G gives a perfect isometry. This would be a consequence of D_G being induced by a self-equivalence of the derived category $D^b(\mathrm{mod}(\mathcal{O}G))$ for a complete discrete valuation ring $\mathcal{O}$ with residue field of characteristic different from p, and Broué conjectured that this was the case.

In this paper, we shall show that this conjecture of Broué is correct, and in fact the equivalence of derived categories is defined over any commutative coefficient ring $\mathcal{O}$ in which p is invertible. Our proof uses a complex, or more precisely a coefficient system, of $\mathbb{Z}[1/p]G$-bimodules which gives a version over $\mathbb{Z}[1/p]$ of the complex of $\mathbb{C}G$-modules introduced by Curtis in [3]. As in the approach by Deligne and Lusztig to the character-theoretic duality in [5], the main step of our proof consists of determining the image of a module which is the Harish-Chandra induction of a cuspidal module (see Theorem 3.1). For that purpose, an essential ingredient is the fact that the Harish-Chandra induction and truncation functors are independent of the choice of a parabolic subgroup ([8], [6]) and the related statements on the bimodule inducing these functors (see Lemma 3.4).

We also prove the structural analogue over $\mathbb{Z}[1/p]$ of the character-theoretic theorem of Curtis stating that Alvis–Curtis duality commutes with Harish-Chandra induction (see for instance [7], Theorem 8.11).

We shall use standard notation involving derived categories and chain homotopy categories for which Weibel's book [11] is a good reference. For background concerning equivalences of derived categories in representation theory, see the lecture notes of König and Zimmermann [9].

2. Preliminaries

2.1. Notation

Let us start by fixing some notation. Throughout this paper, $\boldsymbol{G}$ will be a connected reductive group defined over a finite field $\mathbb{F}_q$, where q is a power of the prime p, and F will be the Frobenius endomorphism of $\boldsymbol{G}$ corresponding to the given $\mathbb{F}_q$-structure. By G we shall denote the finite group $\boldsymbol{G}^F$ of $\mathbb{F}_q$-rational points of $\boldsymbol{G}$.

We shall fix a quasi-split rational maximal torus $\boldsymbol{T}$ of $\boldsymbol{G}$ and a rational Borel subgroup $\boldsymbol{B}$ containing $\boldsymbol{T}$. We denote $W = W(\boldsymbol{G}, \boldsymbol{T})^F$ and let $S \subseteq W$ be the set of generators associated with the choice of $\boldsymbol{B}$. Recall that S is in bijection with the set of orbits of the Frobenius map F on the set of simple reflections of $W(\boldsymbol{G}, \boldsymbol{T})$, and that the conjugacy classes of rational parabolic subgroups of $\boldsymbol{G}$ are parametrized by subsets of S; given $I \subseteq S$, there is a unique member $\boldsymbol{P}_I$ of the corresponding class of parabolics that contains $\boldsymbol{B}$. Let $\boldsymbol{U}_I$ be the unipotent radical of $\boldsymbol{P}_I$ and let $\boldsymbol{L}_I$ be the Levi complement that contains $\boldsymbol{T}$. Thus $\boldsymbol{P}_\emptyset = \boldsymbol{B}$, and $\boldsymbol{P}_S = \boldsymbol{G}$.

To refer to the finite groups of $\mathbb{F}_q$-rational points of these various subgroups of $\boldsymbol{G}$, we shall use the same notation, but in ordinary rather than in bold type. Thus $B = \boldsymbol{B}^F$, $L_I = \boldsymbol{L}_I^F$, and so on. We shall be working mainly with the split BN-pair of G defined by B, $N = N_G(\boldsymbol{T})^F$, and S ([4], Section 69).

We shall be considering representations of these finite groups over a commutative coefficient ring which contains a multiplicative inverse to p. Throughout the paper, $\mathcal{O}$ will usually denote such a ring. Often it will be sufficient to deal with the coefficient ring $\mathbb{Z}[1/p]$, which we shall denote by R.

For A any p-subgroup of G, denote by $[A]$ the idempotent $|A|^{-1} \sum_{a \in A} a \in RG$.

In Sections 3 and 4, we shall restrict to the case where $\mathcal{O}$ is a field with characteristic different from p, which we shall denote by k.

We shall need to consider both left and right modules for group algebras, but we shall adopt the convention that when we do not specify which side of a module M the action is on, we intend M to be a left module. By $\mathrm{Mod}(\mathcal{O}H)$ we mean the abelian category of left $\mathcal{O}H$-modules for a finite group H. In Section 7.1 we shall need to consider the abelian category $\mathrm{mod}(\mathcal{O}H)$ of finitely generated left $\mathcal{O}H$-modules for a commutative noetherian ring $\mathcal{O}$.

2.2. Harish-Chandra induction functors and associated bimodules

Let $\boldsymbol{P}$ be a rational parabolic subgroup of $\boldsymbol{G}$ with unipotent radical $\boldsymbol{U}$ and rational Levi complement $\boldsymbol{L}$, and let P, U and L be the corresponding subgroups of the finite group G. One denotes by

$$\mathrm{R}_L^G : \mathrm{Mod}(\mathcal{O}L) \longrightarrow \mathrm{Mod}(\mathcal{O}G)$$

the functor $\mathrm{R}_L^G = \mathrm{Ind}_P^G \mathrm{Infl}_L^P$ where Infl_L^P is the inflation functor making $\mathcal{O}L$-modules into U-trivial $\mathcal{O}P$-modules. This functor is called *Harish-Chandra induction*. Since p is invertible in $\mathcal{O}$, one may identify R_L^G with the tensor product functor $\mathcal{O}G[U] \otimes_{\mathcal{O}L} -$ ([8], Section 1), where $\mathcal{O}G[U]$ is considered as an $\mathcal{O}G$-$\mathcal{O}L$-bimodule. In the same way, the *Harish-Chandra truncation* functor

$$^*\mathrm{R}_L^G : \mathrm{Mod}(\mathcal{O}L) \longrightarrow \mathrm{Mod}(\mathcal{O}G)$$

is identified with the tensor product functor $[U].\mathcal{O}G \otimes_{\mathcal{O}G} -$. Recall that, if M is an $\mathcal{O}G$-module, then $^*\mathrm{R}_L^G M$ can be seen as the $\mathcal{O}L$-module M^U of fixed points of M under the action of U.

The two functors are induced by bimodules that are projective as right (and as left) modules, so they are exact. They are moreover both left and right adjoint to each other.

The notation R_L^G and $^*\mathrm{R}_L^G$ originates from the case $\mathcal{O} = \mathbb{C}$ where the functors do not depend on the choice of P (see for instance [7], Proposition 6.1). In the general case of a ring $\mathcal{O}$ where p is invertible, Howlett–Lehrer and Dipper–Du proved independently that the same result holds: different parabolic subgroups with the same Levi complement L give isomorphic functors ([8], Theorem 1.1 or [6], Theorem 5.2). We shall be mostly interested in the case $\boldsymbol{L} = \boldsymbol{L}_I$, in which case we shall, to be definite, use the parabolic subgroup $\boldsymbol{P}_I$ to define the functors.

2.3. Alvis–Curtis duality and an associated coefficient system

We define our functor $X_{\mathcal{O}} \otimes_{\mathcal{O}G} -$ in terms of homology of local coefficient systems. For the basic definitions and terminology, see, for example, Benson's book [1], Chapter 7.

Let Δ/G be the set of parabolic subgroups of G containing B, partially ordered by the *opposite* relation to inclusion of subgroups (this coincides with the quotient of the building by the conjugation action of G, but this interpretation is not essential for our purpose). Since the elements of Δ/G are in bijection with subsets of S by $I \mapsto P_I$, it can be considered as a simplicial complex where the i-dimensional simplices are the parabolics P_I such that $|S \setminus I| = i+1$. We shall regard G itself as a (-1)-dimensional simplex.

Recall that $R = \mathbb{Z}[1/p]$.

Let $\mathcal{D}$ be the coefficient system on Δ/G which associates with P_I the RG-RG-bimodule $\mathcal{D}_I = RG[U_I] \otimes_{RP_I} [U_I].RG$. If one has an inclusion in Δ/G (i.e., a reverse inclusion of parabolic subgroups containing B), then one defines a restriction map as follows. For $I \subseteq J \subseteq S$, define

$$\phi_{IJ} : \mathcal{D}_I \to \mathcal{D}_J$$

by

$$x \otimes_{RP_I} y \mapsto x \otimes_{RP_J} y$$

for $x \in RG[U_I]$, $y \in [U_I].RG$.

It is easily checked that this is a map of RG-RG-bimodules and that $\phi_{JK}\phi_{IJ} = \phi_{IK}$ whenever $I \subseteq J \subseteq K \subseteq S$.

We now define a complex X_R as $\widetilde{C}_*(\Delta/G, \mathcal{D})$, the reduced chain complex with coefficients in $\mathcal{D}$, with degrees renumbered so as to produce a cochain complex with degrees varying from 0 to $|S|$. If $\mathcal{O}$ is any ring where p is invertible, we set $X_{\mathcal{O}} = X_R \otimes_R \mathcal{O}$. Then $X_{\mathcal{O}}$ is a cochain complex of $\mathcal{O}G$-$\mathcal{O}G$-bimodules

$$\cdots \to 0 \to X_{\mathcal{O}}^0 \to X_{\mathcal{O}}^1 \to \cdots \to X_{\mathcal{O}}^{|S|} \to 0 \to \cdots$$

with $X_{\mathcal{O}}^i = \bigoplus_I \mathcal{O}G[U_I] \otimes_{\mathcal{O}P_I} [U_I].\mathcal{O}G$ where I ranges over the subsets of S of cardinality i. In particular, $X_{\mathcal{O}}^{|S|} = \mathcal{O}G$.

Note first that the complex above is defined in an intrinsic way from the subgroup B, since the P_I's are just the subgroups of G containing B, $|I|$ is determined by the height of P_I in this poset, and U_I is just the maximal normal p-subgroup of P_I. Since the outcome would be the same with a G-conjugate of B, and since the conjugates of B may be characterized as the normalizers of Sylow p-subgroups of G, we see that X_R is defined in an intrinsic way from the abstract structure of G.

The functor

$$X_{\mathcal{O}} \otimes_{\mathcal{O}G} - : \mathrm{Mod}(\mathcal{O}G) \longrightarrow C^b(\mathrm{Mod}(\mathcal{O}G))$$

extends to an endofunctor of $C^b(\mathrm{Mod}(\mathcal{O}G))$, also given by taking the tensor product with $X_{\mathcal{O}}$. This functor preserves contractibility of complexes, so it induces an endofunctor of the chain homotopy category $K^b(\mathrm{Mod}(\mathcal{O}G))$. Finally, since the terms of $X_{\mathcal{O}}$ are projective as right $\mathcal{O}G$-modules, the functor $X_{\mathcal{O}} \otimes_{\mathcal{O}G} -$ is exact, and so it induces an endofunctor of the derived category $D^b(\mathrm{Mod}(\mathcal{O}G))$. We shall prove in Section 5 that this functor

$$X_{\mathcal{O}} \otimes_{\mathcal{O}G} - : D^b(\mathrm{Mod}(\mathcal{O}G)) \longrightarrow D^b(\mathrm{Mod}(\mathcal{O}G))$$

is a self-equivalence.

The following property is clear from the definition of $\mathcal{D}$.

Lemma 2.1. *If we consider $\mathcal{D}$ as a coefficient system of $\mathcal{O}[G \times G]$-modules in the usual way, via $(g, g')x = gxg'^{-1}$, it is stable under the automorphism of $G \times G$ that interchanges the two factors. Hence $X_{\mathcal{O}}$ is also stable under this automorphism.*

As an easy consequence of this, we get the following result, which will mean that we can understand the functor given by tensoring with the *dual* of $X_{\mathcal{O}}$ (i.e., the adjoint functor to the one described above) in terms of the original functor. For clarity, let us make a distinction between two kinds of dual: if M is a left $\mathcal{O}$-free $\mathcal{O}G$-module, M^* will be the $\mathcal{O}$-linear dual of M, regarded as a *left* $\mathcal{O}G$-module via the Hopf algebra structure of $\mathcal{O}G$; on the other hand, if A and B are $\mathcal{O}$-algebras and N is an $\mathcal{O}$-free A-B-bimodule, then $N^\vee$ will be the $\mathcal{O}$-linear dual of N, regarded as a B-A-bimodule.

Proposition 2.2. *If M is any $\mathcal{O}G$-module, then there is an isomorphism*

$$(X_{\mathcal{O}}^\vee \otimes_{\mathcal{O}G} M)^* \cong X_{\mathcal{O}} \otimes_{\mathcal{O}G} M^*$$

of complexes of $\mathcal{O}G$-modules. If M is finitely generated and free as a $\mathcal{O}$-module, then

$$X_{\mathcal{O}}^\vee \otimes_{\mathcal{O}G} M \cong (X_{\mathcal{O}} \otimes_{\mathcal{O}G} M^*)^*.$$

Let us now relate our tensor product functor to the usual duality operation D_G. If M in an $\mathcal{O}G$-module, the complex $X_{\mathcal{O}} \otimes_{\mathcal{O}G} M$ is a cochain complex concentrated in degrees 0 to $|S|$, with $(X_{\mathcal{O}} \otimes_{\mathcal{O}G} M)^i = \bigoplus_I \mathcal{O}G[U_I] \otimes_{\mathcal{O}P_I} [U_I].\mathcal{O}G \otimes_{\mathcal{O}G} M$ where the sum is over $I \subseteq S$ of cardinality i. We clearly have

$$\mathcal{O}G[U_I] \otimes_{\mathcal{O}P_I} [U_I].\mathcal{O}G \cong \mathcal{O}G[U_I] \otimes_{\mathcal{O}L_I} [U_I].\mathcal{O}G$$

as bimodules, so $\mathcal{O}G[U_I] \otimes_{\mathcal{O}P_I} [U_I].\mathcal{O}G \otimes_{\mathcal{O}G} M \cong \mathrm{R}_{L_I}^G {}^*\mathrm{R}_{L_I}^G M$. If $\mathcal{O}$ is a domain of characteristic zero and M is $\mathcal{O}$-free and finitely generated, then the trace character of $X_{\mathcal{O}} \otimes_{\mathcal{O}G} M$ equals $\sum_{I \subseteq S} (-1)^{|I|} \mathrm{R}_{L_I}^G \circ {}^*\mathrm{R}_{L_I}^G(\chi_M)$ where χ_M is the trace character of M. This is the formula defining the Alvis–Curtis duality operation on characters ([7], Definition 8.8).

Finally, let us relate the functor $X_{\mathcal{O}} \otimes_{\mathcal{O}G} -$ to Curtis' construction [3]. The building Δ of G is a simplicial complex whose poset of simplices is opposite to the poset of proper parabolic subgroups of G, and we shall regard G itself as corresponding to a (-1)-dimensional simplex. Given an $\mathcal{O}G$-module M, a G-equivariant coefficient system $\mathcal{F}_M$ of $\mathcal{O}$-modules can be constructed on the building by attaching the $R_u(P)$-fixed points $M^{R_u(P)}$ to the simplex corresponding to a parabolic subgroup $P \le G$ with unipotent radical $R_u(P)$, with the natural inclusion $M^{R_u(P)} \longrightarrow M^{R_u(P')}$ as the restriction map when $P \le P'$. We also attach M itself to the (-1)-dimensional simplex corresponding to G. Since one has $N_G(P) = N_G(R_u(P)) = P$ for any parabolic subgroup, it is clear that our $X_{\mathcal{O}} \otimes_{\mathcal{O}G} M$ is just the reduced chain complex $\tilde{C}_*(\Delta, \mathcal{F}_M)$ of the building with coefficients in $\mathcal{F}_M$, with the same re-indexing as before to get a cochain complex with degrees increasing from 0 to $|S|$, and viewed as a complex of $\mathcal{O}$-modules with a natural G-action induced by the conjugation action of G on the set of parabolic subgroups ([1], Definition 7.3.1).

In particular, if $M = \mathcal{O}$ is the trivial module, then $X_{\mathcal{O}} \otimes_{\mathcal{O}G} M$ is related to the reduced simplicial chain complex of Δ with constant coefficients $\mathcal{O}$, and therefore has a single non-zero cohomology group, since Δ has the homotopy type of a bouquet of spheres of the same dimension, by the Solomon–Tits theorem ([4], Theorem 66.33).

3. Harish-Chandra inductions of cuspidal modules

We keep the notation of the previous section, but we will now work with a coefficient field k of characteristic different from p.

For any connected reductive group $\boldsymbol{H}$ defined over $\mathbb{F}_q$ (so $\boldsymbol{H}$ might be one of the Levi subgroups $\boldsymbol{L}_I$ of $\boldsymbol{G}$), we say that a kH-module V is *cuspidal* if the Harish-Chandra truncation ${}^*\mathrm{R}_L^H(V)$ is zero for the rational Levi complements $\boldsymbol{L}$ of all proper rational parabolic subgroups $\boldsymbol{P} < \boldsymbol{H}$. In other words, this says that $V^U = 0$ for the unipotent radical $\boldsymbol{U}$ of every proper rational parabolic subgroup $\boldsymbol{P}$ of $\boldsymbol{H}$. This section is devoted to the proof of the following theorem.

Theorem 3.1. *Let k be a field of characteristic different from p, let I_0 be a subset of S, let V be a cuspidal kL_{I_0}-module, and let $E = \mathrm{R}_{L_{I_0}}^G(V)$. Then*

$$X_k \otimes_{kG} E \cong E[-|I_0|]$$

and

$$X_k^\vee \otimes_{kG} E \cong E[|I_0|]$$

in the chain homotopy category $K^b(\mathrm{Mod}(kG))$.

Let us state what we shall prove in terms of coefficient systems.

Let $\mathcal{Y}$ be the coefficient system on Δ/G associating with P_I the set

$$\mathcal{Y}_I = \{P_I g P_{I_0} \ :\ (P_I^g \cap P_{I_0})U_{I_0} = P_{I_0}\}.$$

We define $\psi_{IJ}(P_I g P_{I_0}) = P_J g P_{I_0}$ when $I \subseteq J \subseteq S$. This is a coefficient system taking values in the category of finite sets. We make it into a coefficient system $\mathbb{Z}\mathcal{Y}$ with values in the category of abelian groups in the standard way, replacing each finite set $\mathcal{Y}_I$ with the free abelian group $\mathbb{Z}\mathcal{Y}_I$. Setting $\mathcal{D}_k = \mathcal{D} \otimes_R k$, we shall prove

Lemma 3.2. *$\mathcal{D}_k \otimes_{kG} E \cong E \otimes_{\mathbb{Z}} \mathbb{Z}\mathcal{Y}$ as coefficient systems on Δ/G with values in* $\mathrm{Mod}(kG)$.

Let us first say how this implies Theorem 3.1.

Using the reflection representation space M of W, the fundamental chamber $\mathcal{C} \subseteq M$ and its facets $\mathcal{C}_I$ ($I \subseteq S$) of dimension $|S|-|I|$ ([4], Section 66), it is straightforward to check that

$$w^{-1}\mathcal{C}_I \mapsto P_I w P_{I_0} \qquad (w \in W,\ I \subseteq S)$$

induces a bijection between $\mathcal{Y}$ and the $w^{-1}\mathcal{C}_I$'s that are included in the vector space M_0 spanned by $\mathcal{C}_{I_0}$. The restriction map ψ_{IJ} becomes the map sending $w^{-1}\mathcal{C}_I$ to $w^{-1}\mathcal{C}_J$, so those $w^{-1}\mathcal{C}_I$ are clearly just the simplices in a triangulation of the unit sphere of M_0.

The (reduced) chain complex associated to the coefficient system $\mathbb{Z}\mathcal{Y}$ on Δ/G is thus isomorphic to the (reduced) chain complex, with constant coefficients $\mathbb{Z}$, of the sphere, and so its homology is concentrated in degree $|S| - |I_0| - 1$ by classical results on the (reduced) homology of spheres. Since the chain complex associated with $\mathbb{Z}\mathcal{Y}$ is a complex of free abelian groups, it is actually homotopy equivalent to $\mathbb{Z}[-(|S|-|I_0|-1)]$. The first isomorphism of Theorem 3.1 follows once we re-index as mentioned in Section 2.3 to construct X_k from $\mathcal{D}$ as a cochain complex. The second isomorphism then follows using Proposition 2.2.

The proof of Lemma 3.2 uses two further lemmas.

Let us fix $I \subseteq S$ and $g \in G$. Consider the cuspidal kL_{I_0}-module V as a kP_{I_0}-module with U_{I_0} acting trivially.

Lemma 3.3. *If $P_I g P_{I_0} \notin \mathcal{Y}_I$, then $[U_I]g.kP_{I_0} \otimes_{P_{I_0}} V = 0$.*

Lemma 3.4. *If $P_I g P_{I_0} \in \mathcal{Y}_I$, then the map*

$$kG \otimes_{P_I} kP_I[U_I]g.kP_{I_0}[U_{I_0}] \to kG[U_{I_0}]$$

such that

$$x \otimes y \mapsto xy$$

is an isomorphism.

Proof of Lemmas 3.3 *and* 3.4. We may assume $g \in N$ and that its class modulo T is in the set D_{I,I_0} of I-reduced-I_0 elements of W ([4], Definition 64.39 and Theorem 65.21). Let $I' = I^g \cap I_0$. Then $(P_I^g \cap P_{I_0})U_{I_0} = P_{I'}$ ([4], Theorem 69.16.(i)). So the condition $P_I g P_{I_0} \in \mathcal{Y}_I$ is equivalent to $I_0 \subseteq I^g$ (or $I' = I_0$).

Assume that $P_I g P_{I_0} \notin \mathcal{Y}_I$ and $[U_I]g.kP_{I_0} \otimes_{P_{I_0}} V \neq 0$. Since

$$[U_I^g].kP_{I_0} \otimes_{P_{I_0}} V = [U_I^g][U_I^g \cap P_{I_0}].kP_{I_0} \otimes_{P_{I_0}} V,$$

$[U_I^g].kP_{I_0} \otimes_{P_{I_0}} V \neq 0$ implies that $[U_I^g \cap P_{I_0}].kP_{I_0} \otimes_{P_{I_0}} V \neq 0$, or in other words that $[U_I^g \cap P_{I_0}]V \neq 0$. Since U_{I_0} acts trivially on V, one has $[(U_I^g \cap P_{I_0})U_{I_0}]V \neq 0$. We have $(U_I^g \cap P_{I_0})U_{I_0} = U_{I'}$ ([4], Theorem 69.16.(ii)). Since we are assuming that $P_I g P_{I_0} \notin \mathcal{Y}_I$, we know that I' is properly contained in I_0 and therefore $[U_{I'}]V = 0$ by the hypothesis of Theorem 3.1. This gives Lemma 3.3.

Assume $P_I g P_{I_0} \in \mathcal{Y}_I$. The map of Lemma 3.4 is clearly well-defined (and is a morphism of kG-kP_{I_0}-bimodules). Set $P = (P_I)^g$.

Let us check first that the dimension of $kG \otimes_{kP_I} kP_I[U_I]g.kP_{I_0}[U_{I_0}]$, or equivalently $kG \otimes_{kP} kP[U_I]^g.kP_{I_0}[U_{I_0}]$, is less than or equal to that of $kG[U_{I_0}]$, which is

$|G : U_{I_0}|$. Since $P_{I_0} = (P \cap P_{I_0})U_{I_0}$, we have

$$\begin{aligned} kG \otimes_{kP} kP[U_I]^g.kP_{I_0}[U_{I_0}] &= kG \otimes_{kP} kP[U_I]^g[U_{I_0}] \\ &= kG \otimes_{kP} kP[U_I]^g[U_{I_0} \cap P][U_{I_0}], \end{aligned}$$

and this is clearly equal to the subspace $\{x \otimes_{kP} [U_{I_0}] \; : \; x \in kG[U_I^g(U_{I_0} \cap P)]\}$, so its dimension is less than or equal to the dimension of $kG[U_I^g(U_{I_0} \cap P_I^g)]$, which is just $|G : U_I^g(U_{I_0} \cap P_I^g)|$. By [4], Theorem 69.16.(ii), $U_I({}^gU_{I_0} \cap P_I) = U_{{}^gI'}$ which in turn equals $U_{{}^gI_0}$ since $P_IgP_{I_0} \in \mathcal{Y}_I$. So, to check our claim it suffices to check that $|U_{I_0}| = |U_{{}^gI_0}|$. But this is clear since $L_{{}^gI_0} = {}^gL_{I_0}$ and, for any $J \subseteq S$, U/U_J is a Sylow p-subgroup of $P_J/U_J \cong L_J$ ([4], Proposition 69.9 and Theorem 69.10).

It remains to check the surjectivity of the map defined in Lemma 3.4. Its image is $kG[U_I]g.kP_{I_0}[U_{I_0}] = kG[U_I]g[U_{I_0}].kP_{I_0}$. So it suffices to check that

$$kG[U_I]g[U_{I_0}] \supseteq kG[U_{I_0}]. \tag{i}$$

We now refer to [8], Theorem 2.4. This tells us that

$$kG[U_{I_0}] = kG[U_{{}^gI_0}]g[U_{I_0}]. \tag{ii}$$

On the other hand, our hypothesis implies $I \supseteq {}^gI' = {}^gI_0$, so $kG[U_I] \supseteq kG[U_{{}^gI_0}]$. Then (ii) above implies (i). □

Proof of Lemma 3.2. Let $I \subseteq J \subseteq S$. As a kP_I-kP_{I_0}-bimodule,

$$kG = \bigoplus_{P_IgP_{I_0}} kP_Ig.kP_{I_0},$$

so

$$[U_I]E = [U_I].kG \otimes_{kP_{I_0}} V = \bigoplus_{P_IgP_{I_0}} kP_I[U_I]g.kP_{I_0} \otimes_{kP_{I_0}} V.$$

By Lemma 3.3, this is the same as

$$\bigoplus_{P_IgP_{I_0} \in \mathcal{Y}_I} kP_I[U_I]g.kP_{I_0} \otimes_{kP_{I_0}} V.$$

Then

$$\mathcal{D}_I \otimes_{kG} E = \bigoplus_{P_IgP_{I_0} \in \mathcal{Y}_I} kG \otimes_{kP_I} [U_I].kP_Ig.kP_{I_0} \otimes_{kP_{I_0}} V.$$

By Lemma 3.4, for every g the summand $kG \otimes_{kP_I} [U_I].kP_Ig.kP_{I_0} \otimes_{kP_{I_0}} V$ is isomorphic to $E = kG[U_{I_0}] \otimes_{kP_{I_0}} V$ by the map

$$x \otimes_{kP_I} [U_I]a \otimes_{kP_{I_0}} v \mapsto x[U_I]a \otimes_{kP_{I_0}} v$$

for $x \in kG, a \in P_IgP_{I_0}, v \in V$. Then $\mathcal{D}_I \otimes_{kG} E \cong E \otimes_{\mathbb{Z}} \mathbb{Z}\mathcal{Y}_I$ by the map

$$i\colon x \otimes_{kP_I} [U_I]a \otimes_{kP_{I_0}} v \mapsto (x[U_I]a \otimes_{kP_{I_0}} v) \otimes_{\mathbb{Z}} (P_IaP_{I_0})$$

for $x \in kG$, $a \in \bigcup_{P_I g P_{I_0} \in \mathcal{Y}_I} P_I g P_{I_0}$, $v \in V$.

In the same way, we get an isomorphism $\mathcal{D}_J \otimes_{kG} E \cong E \otimes_{\mathbb{Z}} \mathbb{Z}\mathcal{Y}_J$ given by the map

$$j \colon y \otimes_{kP_J} [U_J]b \otimes_{kP_{I_0}} v \mapsto (y[U_J]b \otimes_{kP_{I_0}} v) \otimes_{\mathbb{Z}} (P_J b P_{I_0})$$

for $y \in kG$, $b \in \bigcup_{P_J g P_{I_0} \in \mathcal{Y}_J} P_J g P_{I_0}$, $v \in V$.

It remains to check that $j \circ \phi_{IJ} = \psi_{IJ} \circ i$. Taking $x \in kG$, $a \in P_I g P_{I_0} \in \mathcal{Y}_I$, $v \in V$, we have

$$\begin{aligned}(\psi_{IJ} \circ i)(x \otimes_{kP_I} [U_I]a \otimes_{kP_{I_0}} v) &= \psi_{IJ}\big((x[U_I]a \otimes_{kP_{I_0}} v) \otimes_{\mathbb{Z}} (P_I a P_{I_0})\big) \\ &= (x[U_I]a \otimes_{kP_{I_0}} v) \otimes_{\mathbb{Z}} (P_J a P_{I_0}).\end{aligned}$$

On the other hand,

$$\phi_{IJ}(x \otimes_{kP_I} [U_I]a \otimes_{kP_{I_0}} v) = x \otimes_{kP_J} [U_I]a \otimes_{kP_{I_0}} v.$$

The latter is also equal to

$$x \otimes_{kP_J} [U_J][U_I]a \otimes_{kP_{I_0}} v = |U_I|^{-1} \sum_{u \in U_I} x \otimes_{kP_J} [U_J]ua \otimes_{kP_{I_0}} v$$

with $P_J ua P_{I_0} = P_J a P_{I_0} \in \mathcal{Y}_J$ for each u. So the image by j is

$$\Big(|U_I|^{-1} \sum_{u \in U_I} x[U_J]ua \otimes_{kP_{I_0}} v\Big) \otimes_{\mathbb{Z}} (P_J a P_{I_0}) = (x[U_I]a \otimes_{kP_{I_0}} v) \otimes_{\mathbb{Z}} (P_J a P_{I_0}).$$

This finishes the proof. □

Defining $\mathcal{D}$ for P_{I_0} instead of G gives a coefficient system of RP_{I_0}-RP_{I_0}-bimodules on the subsets of I_0. We can extend it trivially to a coefficient system on the whole of Δ/G which we denote by $\mathcal{D}(P_{I_0})$. We denote the restriction maps of $\mathcal{D}(P_{I_0})$ by φ_{IJ} for $I \subseteq J \subseteq S$.

Proposition 3.5. *There is a surjective map*

$$\pi : \mathcal{D} \otimes_{RG} RG[U_{I_0}] \to RG[U_{I_0}] \otimes_{RP_{I_0}} \mathcal{D}(P_{I_0})$$

of coefficient systems of RG-RP_{I_0}-bimodules on Δ/G.

Proof. We shall sometimes abbreviate $P_{I_0} = P_0$.

Let us define the map π as follows. Let $I \subseteq S$, and define $e = [U_I]$. Let $e_0 = [U_{I_0}]$.

If $I \nsubseteq I_0$, let $\pi_I = 0$.

Assume $I \subseteq I_0$. Then

$$RGe_0 \otimes_{RP_0} \mathcal{D}(P_0) = RGe_0 \otimes_{RP_0} RP_0 e \otimes_{RP_I} e.RP_0.$$

In

$$\mathcal{D}_I \otimes_{RG} RGe_0 = RGe \otimes_{RP_I} e.RG \otimes_{RG} RGe_0,$$

let us consider the last tensor product. By the obvious maps of RP_I-RP_0-bimodules, we have

$$e.RG \otimes_{RG} RGe_0 \cong e.RGe_0 = \bigoplus_{P_I g P_0 \subseteq G} e.RP_I g.RP_0 e_0,$$

where the summand for $g = 1$ gives $e.RP_0e_0 = ee_0.RP_0 = e.RP_0$. So $\mathcal{D}_I \otimes_{RG} RGe_0$ has a summand isomorphic to

$$RGe \otimes_{RP_I} e.RP_0 \cong RGe_0 \otimes_{RP_0} RP_0e \otimes_{RP_I} e.RP_0.$$

This provides us with a surjective map π_I defined as follows for $x \in RG$, $y, z \in G$.

$$\pi_I(xe \otimes_{RP_I} ey \otimes_{RG} ze_0) = \begin{cases} xe \otimes_{RP_0} e \otimes_{RP_I} eyze_0 & \text{if } yz \in P_0 \\ 0 & \text{if } yz \notin P_0. \end{cases}$$

Let $J \supseteq I$. We must verify the equality of compositions of maps

$$(RGe_0 \otimes_{RP_0} \varphi_{IJ}) \circ \pi_I = \pi_J \circ (\phi_{IJ} \otimes_{RG} RGe_0). \qquad (*)$$

We may assume that $I \subseteq J \subseteq I_0$, since otherwise $\mathcal{D}(P_0)_J = 0$. Set $f = [U_J]$. Then $e = ef = fe$ and $f = fe_0 = e_0f$. Take $x \in RG$, $y, z \in G$ and consider the element

$$xe \otimes_{RP_I} ey \otimes_{RG} ze_0 \in \mathcal{D}_I \otimes_{RG} RGe_0 = RGe \otimes_{RP_I} e.RG \otimes_{RG} RGe_0.$$

If $yz \in P_0$, then the effects of the two compositions of maps $(*)$ on this element are respectively

$$xe \otimes_{RP_I} ey \otimes_{RG} ze_0 \mapsto xe \otimes_{RP_0} e \otimes_{RP_I} eyze_0 \mapsto xe \otimes_{RP_0} e \otimes_{RP_J} eyze_0$$

and

$$\begin{aligned} xe \otimes_{RP_I} ey \otimes_{RG} ze_0 &\mapsto xe \otimes_{RP_J} ey \otimes_{RG} ze_0 = xef \otimes_{RP_J} fey \otimes_{RG} ze_0 \\ &\mapsto xef \otimes_{RP_0} f \otimes_{RP_J} feyze_0 = xe \otimes_{RP_0} f \otimes_{RP_J} eyze_0, \end{aligned}$$

since $eyz \in RP_0$. But $e \in RP_J$, so $xe \otimes_{RP_0} f \otimes_{RP_J} eyze_0 = xe \otimes_{RP_0} e \otimes_{RP_J} eyze_0$.

If $yz \notin P_0$, then we get

$$xe \otimes_{RP_I} ey \otimes_{RG} ze_0 \mapsto 0 \mapsto 0$$

and

$$xe \otimes_{RP_I} ey \otimes_{RG} ze_0 \mapsto xe \otimes_{RP_J} ey \otimes_{RG} ze_0 = xef \otimes_{RP_J} fy \otimes_{RG} ze_0 \mapsto 0.$$

Thus the two compositions are indeed equal. □

4. The main theorem over a field of coefficients

As in the previous section, we shall take $\mathcal{O}$ to be a field k of characteristic different from p.

Since X_k is a complex of finitely generated kG-kG-bimodules that are projective as left and as right modules, and since kG is a symmetric algebra, it follows (see, for example, [9], Corollary 9.2.5) that the k-linear dual $X_k^\vee$ of X_k induces a functor $X_k^\vee \otimes_{kG} -$ that is right adjoint (and also left adjoint) to $X_k \otimes_{kG} -$. This is true whether we regard these as endofunctors of the category $C^b(\mathrm{Mod}(kG))$ of chain complexes, the chain homotopy category $K^b(\mathrm{Mod}(kG))$, or the derived category $D^b(\mathrm{Mod}(kG))$. Hence we get a natural map of complexes of kG-kG-bimodules

$$\epsilon : X_k \otimes_{kG} X_k^\vee \longrightarrow kG$$

that induces the counit of this adjunction.

We shall show that ϵ is an isomorphism in the derived category of kG-kG-bimodules.

The degree $|S|$ term of X_k is kG, and so the degree zero term of $X_k \otimes_{kG} X_k^\vee$ has a summand isomorphic to kG that is mapped isomorphically by ϵ to kG. Thus we have a short exact sequence

$$0 \longrightarrow K_k \longrightarrow X_k \otimes_{kG} X_k^\vee \stackrel{\epsilon}{\longrightarrow} kG \longrightarrow 0$$

of complexes that is split in each degree, where K_k is just the kernel of ϵ.

Lemma 4.1. *If E is a kG-module Harish-Chandra induced from a finite-dimensional cuspidal module for some Levi subgroup L_I, then $K_k \otimes_{kG} E$ is chain contractible.*

Proof. By standard properties of adjoint functors, the map

$$\epsilon \otimes_{kG} X_k : X_k \otimes_{kG} X_k^\vee \otimes_{kG} X_k \longrightarrow X_k$$

is split surjective, a splitting map being given by $X_k \otimes_{kG} \eta$, where

$$\eta : kG \longrightarrow X_k^\vee \otimes_{kG} X_k$$

is the natural map inducing the unit of the adjunction. Thus

$$\epsilon \otimes_{kG} X_k \otimes_{kG} E[|I|] : X_k \otimes_{kG} X_k^\vee \otimes_{kG} X_k \otimes E[|I|] \longrightarrow X_k \otimes_{kG} E[|I|]$$

is a split surjection between objects of $K^b(\mathrm{Mod}(kG))$ that, by Theorem 3.1, are both isomorphic to E. By the Krull–Schmidt theorem, the map must be an isomorphism. Since $X_k \otimes_{kG} E[|I|] \cong E$, the map $\epsilon \otimes_{kG} E$ is also an isomorphism in $K^b(\mathrm{Mod}(kG))$, and hence $K_k \otimes_{kG} E$ is chain contractible. □

We shall also need the following lemma.

Lemma 4.2. *Let K be a bounded complex of projective right A-modules, where A is a finite dimensional algebra over a field. Let $\mathcal{E}$ be a set of (left) A-modules such that* *for every irreducible A-module N there is some $E \in \mathcal{E}$ with $\mathrm{Hom}_A(E, N) \neq 0$. Then K is acyclic if and only if $K \otimes_A E$ is acyclic for every $E \in \mathcal{E}$.*

Proof. Suppose $K \cong K \otimes_A A$ is not acyclic. Let i_0 be the minimal i such that, for all finitely generated left A-modules E, $H^{i+1}(K \otimes_A E) = 0$.

By the projectivity of K, any short exact sequence $0 \to E_3 \to E_2 \to E_1 \to 0$ of modules gives rise to an exact sequence of complexes $0 \to K \otimes_A E_3 \to K \otimes_A E_2 \to K \otimes_A E_1 \to 0$ and then, by the long exact sequence of cohomology, to an exact sequence

$$H^{i_0}(K \otimes_A E_3) \to H^{i_0}(K \otimes_A E_2) \to H^{i_0}(K \otimes_A E_1) \to H^{i_0+1}(K \otimes_A E_3) = 0.$$

If $H^{i_0}(K \otimes_A E_2) \neq 0$, then either $H^{i_0}(K \otimes_A E_1) \neq 0$ or $H^{i_0}(K \otimes_A E_3) \neq 0$. There must therefore be an irreducible A-module N such that $H^{i_0}(K \otimes_A N) \neq 0$. By assumption, we can choose the short exact sequence with $E_1 = N$ and $E_2 \in \mathcal{E}$. Then $H^{i_0}(K \otimes_A E_2) = 0$ by the hypothesis, and the above exact sequence gives $H^{i_0}(K \otimes_A N) = 0$, a contradiction. □

Theorem 4.3. *With the notation above, ϵ and η are isomorphisms in the derived category of kG-kG-bimodules, and hence*

$$X_k \otimes_{kG} - : D^b(\mathrm{Mod}(kG)) \longrightarrow D^b(\mathrm{Mod}(kG))$$

is an equivalence of categories.

Proof. We shall prove that ϵ is an isomorphism by showing that K_k is acyclic; the proof for η is similar, with the cokernel of η playing the role of K_k.

Let N be an irreducible kG-module, and choose $I \subseteq S$ minimal subject to ${}^*\mathrm{R}_{L_I}^G(N) \not\cong 0$ (i.e., $N^{U_I} \neq 0$), so that ${}^*\mathrm{R}_{L_I}^G(N)$ is a cuspidal kL_I-module. By adjointness, the (non-zero) identity map of ${}^*\mathrm{R}_{L_I}^G(N)$ corresponds to a non-zero map $\mathrm{R}_{L_I}^G {}^*\mathrm{R}_{L_I}^G(N) \to N$. So by Lemma 4.1, the conditions of Lemma 4.2 are satisfied for $\mathcal{E}$ the set of kG-modules that are Harish-Chandra induced from finite-dimensional cuspidal modules. Hence the lemma implies that K_k is acyclic.

Now $X_k \otimes_{kG} -$ is an equivalence of categories because we have shown that X_k is a 'two-sided tilting complex' ([9], Definition 6.2.1). To be explicit, a quasi-inverse is given by $X_k^\vee \otimes_{kG} -$, since ϵ and η induce isomorphisms of functors

$$X_k \otimes_{kG} X_k^\vee \otimes_{kG} - \cong kG \otimes_{kG} - \cong \mathrm{id}$$

and

$$\mathrm{id} \cong kG \otimes_{kG} - \cong X_k^\vee \otimes_{kG} X_k \otimes_{kG} -. \qquad \square$$

In fact, more should be true. In [10] the notion of a 'splendid' equivalence is introduced, being an equivalence of derived categories of group algebras or blocks that has various especially good properties. The only aspect of the notion that is not immediate for the equivalence that we have just described is that the chain maps ϵ and

η that we have shown to be isomorphisms in the derived category are actually chain homotopy equivalences.

Conjecture 4.4. *With the notation above, ϵ and η are isomorphisms in the chain homotopy category of complexes of kG-kG-bimodules, and hence*

$$X_k \otimes_{kG} - : K^b(\mathrm{Mod}(kG)) \longrightarrow K^b(\mathrm{Mod}(kG))$$

is an equivalence of categories.

To prove this conjecture, one must show that K_k is contractible, and not merely acyclic.

5. General coefficient rings

We shall now generalize the results of the previous section to a general commutative coefficient ring $\mathcal{O}$ in which p is invertible. Exactly as when we worked over a field k, there is a short exact sequence

$$0 \longrightarrow K_{\mathcal{O}} \longrightarrow X_{\mathcal{O}} \otimes_{\mathcal{O}G} X_{\mathcal{O}}^{\vee} \stackrel{\epsilon}{\longrightarrow} \mathcal{O}G \longrightarrow 0$$

of complexes of $\mathcal{O}G$-$\mathcal{O}G$-bimodules, split in each degree.

Theorem 5.1. *The functor*

$$X_{\mathcal{O}} \otimes_{\mathcal{O}G} - : D^b(\mathrm{Mod}(\mathcal{O}G)) \longrightarrow D^b(\mathrm{Mod}(\mathcal{O}G))$$

is an equivalence of categories.

Proof. As in the proof of Theorem 4.3, we need to show that $K_{\mathcal{O}}$ is acyclic. It is sufficient to do this for the case $\mathcal{O} = R = \mathbb{Z}[1/p]$, since K_R is a bounded complex of free R-modules, and so if it is acyclic then it is contractible as a complex of R-modules, and so, for a general $\mathcal{O}$,

$$K_{\mathcal{O}} \cong K_R \otimes_R \mathcal{O}$$

is contractible as a complex of $\mathcal{O}$-modules, and therefore acyclic.

Since K_R is a complex of finitely generated free R-modules, its homology groups are finitely generated abelian groups with no p-torsion, and since by Theorem 4.3

$$K_{\mathbb{Z}/l\mathbb{Z}} \cong K_R \otimes_{\mathbb{Z}} \mathbb{Z}/l\mathbb{Z}$$

is acyclic for every prime $l \neq p$, it follows from the Universal Coefficient Theorem that K_R is acyclic. □

As in Conjecture 4.4, it should be possible to refine this result.

Conjecture 5.2. *The functor*

$$X_{\mathcal{O}} \otimes_{\mathcal{O}G} - : K^b(\mathrm{Mod}(\mathcal{O}G)) \longrightarrow K^b(\mathrm{Mod}(\mathcal{O}G))$$

is an equivalence of categories.

In fact, this can be reduced to the previous conjecture.

Proposition 5.3. *If Conjecture 4.4 is true, then so is Conjecture 5.2.*

Proof. Suppose that K_k is contractible as a complex of kG-kG-bimodules for every field k with characteristic different from p. We wish to show that $K_{\mathcal{O}}$ is contractible for any coefficient ring $\mathcal{O}$ in which p is invertible. Since $K_{\mathcal{O}}$ is a bounded complex of permutation $\mathcal{O}[G \times G]$-modules, it suffices to prove that the fixed point complex $K_{\mathcal{O}}^H$ is acyclic for every subgroup $H \leq G \times G$ ([10], Lemma 1.1).

Since change of scalars commutes with fixed point functors for permutation modules, it suffices, as in the proof of Theorem 5.1, to do this for $\mathcal{O} = R = \mathbb{Z}[1/p]$, where the same Universal Coefficient Theorem argument works, since

$$K_R^H \otimes_{\mathbb{Z}} \mathbb{Z}/l\mathbb{Z} \cong K_{\mathbb{Z}/l\mathbb{Z}}^H$$

is acyclic by assumption. □

6. A theorem of Curtis type

In this section we check that the equivalence defined above commutes with Harish-Chandra induction.

Let us first recall that, by the remarks following the definition of $X_{\mathcal{O}}$ in Section 2.3, the complex $X_{\mathcal{O}}$ is defined intrinsically in terms of the abstract group G.

When L is a Levi subgroup of G, let us denote by $X_{\mathcal{O}}^{(L)}$ the corresponding complex for L.

The character-theoretic analogue of the following theorem is well-known (see [7], Theorem 8.11, for example).

Theorem 6.1. *Let P be a parabolic subgroup of G, with Levi decomposition $P = LU_P$. Let $\mathcal{O}$ be any commutative ring in which p is invertible. Let R_L^G denote the $\mathcal{O}G$-$\mathcal{O}L$-bimodule $\mathcal{O}G[U_P]$ inducing Harish-Chandra induction of $\mathcal{O}L$-modules. Then*

$$X_{\mathcal{O}}^{(G)} \otimes_{\mathcal{O}G} \mathrm{R}_L^G \cong \mathrm{R}_L^G \otimes_{\mathcal{O}L} X_{\mathcal{O}}^{(L)}$$

in $D^b(\mathrm{Mod}(\mathcal{O}G \otimes \mathcal{O}L^{\mathrm{op}}))$.

Proof. We may choose P containing B, that is $P = P_{I_0}$ for some $I_0 \subseteq S$. Let $e = [U_P]$.

It is sufficient to take $\mathcal{O} = R = \mathbb{Z}[1/p]$, so we need to show that

$$X_R^{(G)} \otimes_{RG} RGe \cong RGe \otimes_{RL} X_R^{(L)}$$

in the derived category $D^b(\mathrm{Mod}(RG \otimes RL^{\mathrm{op}}))$.

The map of Proposition 3.5 gives a surjective chain map

$$X_R^{(G)} \otimes_{RG} RGe \to RGe \otimes_{RL} X_R^{(L)} \to 0.$$

(We may replace $P = P_{I_0}$ by L in the tensor products of Proposition 3.5 since U_P always acts trivially there.) Let K'_R be the kernel of this map. Considered as a map of complexes of RL^{op}-modules, this surjection is split in each degree, since the modules are all projective right RL-modules, so as well as an exact sequence

$$0 \to K'_R \to X_R^{(G)} \otimes_{RG} RGe \to RGe \otimes_{RL} X_R^{(L)} \to 0$$

in $C^b(\mathrm{Mod}(RL^{\mathrm{op}}))$, we have exact sequences

$$0 \to K'_k \otimes_{kL} E \to X_k^{(G)} \otimes_{kG} kGe \otimes_{kL} E \to kGe \otimes_{kL} X_k^{(L)} \otimes_{kL} E \to 0$$

in $C^b(\mathrm{Mod}(kG))$ for any kL-module E, where k is a field of characteristic different from p and $K'_k = K'_R \otimes_R k$.

By Lemma 4.2, to show that K'_k is acyclic it is sufficient to check that $K'_k \otimes E$ is acyclic for any kL-module of the form $E = \mathrm{Ind}_{P_I}^{P}(V)$ for $I \subseteq I_0$ and V a U_I-trivial kP_I-module such that $V^{U_J} = 0$ for every J properly contained in I. Set $\widetilde{E} := kGe \otimes_{kL} E = kG \otimes_{kP} E = \mathrm{Ind}_{P_I}^{G}(V)$. Now Theorem 3.1 implies that

$$X_k^{(G)} \otimes_{kG} kGe \otimes_{kL} E \cong X_k^{(G)} \otimes_{kG} \widetilde{E} \cong \widetilde{E}[-|I|]$$

in $D^b(\mathrm{Mod}(kG))$, and $X_k^{(L)} \otimes_{kL} E \cong E[-|I|]$ in $D^b(\mathrm{Mod}(kL))$. Since kGe is projective, and hence flat, as a right module, this implies that

$$kGe \otimes_{kL} X_k^{(L)} \otimes_{kL} E \cong kGe \otimes_{kL} E[-|I|] = \widetilde{E}[-|I|]$$

in $D^b(\mathrm{Mod}(kG))$. Just as in the proof of Theorem 4.3, this implies that $K'_k \otimes E$ is acyclic.

This shows that K'_k is acyclic for any field k of characteristic different from p. Using the Universal Coefficient Theorem as in the proof of Theorem 5.1, we deduce that K'_R is acyclic. Thus our claim. □

By Lemma 2.1, the corresponding result for Harish-Chandra truncation is also true. As in Conjectures 4.4 and 5.2, it should be possible to refine Theorem 6.1 to the homotopy category.

Conjecture 6.2. *There is an isomorphism*

$$X_{\mathcal{O}}^{(G)} \otimes_{\mathcal{O}G} \mathrm{R}_L^G \cong \mathrm{R}_L^G \otimes_{\mathcal{O}L} X_{\mathcal{O}}^{(L)}$$

in $K^b(\mathrm{Mod}(\mathcal{O}G \otimes \mathcal{O}L^{\mathrm{op}}))$.

7. Concluding remarks

7.1. Contravariant duality

The equivalences of Theorems 4.3 and 5.1 are covariant functors, and so are probably not worthy of the name 'duality', even though they induce an involution on generalized characters. However, it follows from what we have proved and Section 4 of Yekutieli's paper [12] on dualizing complexes that, if the coefficient ring $\mathcal{O}$ is noetherian, the functors

$$\boldsymbol{R}\mathrm{Hom}_{\mathcal{O}G}(-, X_{\mathcal{O}}) : D^b(\mathrm{mod}(\mathcal{O}G))^{\mathrm{op}} \longrightarrow D^b(\mathrm{mod}(\mathcal{O}G^{\mathrm{op}}))$$

and

$$\boldsymbol{R}\mathrm{Hom}_{\mathcal{O}G}(-, X_{\mathcal{O}}) : D^b(\mathrm{mod}(\mathcal{O}G^{\mathrm{op}}))^{\mathrm{op}} \longrightarrow D^b(\mathrm{mod}(\mathcal{O}G))$$

are mutually inverse contravariant equivalences. [Actually, Yekutieli works with a coefficient field k, but his proofs of the results that are relevant here generalize trivially to a general commutative noetherian coefficient ring.]

Identifying $\mathcal{O}G$-modules and $\mathcal{O}G^{\mathrm{op}}$-modules in the usual way, the two functors above may be regarded as contravariant functors from $D^b(\mathrm{mod}(\mathcal{O}G))$ to itself. Lemma 2.1 implies that these two functors are isomorphic, and we shall use the notation

$$D_G : D^b(\mathrm{mod}(\mathcal{O}G))^{\mathrm{op}} \longrightarrow D^b(\mathrm{mod}(\mathcal{O}G))$$

for either of them.

By the remarks above, we have the following.

Theorem 7.1. *Let $\mathcal{O}$ be a commutative noetherian ring in which p is invertible. The functor D_G is an equivalence of triangulated categories with $D_G \circ D_G$ isomorphic to the identity functor.*

There is also a version of Theorem 6.1 for this contravariant duality. We shall use D_L to denote the contravariant duality functor for $D^b(\mathrm{mod}(\mathcal{O}L))$.

Theorem 7.2. *Let P be a parabolic subgroup of G, with Levi decomposition $P = LU_P$. Let $\mathcal{O}$ be a commutative noetherian ring in which p is invertible. Then*

$$D_G \circ \mathrm{R}_L^G \cong \mathrm{R}_L^G \circ D_L$$

as contravariant functors from $D^b(\mathrm{mod}(\mathcal{O}L))$ to $D^b(\mathrm{mod}(\mathcal{O}G))$.

Proof. To calculate the effect of these functors on an object of $D^b(\mathrm{mod}(\mathcal{O}L))$, we apply the functors $\mathrm{Hom}_{\mathcal{O}G}(-\otimes_{\mathcal{O}L}[U_P].\mathcal{O}G, X_{\mathcal{O}}^{(G)})$ and $\mathcal{O}G[U_P]\otimes_{\mathcal{O}L}\mathrm{Hom}_{\mathcal{O}L}(-, X_{\mathcal{O}}^{(L)})$ to a projective resolution P_* of the object, regarded as a *right* $\mathcal{O}L$-module. Since $\mathcal{O}$

is noetherian, P_* is a complex of finitely generated projective right $\mathcal{O}L$-modules, and so

$$\mathrm{Hom}_{\mathcal{O}G}(P_* \otimes_{\mathcal{O}L} [U_P].\mathcal{O}G, X_{\mathcal{O}}^{(G)}) \cong X_{\mathcal{O}}^{(G)} \otimes_{\mathcal{O}G} \mathcal{O}G[U_P] \otimes_{\mathcal{O}L} P_*^{\vee}, \qquad \text{(i)}$$

and

$$\mathcal{O}G[U_P] \otimes_{\mathcal{O}L} \mathrm{Hom}_{\mathcal{O}L}(P_*, X_{\mathcal{O}}^{(L)}) \cong \mathcal{O}G[U_P] \otimes_{\mathcal{O}L} X_{\mathcal{O}}^{(L)} \otimes_{\mathcal{O}L} P_*^{\vee}, \qquad \text{(ii)}$$

where $P_*^{\vee}$ is the $\mathcal{O}$-dual of P_*, regarded as a complex of projective *left* $\mathcal{O}L$-modules.

As shown in the proof of Theorem 6.1, there is a natural chain map

$$X_{\mathcal{O}}^{(G)} \otimes_{\mathcal{O}G} \mathcal{O}G[U_P] \to \mathcal{O}G[U_P] \otimes_{\mathcal{O}L} X_{\mathcal{O}}^{(L)}$$

that is a split epimorphism in each degree and induces an isomorphism in homology. Taking the tensor product with $P_*^{\vee}$, we get a quasi-isomorphism (i.e., an isomorphism in the derived category) between (i) and (ii).

Hence $D_G \circ \mathrm{R}_L^G \cong \mathrm{R}_L^G \circ D_L$ as functors taking values in $D^b(\mathrm{mod}(\mathcal{O}G))$. □

There is of course a corresponding statement about Harish-Chandra truncation ${}^*\mathrm{R}_L^G$.

7.2. The Steinberg module

For a field k, usually of characteristic p or zero, the Steinberg module St_k for kG is defined as the reduced simplicial homology, with coefficients in k, of the building of G ([4], Definition 66.34). Since, by the Solomon–Tits theorem, the building of G has the homotopy type of a bouquet of $(n-1)$-spheres (where $n = |S|$), this reduced homology is concentrated in the single degree $n-1$ ([4], Theorem 66.33). It is consistent to define the Steinberg module $\mathrm{St}_{\mathcal{O}}$ for $\mathcal{O}G$ similarly for a general coefficient ring $\mathcal{O}$.

Applying our functor $X_{\mathcal{O}} \otimes_{\mathcal{O}G} -$ to the trivial module $\mathcal{O}$ amounts, up to a suitable re-indexing, to calculating the reduced homology of the building of G with coefficients in $\mathcal{O}$, so $X_{\mathcal{O}} \otimes_{\mathcal{O}G} \mathcal{O}$ is isomorphic in $D^b(\mathrm{Mod}(\mathcal{O}G))$ to $\mathrm{St}_{\mathcal{O}}$, the complex consisting of the Steinberg module concentrated in degree zero. Since, by Theorem 5.1, $X_{\mathcal{O}} \otimes_{\mathcal{O}G} -$ is an equivalence of derived categories, the following theorem follows immediately.

Theorem 7.3. *There is an isomorphism*

$$\mathrm{Ext}^*_{\mathcal{O}G}(St_{\mathcal{O}}, St_{\mathcal{O}}) \cong H^*(G, \mathcal{O})$$

of graded rings for any commutative ring $\mathcal{O}$ in which p is invertible.

Of course, this is far from true if $\mathcal{O}$ is a field of characteristic p, since then $\mathrm{St}_{\mathcal{O}}$ is a projective $\mathcal{O}G$-module, and so $\mathrm{Ext}^*_{\mathcal{O}G}(\mathrm{St}_{\mathcal{O}}, \mathrm{St}_{\mathcal{O}})$ is zero in non-zero degrees, which is certainly not the case for $H^*(G, \mathcal{O})$.

References

[1] D. J. Benson, Representations and cohomology II, Cambridge Stud. Adv. Math. 31, Cambridge University Press, 1991.

[2] M. Broué, Blocs, isométries parfaites, catégories dérivées, C. R. Acad. Sci. Paris 307–1 (1988), 13–18.

[3] C. W. Curtis, Homology representations of finite groups, Proc. ICRA II (Ottawa, 1979), Lecture Notes in Math. 832, Springer-Verlag, Berlin 1980, 177–194.

[4] C. W. Curtis and I. Reiner, Methods of representation theory with applications to finite groups and orders (Volume II), Wiley-Interscience, 1987.

[5] P. Deligne and G. Lusztig, Duality for representations of a reductive group over a finite field, J. Algebra 74 (1982), 284–291.

[6] R. Dipper and J. Du, Harish-Chandra vertices, J. Reine Angew. Math. 437 (1993), 101–130.

[7] F. Digne and J. Michel, Representations of finite groups of Lie type, London Math. Soc. Stud. Texts 21, Cambridge University Press, 1991.

[8] R. B. Howlett and G. I. Lehrer, On Harish-Chandra induction and restriction for modules of Levi subgroups, J. Algebra 165 (1994), 172–183.

[9] S. König and A. Zimmermann, Derived equivalences for group rings (with contributions by B. Keller, M. Linckelmann, J. Rickard and R. Rouquier), Lecture Notes in Math. 1685, Springer-Verlag, Berlin 1998.

[10] J. Rickard, Splendid equivalences: derived categories and permutation modules, Proc. London Math. Soc. (3) 72 (1996), 331–358.

[11] C. A. Weibel, An introduction to homological algebra, Cambridge Stud. Adv. Math. 38, Cambridge University Press 1994.

[12] A. Yekutieli, Dualizing complexes, Morita equivalence and the derived Picard group of a ring, J. London Math. Soc. (2) 60 (1999), 723–746.

UFR de Mathématiques, Université Paris 7, 2 Place Jussieu, 75251 Paris, France

E-mail: `cabanes@math.jussieu.fr`

`http://www.math.jussieu.fr/~cabanes`

School of Mathematics, University of Bristol, University Walk, Bristol BS8 1TW, England

E-mail: `j.rickard@bristol.ac.uk`

`http://www.maths.bris.ac.uk/~pure/staff/majcr`

Relating the cohomology of general linear groups and symmetric groups

*Stephen R. Doty and Daniel K. Nakano**

1. Introduction

1.1. In their pioneering work Frobenius and Schur demonstrated that there is an intimate connection between the representation theory for the general linear group and complex representations of the symmetric groups Σ_d. During the last twenty years, much effort has been made in relating the modular representation theory of these two groups. For a detailed surveys of results in this area, we refer the reader to [Gr] and [Mar]. Recently, the authors with K. Erdmann discovered a method for relating the cohomology theories of the general linear group and the symmetric group over fields with non-zero characteristic. This approach is based on constructing a Grothendieck spectral sequence with E_2-page involving extensions of modules over the general linear group and converging to extensions of modules over the symmetric group. The construction of this spectral sequence along with the basic results is given in [DEN].

The purpose of this paper is to demonstrate how this new approach in relating the extensions for these two groups can be used to obtain previously known results in the area. Many representation theoretic problems dealing with these groups can be phrased in the language of homological algebra. With an appropriate translation our methods involving the spectral sequence become useful in solving these problems in a systematic way.

1.2. The paper is organized as follows. In Section 2, elementary properties of the Schur functor are reviewed. Furthermore, several of the main results of [DEN] which will be used in this paper are stated. The following section is devoted to showing how the spectral sequence can be used to derive representation theoretic results about the decomposition numbers and blocks for the Schur algebras $S(n, d)$. Old results of James and Donkin are proved by using this new approach. The results in Sections 4

*Research of the second author was supported in part by NSF grant 9800960

and 5 demonstrate the computational power of our approach. We first compute $R^i\mathcal{G}(\mathrm{sgn})$ where $i = 1, 2$. Here $R^i\mathcal{G}(-)$ denotes the ith higher right derived functor of a right adjoint, $\mathcal{G}$, to the Schur functor. This explicit computation allows us to recover the main results of Burichenko, Kleshchev, and Martin [BKM]. These results entail computing the first and second cohomology of Σ_d with coefficients in a dual Specht module. Finally, in the last section, results pertaining to the realization of tilting modules for $\mathrm{GL}_n(k)$ as images of the functor $\mathcal{G}$ are used to gain insight about the structure of tilting modules.

Acknowledgements. This work was completed during the conference on modular representation theory held at the University of Virginia in May 1998. Both authors thank Michael Collins, Brian Parshall and Leonard Scott for their efforts in organizing a productive meeting. The authors also acknowledge Stephen Donkin and Brian Parshall for several useful discussions involving the realization of tilting modules.

2. The Schur functor

2.1. Notation. Let k be an algebraically closed field of characteristic $p > 0$. Let A be a finite-dimensional algebra over k and let $\mathrm{Mod}(A)$ be the category of left modules over A. The subcategory of finite-dimensional modules over A will be denoted by $\mathrm{mod}(A)$. For $M \in \mathrm{mod}(A)$ and S a simple A-module let $[M : S]$ be the multiplicity of S as a composition factor of M.

Let $\mathrm{GL}_n(k)$ be the general linear group over k and $k\Sigma_d$ be the group algebra of the symmetric group on d letters Σ_d. The Schur algebra $S(n, d)$ is the finite-dimensional associative k-algebra $\mathrm{End}_{k\Sigma_d}(V^{\otimes d})$ where V is the natural representation of $\mathrm{GL}_n(k)$. The category of homogeneous polynomial $\mathrm{GL}_n(k)$-modules of a fixed degree $d \geq 0$ is equivalent to the category of modules for $S(n, d)$. The basic references for this section are [Gr] and [Mar].

Throughout the paper we assume that $n \geq d$. Let $e = \zeta_{(1,2,\ldots,d)(1,2,\ldots,d)}$ be the idempotent in $S(n, d)$ described in [Gr, (6.1)]. Then $eS(n, d)e \cong k\Sigma_d$ for $n \geq d$. The Schur functor $\mathcal{F}$ is the covariant exact functor from $\mathrm{Mod}(S(n, d))$ to $\mathrm{Mod}(k\Sigma_d)$ defined by $\mathcal{F}(M) = eM$.

The simple $S(n, d)$-modules are in one-to-one correspondence with the set of partitions of d. We will denote this set by $\Lambda := \Lambda^+(n, d)$ and the corresponding simple $S(n, d)$-module by $L(\lambda)$ for $\lambda \in \Lambda$. The set Λ can be identified as the set of dominant polynomial weights of $\mathrm{GL}_n(k)$ of degree d. The standard dominance order on partitions will be denoted by $\rhd$ [Mar, Definition 1.4.2]. For $\lambda \in \Lambda$, set $H^0(\lambda) = \mathrm{ind}_B^G \lambda$ where B is the lower triangular Borel subgroup of $\mathrm{GL}_n(k)$ and let $V(\lambda)$ be the Weyl module. Furthermore, if $\lambda \in \Lambda$, let $I(\lambda)$ be the injective hull of $L(\lambda)$, $P(\lambda)$ be the projective cover of $L(\lambda)$ and $T(\lambda)$ be the corresponding tilting module. There exists a duality on $\mathrm{mod}(S(n, d))$ which fixes simple modules called the *transposed dual*. This duality will be denoted by τ.

Set $D_\lambda = eL(\lambda)$. Either $D_\lambda = (0)$ or D_λ is a simple $k\Sigma_d$-module. A partition $(\lambda_1, \lambda_2, \dots)$ is called *p-restricted* if $\lambda_i - \lambda_{i+1} \leq p-1$ for all i. A complete set of non-isomorphic simple $k\Sigma_d$-modules is given by $\{D_\lambda \mid \lambda \in \Lambda \text{ is } p\text{-restricted}\}$. Therefore, the simple $k\Sigma_d$-modules will be labelled by the p-restricted partitions $\lambda \in \Lambda$. Let Λ_{reg} denote the set of p-restricted partitions, and Λ_{sing} be the set of partitions λ such that $eL(\lambda) = 0$. We will often say that $L(\lambda)$ is regular if $\lambda \in \Lambda_{\text{reg}}$ and singular otherwise. The Specht module corresponding to a partition $\lambda \in \Lambda$ over $k\Sigma_d$ is denoted by S^λ. For $\lambda = (\lambda_1, \lambda_2, \dots, \lambda_d) \in \Lambda$, let $M^\lambda = \text{ind}_{k\Sigma^\lambda}^{k\Sigma_d} k$ be the permutation module obtained from inducing up the trivial module k from the Young subgroup $\Sigma^\lambda = \Sigma_{\lambda_1} \times \Sigma_{\lambda_2} \times \cdots \times \Sigma_{\lambda_d}$. The Young module Y^λ is the unique indecomposable summand of M^λ which contains S^λ as a submodule [Mar, Definition 4.6.1].

Let sgn be the one-dimensional sign representation of $k\Sigma_d$. Also, if λ is a partition of d, let λ' be the dual partition. The usual linear dual on $\text{mod}(k\Sigma_d)$ will be denoted by $*$. One has following correspondences between $S(n,d)$-modules and $k\Sigma_d$-modules under $\mathcal{F}$ (see [Gr, § 6], [Mar, Lemma 4.6.2] and [Don3, 3.6]):

$$\begin{aligned}
\mathcal{F}(H^0(\lambda)) &= S^\lambda; \\
\mathcal{F}(V(\lambda)) &= (S^\lambda)^*; \\
\mathcal{F}(I(\lambda)) &= \mathcal{F}(P(\lambda)) = Y^\lambda; \\
\mathcal{F}(T(\lambda)) &= Y^{\lambda'} \otimes \text{sgn}.
\end{aligned}$$

2.2. Spectral sequences. The Schur functor $\mathcal{F}$ can be represented as a tensor functor: $\mathcal{F}(M) \cong eM \cong eS(n,d) \otimes_{S(n,d)} M$, because $V^{\otimes d} \cong eS(n,d)$ is a projective right $S(n,d)$-module. The functor $\mathcal{F}$ admits a natural right adjoint functor $\mathcal{G}$ defined as follows:

$$\mathcal{G}(N) = \text{Hom}_{eS(n,d)e}(eS(n,d), N),$$

for $N \in \text{Mod}(eS(n,d)e) \cong \text{Mod}(k\Sigma_d)$. Furthermore, $\mathcal{F}$ is also a left inverse to $\mathcal{G}$, and $\mathcal{G}$ takes injective $k\Sigma_d$-modules to injective $S(n,d)$-modules. One can use these facts to construct a Grothendieck spectral sequence [DEN, Theorem 2.2].

Theorem A. *Let $M \in \text{Mod}(S(n,d))$ and $N \in \text{Mod}(k\Sigma_d)$ with $n \geq d$. There exists a first-quadrant spectral sequence, with E_2-page given by*

$$E_2^{i,j} = \text{Ext}^i_{S(n,d)}(M, \text{Ext}^j_{k\Sigma_d}(V^{\otimes d}, N)) \Rightarrow \text{Ext}^{i+j}_{k\Sigma_d}(eM, N).$$

Let $R^\bullet\mathcal{G}(-) := \text{Ext}^\bullet_{k\Sigma_d}(V^{\otimes d}, -)$ denote the higher right derived functors of $\mathcal{G}$. These functors are of central importance. We will show later that knowledge about $R^\bullet\mathcal{G}(N)$ as an $S(n,d)$-module can be used to directly compute cohomology for the symmetric group. The following result provides information about the composition factors of the higher right derived functors of $\mathcal{G}$ [DEN, Proposition 5.1].

Theorem B. *Let $N \in \text{mod}(k\Sigma_d)$ and $\mu \in \Lambda$.*

(i) $[R^j\mathcal{G}(N) : L(\mu)] = \dim_k \text{Ext}^j_{k\Sigma_d}(Y^\mu, N)$ *for* $j \geq 0$.

(ii) *In particular,* $e(R^j\mathcal{G}(N)) = 0$ *for* $j > 0$.

(iii) $[\mathcal{G}(D_\lambda) : L(\mu)] = \dim_k \text{Hom}_{k\Sigma_d}(Y^\mu, D_\lambda)$.

Finally, we will need some results describing the image of $\mathcal{G}$ on simple $k\Sigma_d$-modules [DEN, Theorem 2.4A, Theorem 2.5A, Corollary 3.3].

Theorem C. *Let D_λ be a simple $k\Sigma_d$-module.*

(i) *If I_λ is the injective hull of D_λ, then $\mathcal{G}(I_\lambda) = I(\lambda)$.*

(ii) *The module $\mathcal{G}(D_\lambda)$ is isomorphic to the largest submodule of $I(\lambda)$ whose socle quotient contains only composition factors of the form $L(\mu)$ where $\mu \in \Lambda_{\text{sing}}$.*

(iii) *If $\lambda, \mu \in \Lambda_{\text{reg}}$, then*

$$\text{Ext}^1_{k\Sigma_d}(D_\lambda, D_\mu) \cong \text{Ext}^1_{S(n,d)}(L(\lambda), \mathcal{G}(D_\mu)).$$

3. Decomposition numbers and blocks

3.1. James' theorem. In this section we will indicate how the spectral sequence in Theorem 2.2 A encodes information about decomposition numbers for $\text{GL}_n(k)$ and $k\Sigma_d$ by proving a well-known result of James.

Theorem. *Let $\lambda \in \Lambda$ and $\mu \in \Lambda_{\text{reg}}$. Then $[H^0(\lambda) : L(\mu)] = [S^\lambda : D_\mu]$.*

Proof. Consider the spectral sequence in Theorem 2.2 A with $M = H^0(\lambda)$ and $N = I_\mu$ where I_μ is the injective hull of D_μ. By using the fact that $E_0 = E^{0,0}$ (or adjointness) and Theorem 2.2 C, we have

$$\begin{aligned}
[H^0(\lambda) : L(\mu)] &= \dim_k \text{Hom}_{S(n,d)}(H^0(\lambda), I(\mu)) \\
&= \dim_k \text{Hom}_{S(n,d)}(H^0(\lambda), \mathcal{G}(I_\mu)) \\
&= \dim_k \text{Hom}_{k\Sigma_d}(S^\lambda, I_\mu) \\
&= [S^\lambda : D_\mu].
\end{aligned}$$

□

3.2. Blocks for $S(n,d)$, $n \geq d$. Let $\sim$ be the equivalence relation on $\Lambda = \Lambda^+(n,d)$ given by $\lambda = (\lambda_1, \lambda_2, \ldots, \lambda_d) \sim \mu = (\mu_1, \mu_2, \ldots, \mu_d)$ if and only if there exists $\sigma \in \Sigma_d$ such that $\lambda_i - i = \mu_{\sigma(i)} - \sigma_i \pmod p$ for $i = 1, 2, \ldots, d$. The following theorem was proved by Donkin in 1987 [Don2, (2.12)]. Donkin constructed a "double Schur functor" to prove this theorem. The proof presented here employs Theorem 2.2 C, the Nakayama Rule and the Linkage Principle.

Theorem. *Let $n \geq d$ and $\lambda, \mu \in \Lambda$. Then the simple modules $L(\lambda)$ and $L(\mu)$ are in the same block of $S(n, d)$ if and only if $\lambda \sim \mu$.*

Proof. Let $L(\lambda)$ and $L(\mu)$ be in the same block of $S(n, d)$. Then there exist $\lambda_0, \lambda_1, \ldots, \lambda_t \in \Lambda$ such that $\lambda = \lambda_0$, $\mu = \lambda_n$ with $\mathrm{Ext}^1_{S(n,d)}(L(\lambda_i), L(\lambda_{i+1})) \neq 0$. By the Linkage Principle, $\lambda_i \sim \lambda_{i+1}$ for all i; thus $\lambda \sim \mu$.

On the other hand, suppose that $\lambda \sim \mu$ and consider the Young modules $Y^\lambda = \mathcal{F}(P(\lambda))$ and $Y^\mu = \mathcal{F}(P(\mu))$. Even if one (or both) of the weights λ or μ is not regular, there exist $\hat{\lambda}$ and $\hat{\mu}$ in Λ_{reg} such that $D_{\hat{\lambda}}$ (resp. $D_{\hat{\mu}}$) is a composition factor of $Y^\lambda = \mathcal{F}(P(\lambda))$ (resp. $Y^\mu = \mathcal{F}(P(\mu))$) because the Young modules are indecomposable and non-zero modules for $k\Sigma_d$. By the exactness of the Schur functor it now follows that $L(\lambda)$ and $L(\hat{\lambda})$ (resp. $L(\mu)$ and $L(\hat{\mu})$) must be $S(n, d)$-composition factors of $I(\lambda)$ (resp. $I(\mu)$). Therefore $L(\lambda)$ (resp. $L(\mu)$) and $L(\hat{\lambda})$ (resp. $L(\hat{\mu})$) lie in the same block for $S(n, d)$. Moreover, note that $\lambda \sim \hat{\lambda}$ and $\mu \sim \hat{\mu}$ by the argument given in the preceding paragraph. Therefore it suffices to show that $L(\hat{\lambda})$ and $L(\hat{\mu})$ are in the same block of $S(n, d)$.

Since $\hat{\lambda} \sim \hat{\mu}$, it follows by the Nakayama Rule [Mar, Theorem 5.1.1] that $D_{\hat{\lambda}}$ and $D_{\hat{\mu}}$ are in the same block of $k\Sigma_d$. Thus there exist $\sigma_0, \sigma_1, \ldots, \sigma_s$ in Λ_{reg} such that $\hat{\lambda} = \sigma_0$, $\hat{\mu} = \sigma_s$ with $\mathrm{Ext}^1_{k\Sigma_d}(D_{\sigma_i}, D_{\sigma_{i+1}}) \neq 0$. According to Theorem 2.3(C),

$$0 \neq \mathrm{Ext}^1_{k\Sigma_d}(D_{\sigma_i}, D_{\sigma_{i+1}}) \cong \mathrm{Ext}^1_{S(n,d)}(L(\sigma_i), \mathcal{G}(D_{\sigma_{i+1}})).$$

We also know that $\mathrm{soc}_{S(n,d)} \mathcal{G}(D_{\sigma_{i+1}}) \cong L(\sigma_{i+1})$, so $\mathcal{G}(D_{\sigma_{i+1}})$ is indecomposable and belongs to the same block containing $L(\sigma_{i+1})$. Therefore $L(\sigma_i)$ and $L(\sigma_{i+1})$ are in the same block for all i, so $L(\hat{\lambda})$ and $L(\hat{\mu})$ are in the same block. □

4. The calculation of first cohomology groups

4.1. In this and the following section, we will take $n = d$. The main purpose of this section will be to calculate $R^1\mathcal{G}(\mathrm{sgn})$ for $p \geq 3$ as an $S(d, d)$-module. This result will then be applied in the next section to compute cohomology for duals of Specht modules over Σ_d.

Theorem. *Let $p \geq 3$.*

$$R^1\mathcal{G}(\mathrm{sgn}) \cong \begin{cases} L(3) & d = 3 \text{ and } p = 3 \\ L(4, 1^{d-4}) & d \geq 4 \text{ and } p = 3 \\ 0 & \text{otherwise.} \end{cases}$$

Proof. Let λ' be the transposed partition of λ. We have

$$\begin{aligned}[R^1\mathcal{G}(\mathrm{sgn}) : L(\lambda')] &= \dim_k \mathrm{Ext}^1_{k\Sigma_d}(Y^{\lambda'}, \mathrm{sgn}) \\ &\leq \dim_k \mathrm{Ext}^1_{k\Sigma_d}(M^{\lambda'}, \mathrm{sgn}) \\ &= \dim_k \mathrm{Ext}^1_{k\Sigma^{\lambda'}}(k, \mathrm{sgn}).\end{aligned}$$

One can express $\lambda' = (\sigma_1, \sigma_2, \ldots, \sigma_t)$. By the Kunneth formula,

$$\mathrm{H}^1(\Sigma^{\lambda'}, \mathrm{sgn}) \cong \bigoplus_{i=1}^{t} \mathrm{H}^0(\Sigma_{\sigma_1}, \mathrm{sgn}) \otimes \cdots \otimes \mathrm{H}^1(\Sigma_{\sigma_i}, \mathrm{sgn}) \otimes \cdots \otimes \mathrm{H}^0(\Sigma_{\sigma_t}, \mathrm{sgn}).$$

An easy computation shows that, for $p \geq 3$,

$$\mathrm{H}^1(\Sigma_d, \mathrm{sgn}) \cong \begin{cases} 0 & d \geq 5 \\ k & d = 3 \text{ or } 4. \end{cases}$$

This shows that $[R^1\mathcal{G}(\mathrm{sgn}) : L(\lambda')] \neq 0$ implies that $\lambda' = (3, 1^{d-3})$ or $(4, 1^{d-4})$ when $p = 3$. In these cases $[R^1\mathcal{G}(\mathrm{sgn}) : L(\lambda')] \leq 1$. Since $(3, 1^{d-3}) \in \Lambda_{\mathrm{reg}}$ for $d \geq 4$, $Y^{(3,1^{d-3})}$ is a projective $k\Sigma_d$-module for $d \geq 4$. In this case

$$[R^1\mathcal{G}(\mathrm{sgn}) : L(3, 1^{d-3})] = \dim_k \mathrm{Ext}^1_{k\Sigma_d}(Y^{(3,1^{d-3})}, \mathrm{sgn}) = 0.$$

On the other hand,

$$[R^1\mathcal{G}(\mathrm{sgn}) : L(3)] = \dim_k \mathrm{Ext}^1_{k\Sigma_3}(k, \mathrm{sgn}) = 1$$

because $Y^{(3)} = eI(3) \cong k$.

Now consider the case $\lambda' = (4, 1^{d-4})$ for $p = 3$. We have $M^{\lambda'} \cong Y^{\lambda'} \oplus N$ where N is a direct sum of Young modules $Y^{\mu'}$ with $\mu' \rhd \lambda'$. Therefore

$$\mathrm{Ext}^1_{k\Sigma_d}(M^{\lambda'}, \mathrm{sgn}) \cong \mathrm{Ext}^1_{k\Sigma_d}(Y^{\lambda'}, \mathrm{sgn}) \oplus \mathrm{Ext}^1_{k\Sigma_d}(N, \mathrm{sgn}).$$

However, $\mathrm{Ext}^1_{k\Sigma_d}(M^{\lambda'}, \mathrm{sgn}) \cong k$ and $\mathrm{Ext}^1_{k\Sigma_d}(N, \mathrm{sgn}) = 0$. Hence

$$[R^1\mathcal{G}(\mathrm{sgn}) : L(4, 1^{d-4})] = \dim_k \mathrm{Ext}^1_{k\Sigma_d}(Y^{(4,1^{d-4})}, \mathrm{sgn}) = 1. \qquad \square$$

4.2. The previous result can be used in conjunction with the spectral sequence to prove a result first obtained by Burichenko, Kleshchev and Martin [BKM, Theorem 2.4].

Corollary. *Let* $p \geq 3$.

$$\mathrm{H}^1(\Sigma_d, (S^\lambda)^*) \cong \begin{cases} k & p = 3 \text{ and } \lambda = (1^3) \text{ or } (d-3, 1^3) \\ 0 & \text{otherwise.} \end{cases}$$

Proof. Let $M = V(\lambda')$ and $N = \mathrm{sgn}$ in Theorem 2.2 A. Observe that

$$\mathrm{Ext}^{\bullet}_{k\Sigma_d}((S^{\lambda'})^*, \mathrm{sgn}) \cong \mathrm{Ext}^{\bullet}_{k\Sigma_d}(k, (S^{\lambda})^*) \cong \mathrm{H}^{\bullet}(\Sigma_d, (S^{\lambda})^*).$$

Consider the five term exact sequence $0 \to E^{1,0} \to E^1 \to E^{0,1} \to E^{2,0} \to E^2$. For $i > 0$ we have

$$E^{i,0} = \mathrm{Ext}^i_{S(d,d)}(V(\lambda'), \delta) = 0.$$

From the five term sequence one obtains the isomorphism

$$\mathrm{H}^1(\Sigma_d, (S^{\lambda})^*) \cong \mathrm{Hom}_{S(d,d)}(V(\lambda'), R^1\mathcal{G}(\mathrm{sgn})).$$

The result follows immediately from Theorem 4.1 and the fact that $V(\lambda')$ has simple head $L(\lambda')$. □

5. The calculation of second cohomology groups

5.1. We will now compute $R^2\mathcal{G}(\mathrm{sgn})$ for $p \geq 3$ as an $S(d,d)$-module. The main result of this section is the following theorem.

Theorem. *Let* $p \geq 3$.

$$R^2\mathcal{G}(\mathrm{sgn}) \cong \begin{cases} L(3) & d = 3 \text{ and } p = 3 \\ L(4) & d = 4 \text{ and } p = 3 \\ L(4,1) & d = 5 \text{ and } p = 3 \\ L(4,1^2) \oplus L(6) & d = 6 \text{ and } p = 3 \\ L(4,1^{d-4}) \oplus L(7,1^{d-7}) & d \geq 7 \text{ and } p = 3 \\ 0 & \text{otherwise.} \end{cases}$$

Proof. Once again let λ' be the transposed partition of λ. We have

$$\begin{aligned} [R^2\mathcal{G}(\mathrm{sgn}) : L(\lambda')] &= \dim_k \mathrm{Ext}^2_{k\Sigma_d}(Y^{\lambda'}, \mathrm{sgn}) \\ &\leq \dim_k \mathrm{Ext}^2_{k\Sigma_d}(M^{\lambda'}, \mathrm{sgn}) \\ &= \dim_k \mathrm{Ext}^2_{k\Sigma^{\lambda'}}(k, \mathrm{sgn}). \end{aligned}$$

By [BKM, Lemma 2.3], $\mathrm{Ext}^2_{k\Sigma^{\lambda'}}(k, \mathrm{sgn}) \cong k$ whenever $\lambda' = (3, 1^{d-3}), (4, 1^{d-4}), (6, 1^{d-6}), (7, 1^{d-7}), (3^2, 1^{d-6}), (4, 3, 1^{d-7}), (4^2, 1^{d-8})$ with $p = 3$ and 0 otherwise. One can now eliminate the cases when $Y^{\lambda'}$ is projective (i.e. if λ' is p-restricted) to conclude that $[R^2\mathcal{G}(\mathrm{sgn}) : L(\lambda')] \neq 0$ implies that $\lambda' = (3), (4, 1^{d-4}), (6, 1^{d-6}), (7, 1^{d-7}), (3^2), (4, 3), (4^2, 1^{d-8})$.

We have $M^{\lambda'} \cong Y^{\lambda'} \oplus N$ where N is a direct sum of Young modules $Y^{\mu'}$ with $\mu' \rhd \lambda'$. Therefore

$$\operatorname{Ext}^2_{k\Sigma_d}(M^{\lambda'}, \operatorname{sgn}) \cong \operatorname{Ext}^2_{k\Sigma_d}(Y^{\lambda'}, \operatorname{sgn}) \oplus \operatorname{Ext}^2_{k\Sigma_d}(N, \operatorname{sgn}).$$

For $\lambda' = (3), (4), (4, 1), (6), (7, 1^{d-7})$, $\operatorname{Ext}^2_{k\Sigma_d}(N, \operatorname{sgn}) = 0$. Therefore

$$[R^2\mathcal{G}(\operatorname{sgn}) : L(\lambda')] = \dim_k \operatorname{Ext}^2_{k\Sigma_d}(Y^{\lambda'}, \operatorname{sgn}) = 1$$

for $\lambda' = (3), (4), (4, 1), (6), (7, 1^{d-7})$.

A partial decomposition of $M^{\lambda'}$ for $\lambda' = (3^2), (4, 3), (4^2, 1^{d-8})$, and $(6, 1, 1^{d-7})$ is given as follows. One can obtain this decomposition by using the fact that the number of times $Y^{\mu'}$ appears in as a direct summand of $M^{\lambda'}$ is the dimension of the λ' weight space of $L(\mu')$ [Mar, p. 126]. The character of $L(\mu')$ can be determined for these cases by using Steinberg's tensor product theorem. We find that

$$\begin{aligned} M^{(3^2)} &\cong Y^{(3^2)} \oplus Y^{(6)} \oplus N \\ M^{(4,3)} &\cong Y^{(4,3)} \oplus Y^{(7)} \oplus N \\ M^{(4^2,1^{d-8})} &\cong Y^{(4^2,1^{d-8})} \oplus Y^{(7,1^{d-7})} \oplus N \\ M^{(6,1,1^{d-7})} &\cong Y^{(6,1,1^{d-7})} \oplus Y^{(7,1^{d-7})} \oplus N. \end{aligned}$$

Recall that $\operatorname{Ext}^2_{k\Sigma_d}(Y^{\lambda'}, \operatorname{sgn}) \cong k$ for $\lambda' = (6), (7, 1^{d-7})$ and $\operatorname{Ext}^2_{k\Sigma_d}(M^{\lambda'}, \operatorname{sgn}) \cong k$ for $\lambda' = (3^2), (4, 3), (4^2, 1^{d-8})$, and $(6, 1, 1^{d-7})$. Hence $\operatorname{Ext}^2_{k\Sigma_d}(Y^{\lambda'}, \operatorname{sgn}) = 0$ for $\lambda' = (3^2), (4, 3), (4^2, 1^{d-8})$, and $(6, 1, 1^{d-7})$.

Finally we need to show that $\operatorname{Ext}^2_{k\Sigma_d}(Y^{\lambda'}, \operatorname{sgn}) \cong k$ for $\lambda' = (4, 1^{d-4})$. One can show by using the fact given above about the decomposition of $M^{\lambda'}$ in terms of Young modules that $Y^{(7,1^{d-7})}$ is not a summand of $M^{(4,1^{d-4})}$ and $Y^{(6)}$ is not a summand of $M^{(4,1^2)}$. Since $M^{\lambda'} \cong Y^{\lambda'} \oplus N$ where N is a direct sum of Young modules $Y^{\mu'}$ with $\mu' \rhd \lambda'$, it follows that

$$[R^2\mathcal{G}(\operatorname{sgn}) : L(\lambda')] = \dim_k \operatorname{Ext}^2_{k\Sigma_d}(Y^{\lambda'}, \operatorname{sgn}) = \dim_k \operatorname{Ext}^2_{k\Sigma_d}(M^{\lambda'}, \operatorname{sgn}) = 1$$

for $\lambda' = (4, 1^{d-4})$. □

5.2. The preceding calculation can be used to recover another result by Burichenko, Kleshchev and Martin [BKM, Theorem 4.1].

Corollary. *Let $p \geq 3$.*

$$\mathrm{H}^2(\Sigma_d, (S^\lambda)^*) \cong \begin{cases} k & p = 3 \text{ and } \lambda = (1^3),\ (2,1),\ (2^2), (1^6), (d-3, 1^3) \\ & (d-3, 2, 1) \text{ or } (d-6, 1^6) \\ 0 & \text{otherwise.} \end{cases}$$

Proof. Set $M = V(\lambda')$ and $N = \text{sgn}$ in Theorem 2.2 A. Then $E^2 = \mathrm{H}^2(\Sigma_d, (S^\lambda)^*)$ and $E^{3,0} = \mathrm{Ext}^3_{S(d,d)}(V(\lambda'), \delta) = 0$. Also, note that $E^{2,0} = \mathrm{Ext}^2_{S(d,d)}(V(\lambda'), \delta) = 0$. Since the differentials d_r in the spectral sequence have bidegree $(r, 1-r)$, it immediately follows that $E^{1,1}$ is a vector space direct summand of E^2.

Consider the term $E^{2,1} = \mathrm{Ext}^2_{S(d,d)}(V(\lambda'), R^1\mathcal{G}(\text{sgn}))$. According to Theorem 4.1, $E^{2,1} = 0$ for $p > 3$. Now let $p = 3$ and, for any $\sigma \in \Lambda$, consider the short exact sequence

$$0 \to L(\sigma) \to H^0(\sigma) \to H^0(\sigma)/L(\sigma) \to 0.$$

Once can apply the long exact sequence in cohomology to conclude that

$$\mathrm{Ext}^2_{S(d,d)}(V(\lambda'), L(\sigma)) \cong \mathrm{Ext}^1_{S(d,d)}(V(\lambda'), H^0(\sigma)/L(\sigma))$$

$$\mathrm{Ext}^1_{S(d,d)}(V(\lambda'), L(\sigma)) \cong \mathrm{Hom}_{S(d,d)}(V(\lambda'), H^0(\sigma)/L(\sigma)).$$

Therefore $E^{2,1} \neq 0$ implies that $\lambda' \lhd \sigma$ where $\sigma = (3),\ (4, 1^{d-4})$. Now let us look at $E^{0,2} = \mathrm{Hom}_{S(d,d)}(V(\lambda'), R^2\mathcal{G}(\text{sgn}))$. By Theorem 5.1, and the preceding observation, if $E^{0,2} \neq 0$ then $E^{2,1} = 0$. This analysis shows that

$$\begin{aligned}\mathrm{H}^2(\Sigma_d, (S^\lambda)^*) &\cong E^{0,2} \oplus E^{1,1}\\ &\cong \mathrm{Hom}_{S(d,d)}(V(\lambda'), R^2\mathcal{G}(\text{sgn})) \oplus \mathrm{Ext}^1_{S(d,d)}(V(\lambda'), R^1\mathcal{G}(\text{sgn})).\end{aligned}$$

Next observe that $E^{1,1} = 0$ for $p > 3$. For $p = 3$, $E^{1,1} = \mathrm{Ext}^1_{S(d,d)}(V(\lambda'), L(\sigma))$ where $\sigma = (3)$ or $(4, 1^{d-4})$. The decomposition of $H^0(\sigma)$ for $\sigma = (3)$ and $(4, 1^{d-4})$ is given by

$$H^0(3):\ \begin{matrix} L(2,1)\\ L(3)\end{matrix} \qquad H^0(4):\ \begin{matrix} L(2^2)\\ L(4)\end{matrix} \qquad H^0(4,1^{d-4}):\ \begin{matrix} L(3,2,1^{d-5})\\ L(4,1^{d-4})\,.\end{matrix}$$

Therefore

$$E^{1,1} = \mathrm{Ext}^1_{S(d,d)}(V(\lambda'), R^1\mathcal{G}(\text{sgn})) \cong \begin{cases} 0 & p > 3\\ k & p = 3,\ \lambda' = (2,1),\ (2^2),\ (3,2,1^{d-5}).\end{cases}$$

The result now follows by Theorem 5.1. □

6. Tilting modules

6.1. Let $A = S(n, d)$. For an A-module M, let $S(M) = \{m \in M : eAm = 0\}$. It is easy to see that $S(M)$ is the unique largest A-submodule of M containing composition factors of the form $L(\mu)$ such that $\mu \in \Lambda_{\text{sing}}$. In this section we will first look at conditions when $\mathcal{G}(Y^\lambda) = I(\lambda)$. Our first result shows that this happens only when $\lambda \in \Lambda_{\text{reg}}$.

Proposition. *Let $\lambda \in \Lambda$ and $n \geq d$. The following statements are equivalent.*

(a) $S(I(\lambda)) = 0$.

(b) $\mathrm{soc}_{S(n,d)} I(\lambda) = L(\lambda)$ *is regular.*

(c) $\mathcal{G}(Y^\lambda) = I(\lambda)$.

Proof. (a) $\Leftrightarrow$ (b) . This follows by the definition of $S(-)$.

(b) $\Rightarrow$ (c). Suppose that $\mathrm{soc}_{S(n,d)} I(\lambda) = L(\lambda)$ is regular. Then $Y^\lambda = eI(\lambda)$ is injective, and by Theorem 2.2 C, $I(\lambda) \cong \mathcal{G}(Y^\lambda)$.

(c) $\Rightarrow$ (b). Suppose that $\mathcal{G}(Y^\lambda) \cong I(\lambda)$. Then by adjointness,

$$\begin{aligned} \mathrm{Hom}_{S(n,d)}(L(\mu), I(\lambda)) &\cong \mathrm{Hom}_{S(n,d)}(L(\mu), \mathcal{G}(Y^\lambda)) \\ &\cong \mathrm{Hom}_{k\Sigma_d}(eL(\mu), Y^\lambda). \end{aligned}$$

Therefore, $\mathrm{soc}_{S(n,d)} I(\lambda)$ is regular. □

6.2. Unlike the injective modules, the tilting modules can often be realized as the image of "twisted Young modules" under the functor $\mathcal{G}$. The following result is dual to the one proved by Cline, Parshall and Scott [CPS, Theorem 5.2.4].

Theorem. *Let $p \geq 3$. For all $\lambda \in \Lambda$ one has $\mathcal{G}(Y^{\lambda'} \otimes \mathrm{sgn}) = T(\lambda)$.*

Proof. Let $\lambda \in \Lambda$. We have $T(\lambda) \cong \mathrm{Hom}_{S(n,d)}(S(n,d), T(\lambda))$. Since $eS(n,d) \cong V^{\otimes d}$ and $eT(\lambda) \cong Y^{\lambda'} \otimes \mathrm{sgn}$, there exists a map

$$e : \mathrm{Hom}_{S(n,d)}(S(n,d), T(\lambda)) \to \mathrm{Hom}_{k\Sigma_d}(V^{\otimes d}, Y^{\lambda'} \otimes \mathrm{sgn})$$

induced by the Schur functor.

Let M and N be in $\mathrm{mod}(S(n,d))$. We have the natural vector space isomorphisms

$$\begin{aligned} f &: \mathrm{Hom}_{S(n,d)}(M, N) \to \mathrm{Hom}_{S(n,d)}(N^\tau, M^\tau) \\ g &: \mathrm{Hom}_{k\Sigma_d}(eM, eN) \to \mathrm{Hom}_{k\Sigma_d}((eN)^*, (eM)^*). \end{aligned}$$

Furthermore, by using the fact that $eM^\tau \cong (eM)^*$ and $eN^\tau \cong (eN)^*$, we have the following maps induced by the Schur functor

$$\begin{aligned} e_1 &: \mathrm{Hom}_{S(n,d)}(M, N) \to \mathrm{Hom}_{k\Sigma_d}(eM, eN) \\ e_2 &: \mathrm{Hom}_{S(n,d)}(N^\tau, M^\tau) \to \mathrm{Hom}_{k\Sigma_d}((eN)^*, (eM)^*)). \end{aligned}$$

These maps are compatible in the sense that $e_2 \circ f = g \circ e_1$.

Now set $M = S(n,d)$ and $N = T(\lambda)$. Then M^τ and N^τ admit good filtrations so, by [Mar, 4.5.6], e_2 is injective. Therefore e_1 must also be injective. Hence $e = e_1$

gives an embedding $T(\lambda) \hookrightarrow \mathcal{G}(Y^{\lambda'} \otimes \mathrm{sgn})$. Finally, we have

$$\begin{aligned}\dim_k \mathcal{G}(Y^{\lambda'} \otimes \mathrm{sgn}) &= \dim_k \mathrm{Hom}_{k\Sigma_d}(V^{\otimes d}, Y^{\lambda'} \otimes \mathrm{sgn}) \\ &= \dim_k \mathrm{Hom}_{k\Sigma_d}((Y^{\lambda'} \otimes \mathrm{sgn})^*, (V^{\otimes d})^*) \\ &= \dim_k \mathrm{Hom}_{k\Sigma_d}(Y^{\lambda'} \otimes \mathrm{sgn}, V^{\otimes d}) \\ &= \dim_k T(\lambda)\end{aligned}$$

by [CPS, Theorem 5.2.4]. Hence $T(\lambda) \cong \mathcal{G}(Y^{\lambda'} \otimes \mathrm{sgn})$ as desired. □

Corollary A. *Let $p \geq 3$ and $\lambda, \mu \in \Lambda$.*

(i) $[T(\lambda) : L(\mu)] = \dim_k \mathrm{Hom}_{k\Sigma_d}(Y^{\mu}, Y^{\lambda'} \otimes \mathrm{sgn})$.

(ii) $[\mathrm{soc}_{S(n,d)} T(\lambda) : L(\mu)] = \begin{cases} [\mathrm{soc}_{k\Sigma_d} Y^{\lambda'} \otimes \mathrm{sgn} : D_\mu] & \mu \in \Lambda_{\mathrm{reg}} \\ 0 & \mu \in \Lambda_{\mathrm{sing}}. \end{cases}$

Proof. (i) Since $p \geq 3$, $T(\lambda) \cong \mathcal{G}(Y^{\lambda'} \otimes \mathrm{sgn})$. Then, by Theorem 2.2 B, we have

$$[T(\lambda) : L(\mu)] = \dim_k \mathrm{Hom}_{k\Sigma_d}(Y^{\mu}, Y^{\lambda'} \otimes \mathrm{sgn}).$$

(ii) From adjointness, one has

$$\begin{aligned}[\mathrm{soc}_{S(n,d)} T(\lambda) : L(\mu)] &\cong \dim_k \mathrm{Hom}_{S(n,d)}(L(\mu), T(\lambda)) \\ &\cong \dim_k \mathrm{Hom}_{S(n,d)}(L(\mu), \mathcal{G}(Y^{\lambda'} \otimes \mathrm{sgn})) \\ &\cong \dim_k \mathrm{Hom}_{k\Sigma_d}(eL(\mu), Y^{\lambda'} \otimes \mathrm{sgn}).\end{aligned}$$

Observe that, if $eL(\mu) = 0$, then the last quantity is zero; otherwise $eL(\mu) = D_\mu$. □

Corollary B. *Let $p \geq 3$ and $\lambda \in \Lambda$.*

(i) $\mathrm{soc}_{S(n,d)} T(\lambda) = \mathrm{hd}_{S(n,d)} T(\lambda)$ *has only regular factors.*

(ii) $S(T(\lambda)) = 0$.

Proof. By Corollary 6.2 A, the socle of $T(\lambda)$ contains only regular composition factors. Since $T(\lambda)$ is self-dual, it follows that $\mathrm{hd}_{S(n,d)} T(\lambda)$ has only regular factors. Once again by the definition of $S(-)$, we have $S(T(\lambda)) = 0$ if and only if $\mathrm{soc}_{S(n,d)} T(\lambda)$ has composition factors with regular highest weights. □

6.3. We can now look at the question of computing extensions between tilting modules and simple modules for $S(n,d)$.

Theorem. *Let $p \geq 3$ and $\lambda, \mu \in \Lambda$.*

(a) $\operatorname{Ext}^1_{S(n,d)}(L(\lambda), T(\mu)) \cong \begin{cases} \operatorname{Ext}^1_{k\Sigma_d}(D_\lambda, Y^{\mu'} \otimes \mathrm{sgn}) & \lambda \in \Lambda_{\mathrm{reg}} \\ 0 & \lambda \in \Lambda_{\mathrm{sing}}. \end{cases}$

(b) $\operatorname{Ext}^1_{k\Sigma_d}(Y^\lambda, Y^\mu) = 0.$

Proof. (a) If λ is in Λ_{reg}, then by [DEN, Theorem 2.4A]

$$\begin{aligned}\operatorname{Ext}^1_{S(n,d)}(L(\lambda), T(\mu)) &\cong \operatorname{Ext}^1_{S(n,d)}(L(\lambda), \mathcal{G}(Y^{\mu'} \otimes \mathrm{sgn})) \\ &\cong \operatorname{Ext}^1_{k\Sigma_d}(D_\lambda, Y^{\mu'} \otimes \mathrm{sgn}).\end{aligned}$$

On the other hand, by using the five term exact sequence obtained from the spectral sequence, we have an embedding

$$\operatorname{Ext}^1_{S(n,d)}(L(\lambda), T(\mu)) \hookrightarrow \operatorname{Ext}^1_{k\Sigma_d}(eL(\lambda), Y^{\mu'} \otimes \mathrm{sgn}) \to \cdots$$

for all $\lambda \in \Lambda$. In particular, if $\lambda \in \Lambda_{\mathrm{sing}}$, then $\operatorname{Ext}^1_{S(n,d)}(L(\lambda), T(\mu)) = 0$.

(b) The tilting modules admit both a good and Weyl filtration. Moreover, the tilting modules when $p \geq 3$ have only regular composition factors in the head. Therefore, by Theorem 6.2 and [DEN, Theorem 2.4A], we have

$$\begin{aligned}0 &= \operatorname{Ext}^1_{S(n,d)}(T(\lambda'), T(\mu')) \\ &\cong \operatorname{Ext}^1_{k\Sigma_d}(Y^\lambda \otimes \mathrm{sgn}, Y^\mu \otimes \mathrm{sgn}) \\ &\cong \operatorname{Ext}^1_{k\Sigma_d}(Y^\lambda, Y^\mu).\end{aligned}$$ □

The preceding result can be used to provide an alternative proof of the following well-known result (see [CPS, (4.6.1)]).

Corollary. *Let $p \geq 3$. Then* $\operatorname{Ext}^1_{k\Sigma_d}(V^{\otimes d}, V^{\otimes d}) = 0$.

Proof. The module $V^{\otimes d}$ is a direct sum of permutation modules each of which is a direct sum of Young modules. The result now follows by Theorem 6.3(b). □

References

[BKM] V. P. Burichenko, A. S. Kleshchev, S. Martin, On cohomology of dual Specht modules, J. Pure Appl. Alg. 112 (1996), 440–480.

[CPS] E. Cline, B. J. Parshall, L. L. Scott, Stratifying Endomorphism Algebras, Mem. Amer. Math. Soc. 571, Amer. Math. Soc., Providence, RI, 1996.

[Don1] S. Donkin, On Schur algebras and related algebras I, J. Algebra 104 (1986), 310–328.

[Don2] S. Donkin, On Schur algebras and related algebras II, J. Algebra 111 (1987), 354–364.

[Don3] S. Donkin, On tilting modules for algebraic groups, Math. Z. 212 (1993), 39–60.

[DEN] S. R. Doty, K. Erdmann, D. K. Nakano, Extensions of modules over Schur algebras, symmetric groups and Hecke algebras, Algebr. Represent. Theory, to appear.

[E] K. Erdmann, Decomposition numbers for symmetric groups and composition factors of Weyl modules, J. Algebra 180 (1996), 316–320.

[Gr] J. A. Green, Polynomial Representations of GL_n, Lecture Notes in Math. 830, Springer-Verlag, Berlin 1980.

[Ja1] G. D. James, On the decomposition matrices of the symmetric groups, II, J. Algebra 43 (1976), 45–54.

[Ja2] G. D. James, The representation theory of the symmetric group, Lecture Notes in Math. 682, Springer-Verlag, Berlin 1978.

[Ja3] G. D. James, The decomposition of tensors over fields of prime characteristic, Math. Z. 172 (1980), 161–178.

[Jan] J. C. Jantzen, Representations of Algebraic Groups, Academic Press, Orlando 1987.

[KN] A. S. Kleshchev, D. K. Nakano, On comparing the cohomology of general linear groups and symmetric groups, Pacific J. Math., to appear.

[Mar] S. Martin, Schur Algebras and Representation Theory, Cambridge Univ. Press, 1993.

Mathematical Sciences, Loyola University, Chicago, IL 60626, U.S.A.

E-mail: doty@math.luc.edu

Department of Mathematics, University of Georgia, Athens, GA 30602, U.S.A.

E-mail: nakano@math.uga.edu

Generalized q-Schur algebras and the ways to approach them

*Jie Du**

1. Introduction

Generalized q-Schur algebras are certain endomorphism algebras arising from representations of finite groups G of Lie type and their associated (Iwahori–)Hecke algebras, or sometimes from those of quantum groups. When G is of type A, these algebras are the usual q-Schur algebras which play a central rôle in linking the representations of finite general linear groups, quantum general linear groups and type A Hecke algebras [DJ], [D1], [Du3]. If G is of type other than A, no algebras found so far can play such a rôle as the q-Schur algebras do. However, various versions of similar algebras have been introduced in different contexts. For example, by mimicking the relation between quantum groups $U_q(\mathfrak{gl}_n)$ and q-Schur algebras, a version of generalized q-Schur algebras was introduced and intensively studied by Du–Scott [DS1]. Using Iwahori's pioneering work, generalized q-Schur algebras of parabolic type can be introduced naturally (see, e.g., [Du2], [DPS1]). More generally, Howlett–Lehrer theory and its generalization in the non-describing characteristic theory allow us to consider generalized q-Schur algebras associated with cuspidal modules [GHM]. In the cases of classical types, certain generalized q-Schur algebras of Young type have also been introduced and investigated by Geck, Gruber and Hiss [GrH], [GH]. Some of these versions can also be turned into endomorphism algebras of corresponding Hecke algebra modules.

Generally speaking, the structure of these generalized q-Schur algebras is not as good as those arising from quantum groups. For example, the latter are always quasi-hereditary, but the former are not. To have a better understanding of their structure, the 'enlargement' idea has recently been emerged in the work [DPS1] and [DS2] (compare also [DJM]). In these, we aimed to find a larger and nicer endomorphism algebra which is, say, quasi-hereditary or stratified, such that the given one is a centralizer subalgebra of the larger one. The significance of this enlargement idea can be seen from the following simple result [DS2, Theorem (6.3.2)].

Let $(\mathcal{O}, K, k)$ be a local triple with $\mathcal{O}$ regular. (A *local triple* is a triple $(\mathcal{O}, K, k)$ consisting of a commutative, Noetherian local domain $\mathcal{O}$ having fraction field K and residue field $k = \mathcal{O}/\mathfrak{m}$.)

*Research was supported by the Australian Research Council

Lemma 1.1. *Let A be a finite $\mathcal{O}$-free algebra and $e \in A$ an idempotent such that eAe is $\mathcal{O}$-free. Put $B = eAe$. Then the decomposition matrix of B is a submatrix of the decomposition matrix of A. In particular, if A has a unitriangular decomposition matrix, so does B.*

The following result, which is used in turning an endomorphism algebra of a certain G-module into that of a Hecke algebra module, is proved in [C], attributed to Dipper.

Lemma 1.2. *Let G be a finite group and $\mathcal{O}$ a principal ideal domain. Let M be an $\mathcal{O}$-free $\mathcal{O}G$-module. For a submodule N of M, let $\sqrt{N}$ denote the smallest $\mathcal{O}$-pure submodule of M containing N. Suppose $H = \mathrm{End}_{\mathcal{O}G}(M)$ is symmetric.*

(a) *If I is an $\mathcal{O}$-pure right ideal of H, then $\mathrm{Hom}_{\mathcal{O}G}(M, \sqrt{IM}) = I$.*

(b) *If I, J are $\mathcal{O}$-pure right ideals of H, then $\mathrm{Hom}_{\mathcal{O}G}(M, -)$ induces a natural isomorphism*

$$\eta : \mathrm{Hom}_{\mathcal{O}G}(\sqrt{IM}, \sqrt{JM}) \xrightarrow{\sim} \mathrm{Hom}_H(I, J).$$

Let (W, S) be a finite Coxeter group with generating system S. For any $s \in S$, let v_s be an indeterminate such that $v_s = v_t$ if s and t are W-conjugate and let $\mathcal{Z} = \mathbb{Z}[v_s, v_s^{-1}]_{s\in S}$. The (generic) Hecke algebra $\mathcal{H} = \mathcal{H}(W)$ over $\mathcal{Z}$ associated to W is an associative algebra with free $\mathcal{Z}$-basis $\{\tau_w\}_{w\in W}$ satisfying the following multiplicative rules:

$$\begin{cases} (\tau_s - v_s^2)(\tau_s + 1) = 0, & \text{if } s \in S, \\ \tau_y\tau_w = \tau_{yw}, & \text{if } \ell(yw) = \ell(y) + \ell(w). \end{cases} \tag{1.3}$$

Here ℓ is the length function on W.

Two specific linear characters will often be used. These are the "trivial" representation IND and the "sign" representation SGN, defined on generators by:

$$\mathrm{IND} : \mathcal{H} \to \mathcal{Z},\ \tau_s \mapsto v_s^2, \qquad \mathrm{SGN} : \mathcal{H} \to \mathcal{Z},\ \tau_s \mapsto -1 \quad (s \in S). \tag{1.4}$$

For $w \in W$, let $v_w^2 = \mathrm{IND}(\tau_w)$. If W_λ is a parabolic subgroup of W, put $x_\lambda = \sum_{w\in W_\lambda} \tau_w$ and $y_\lambda = \sum_{w\in W_\lambda} (-1)^{\ell(w)} v_w^{-2}\tau_w$; then the right $\mathcal{H}$-modules $x_\lambda\mathcal{H}$ and $y_\lambda\mathcal{H}$ are the "induced" modules from IND_λ and SGN_λ, respectively, to $\mathcal{H}$.

Throughout, $(\mathcal{O}, K, k)$ is a fixed local triple with $\mathcal{O}$ a DVR. For a commutative $\mathcal{Z}$-algebra $\mathcal{Z}'$ and a $\mathcal{Z}$-module M, we denote by $M_{\mathcal{Z}'}$ the $\mathcal{Z}'$-module $M \otimes_{\mathcal{Z}} \mathcal{Z}'$ obtained by base change. In particular, $\mathcal{H}_{\mathcal{Z}'}$ is the (specialized) Hecke algebra obtained by base change from $\mathcal{Z}$ to $\mathcal{Z}'$. All modules are right modules unless otherwise stated.

Acknowledgement. The paper is based on a talk which I gave at the Symposium on the modular representation theory of finite groups, Charlottesville, Virginia, May 1998. I would like to thank the organizers of the meeting for the opportunity to attend this very interesting conference.

2. Background and examples

Let $\boldsymbol{G}$ be a connected reductive algebraic group defined over the finite field $\mathbb{F}_q$ of positive characteristic with Frobenius morphism F and let G be the subgroup of fixed points $\boldsymbol{G}^F$. Let W be the associated Weyl group of G with generating system S. Given a subset J of S, let P_J be the corresponding standard parabolic subgroup with the Levi decomposition $L_J U_J$. For any Levi subgroup L of G we denote the corresponding Harish-Chandra induction and truncation by R_L^G and T_L^G respectively. Thus, for $X \in$ **mod**-kL, $Y \in$ **mod**-kG and a parabolic subgroup P with Levi decomposition $P = LU$, $R_L^G(X)$ is defined to be the kG-module obtained by first lifting X to kP with trivial U-action and then inducing this module from P to G, while $T_L^G(Y)$ is the kL-module obtained by taking fixed points of U on Y, on which L acts since U is normal in P. If k is of characteristic $p \geq 0$ and p is not the describing characteristic of G, then these operations are independent of the choice of P (see [HL1], [DD] and [HL2]).

We shall always assume that $p \nmid q$ and k is large enough. A kG-module C is called *cuspidal* if $T_L^G(C) = 0$ for all proper Levi subgroups of G.

Theorem 2.1 ([GHM]). *Let L be a Levi subgroup of G and C an irreducible cuspidal kL-module. Let $W(L, C)$ be the ramification group of $R_L^G(C)$. Then*

$$H = H(L, C) = \mathrm{End}_{kG}(R_L^G(C))$$

is a Hecke algebra associated with $W(L, C)$ with respect to certain powers of q, possibly twisted by a 2-cocycle. Moreover, H is a symmetric algebra ([GH, (2.7)]).

Without loss of generality, we may assume that $L = L_J$ is a standard Levi subgroup. Thus, the ramification group of $R_L^G(C)$ is given by

$$W(L, C) = \{w \in W \mid w(J) = J, C^w = C\}$$

(see [GHM, §3]). As shown in [HL1], this group is a semi-direct product of a reflection group by a group which leaves invariant the positive root system associated with the reflection group.

It is known ([Cr1, §9.2], [H1]) that the set $\mathrm{Irr}_k(G)$ (write $\mathrm{Irr}(G)$ for $\mathrm{Irr}_k(G)$ if $\mathrm{char}\, k = 0$) of irreducible characters of G can be classified as follows. First, they are divided into *Harish-Chandra series* $\mathrm{Irr}_k(G; (L, C))$ indexed by the associate classes of pairs (L, C), where C is a cuspidal kL-module; second, the irreducible characters in the series corresponding to (L, C) are indexed by the irreducible modules of $H(L, C)$. So, to achieve this classification, we need first classify the irreducible cuspidal kL-modules for all Levi subgroups of G, and then determine the irreducible characters of the associated endomorphism algebra. See the examples below.

When $\mathrm{char}\, k = 0$, the set $\mathrm{Irr}(G)$ is also decomposed into *geometric conjugacy classes* $\mathcal{E}(G, s)$ by using Deligne–Lusztig characters $R_{\boldsymbol{T},\theta}^{\boldsymbol{G}}$, where $\boldsymbol{T}$ is an F-stable torus of $\boldsymbol{G}$, and θ an irreducible character of $\boldsymbol{T}^F$. The set $\mathcal{E}(G, s)$ is defined as follows. Since $(\boldsymbol{T}, \theta)$ are partitioned into geometric conjugacy classes (see [Cr1, (4.1.3)]) indexed by the G^*-conjugacy classes of semi-simple elements in G^*, where $G^* = \boldsymbol{G}^{*F^*}$ with $(\boldsymbol{G}^*, F^*)$ dual to $(\boldsymbol{G}, F)$, then, for any semi-simple element $s \in G^*$, $\mathcal{E}(G, s)$ is the subset of $\mathrm{Irr}(G)$ consisting of all irreducible constituents of $R^{\boldsymbol{G}}_{\boldsymbol{T},\theta}$, where $(\boldsymbol{T}, \theta)$ runs through the geometric conjugacy class indexed by s. So we have

$$\mathrm{Irr}(G) = \bigcup_{s \in G^* \text{semi-simple}} \mathcal{E}(G, s).$$

An ordinary irreducible character of G is called *unipotent* if it is a constituent of a Deligne–Lusztig character $R^{\boldsymbol{G}}_{\boldsymbol{T},1}$ for some F-stable torus $\boldsymbol{T}$ of $\boldsymbol{G}$.

On the other hand, from general theory, the set $\mathrm{Irr}_k(G)$ is also partitioned into *blocks*. Broué and Michel [BM] have proved that, for a p-regular semi-simple element $s \in G^*$, the set

$$\mathcal{E}_p(G, s) = \cup_{t \in C_{G^*}(s)_p^{F^*}} \mathcal{E}(G, st) \tag{2.2}$$

is a union of p-blocks. Furthermore, Carter [Cr2] proved that this is the smallest set of characters which is at the same time unions of geometric conjugacy classes and p-blocks. Each $\mathcal{E}_p(G, s)$ is called a *geometric conjugacy class of p-blocks.*

We remark that the modular irreducible characters in a geometric conjugacy class of p-blocks are sometimes parameterized by the irreducible characters of a certain generalized q-Schur algebra, and the corresponding decomposition numbers are determined by that of the algebra.

We now present some examples showing how generalized q-Schur algebras enter into the picture.

Example 2.3. Let $G = \mathrm{GL}_n(q)$ and $s \in G$ a semi-simple element. Decompose the characteristic polynomial $f_s(x)$ of s over $\mathbb{F}_q$ into a product of irreducible polynomials $f_s(x) = p_1(x)^{m_1} \dots p_a(x)^{m_a}$, where $p_i(x)$ are distinct and $\deg p_i(x) = d_i$ are weakly increasing. Thus

$$L_s = \mathrm{GL}_{d_1}(q)^{\times m_1} \times \cdots \times \mathrm{GL}_{d_a}(q)^{\times m_a}$$

is the minimal Levi subgroup containing s and $s = s_1^{\times m_1} \dots s_a^{\times m_a}$ with $s_i \in \mathrm{GL}_{d_i}(q)$. Each s_i defines a cuspidal representation $C_{s_i,K}$ since the characteristic polynomial of s_i agrees with the minimal polynomial of s_i (see [G, p. 437]), and so $C_{s,K} = C_{s_1,K}^{\otimes m_1} \otimes \cdots \otimes C_{s_a,K}^{\otimes m_a}$ is a cuspidal representation of L_s which possesses an $\mathcal{O}L_s$-lattice $C_{s,\mathcal{O}}$, and for $\mathcal{Z}' \in \{K, \mathcal{O}, k\}$ we have

$$H(L_s, C_{s,\mathcal{Z}'}) = \mathrm{End}_{\mathcal{Z}'G}(R^G_{L_s}(C_{s,\mathcal{Z}'})) \cong \mathcal{H}_{\mathcal{Z}',q^{d_1}}(\mathfrak{S}_{m_1}) \otimes \cdots \otimes \mathcal{H}_{\mathcal{Z}',q^{d_a}}(\mathfrak{S}_{m_a}).$$

If $f_s(x) = (x-1)^n$, then $L_s = T$ is a maximal torus of G. In this case, $C_{s,\mathcal{Z}'} \cong \mathcal{Z}'$ and $\mathrm{End}_{\mathcal{Z}'G}(R^G_T(\mathcal{Z}')) \cong \mathcal{H}_{\mathcal{Z}'}(\mathfrak{S}_n))$ for all $\mathcal{Z}' \in \{K, \mathcal{O}, k\}$.

For $\lambda \vdash n$, let $L_\lambda = \mathrm{GL}_{\lambda_1}(q) \times \mathrm{GL}_{\lambda_2}(q) \times \cdots$ be the corresponding Levi subgroup of G and let $\mathrm{St}_{\lambda,K}$ be the Steinberg module of L_λ. It has an $\mathcal{O}L_\lambda$-lattice $\mathrm{St}_{\lambda,\mathcal{O}}$. Then we have (see [DJ]), for $\mathcal{Z}' \in \{K, \mathcal{O}, k\}$, $\sqrt{y_\lambda R^G_T(\mathcal{Z}')} \cong R^G_{L_\lambda}(\mathrm{St}_{\lambda,\mathcal{Z}'})$ and (compare

Lemma 1.2)

$$\mathcal{S}_q(n) = \mathrm{End}_{\mathcal{Z}'G}(\oplus_{\lambda\vdash n} R^G_{L_\lambda}(\mathrm{St}_{\lambda,\mathcal{Z}'})) \cong \mathrm{End}_{\mathcal{H}_{\mathcal{Z}'}}(\oplus_{\lambda\vdash n} y_\lambda \mathcal{H}_{\mathcal{Z}'}).$$

This algebra is called a *q-Schur algebra* whose decomposition matrix gives that part of the decomposition matrix of G corresponding to unipotent characters (see [DJ]).

In general, if s is a semi-simple p-regular element, one may define an endomorphism algebra $\mathcal{S}_q(s)$, which is a tensor product of q-Schur algebras. The decomposition is then the same as that of $\mathcal{E}_p(G, s)$. For details, see [DJ] and [DD].

Example 2.4. Let $G = G_n(q)$ be one of the following groups

$$\mathrm{GU}_n(q), \quad \mathrm{Sp}_n(q)\ (n \text{ even}), \quad \mathrm{SO}_n(q)\ (q \text{ odd}, n \text{ odd})$$

(see [GH, (3.5)]), and assume that p is linear, i.e., $q^{\delta m-1}+1$ is not divisible by p for all positive integer m, where

$$\delta = \delta(G) = \begin{cases} 2, & \text{if } G = \mathrm{GU}_n(q) \\ 1, & \text{otherwise.} \end{cases}$$

Every (standard) Levi subgroup L of G is of the form $L_{a,\lambda} = G_a \times L_\lambda$ where L_λ for a partition $\lambda = (\lambda_1, \ldots, \lambda_b)$ of m ($\lambda \vdash m$) is the Levi subgroup of $\mathrm{GL}_m(q^\delta)$ isomorphic to $\mathrm{GL}_{\lambda_1}(q^\delta) \times \cdots \times \mathrm{GL}_{\lambda_b}(q^\delta)$ (thus, $n = a + 2m$). Fix a Levi subgroup of the form $L_a \cong G_a(q) \times \mathrm{GL}_1(q^\delta)^m$ such that G_a has an irreducible cuspidal character. This is a defect 0 representation for $G_a/Z(G_a)$, and we let C_a denote the $\mathcal{O}G_a$-lattice affording it. Let St_λ be the Steinberg lattice of L_λ and put $C_{a,\lambda} = C_a \otimes_{\mathcal{O}} \mathrm{St}_\lambda$. Then the endomorphism algebra $H(L_a, C_{a,(1^m)})$ of $Z = R^G_{L_a}(C_{a,(1^m)})$ is isomorphic to the specialized Hecke algebra $\mathcal{H}_{\mathcal{O}}(B_m)$ of type B_m with parameters Q_a and q^δ, where Q_a depends on a and the type of G, and $\sqrt{y_\lambda Z} \cong R^G_{L_{a,\lambda}}(C_{a,\lambda})$. By [GH, 4.11], we have an isomorphism of the following endomorphism algebras (cf. Lemma 1.2)

$$\mathrm{End}_{\mathcal{O}G}(\oplus_{\lambda\vdash m} R^G_{L_{a,\lambda}}(C_{a,\lambda})) \cong \mathrm{End}_{\mathcal{H}_{\mathcal{O}}(B_m)}(\oplus_{\lambda\vdash m} y_\lambda \mathcal{H}_{\mathcal{O}}(B_m)).$$

These generalized q-Schur algebras are called the q-Schur^2 algebras of *Young* type in [DS2]. It is proved (see [GH, (4.7)]) that, in the linear prime case, the decomposition matrix of the unipotent characters of G is a block diagonal matrix in which each diagonal block is the decomposition matrix of such a generalized q-Schur algebra.

Example 2.5. Let $G = G^F$ be as before, and B a Borel subgroup of G containing the maximal torus T. Let $C = k$ be the trivial kT-module. If (W, S) is the associated Coxeter system of G, then there exist index parameters $c_s \in \mathbb{N}$, $s \in S$, defined by $[B : B^s \cap B] = q^{c_s}$ such that the specialization $\mathcal{Z} \to k$ with $v_s^{2c_s} \mapsto q^{c_s}$ induces the isomorphism $H(T, k)(:= \mathrm{End}_{kG}(R^G_T(k))) \cong \mathcal{H}_k$. Thus we may define a generalized q-Schur algebra

$$S(G, k) = \mathrm{End}_{kG}\Big(\bigoplus_{P\supseteq B} \mathrm{ind}^G_P k\Big).$$

Note that $S(G,k) \cong \mathrm{End}_{\mathcal{H}_k}(\oplus_{\lambda\subseteq S} x_\lambda \mathcal{H}_k)$. These algebras are called Hecke endomorphism algebras in [DPS1] or the generalized q-Schur algebras of *parabolic* type. We shall also call the endomorphism algebra $\mathrm{End}_{\mathcal{H}_k}(\oplus_{\lambda\subseteq S} y_\lambda \mathcal{H}_k)$ a *twisted* Hecke endomorphism algebra. Note that if G is the finite general linear group $\mathrm{GL}_n(q)$, $S(G,k)$ is (Morita equivalent to) the q-Schur algebra.

3. q-Schur^2 algebras

The structure of the generalized q-Schur algebras in Examples 2.4 and 2.5 is not clear except for type A. However, if W is of type B or C, it can be embedded into a quasi-hereditary algebra, the so-called q-Schur^2 algebra, as a centralizer subalgebra (see [DS2]). In this section, we first look at how such a natural enlargement is done in this case.

In this section, we assume that W is the Weyl group of type B_r (or C_r). As a Coxeter group, we denote the set of Coxeter generators of W by $S = \{s_0, s_1, \ldots, s_{r-1}\}$ with relations described in the following Coxeter diagram

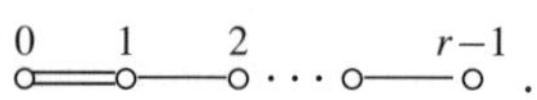

Let B be the subgroup generated by $t_1, \ldots, t_r$, where $t_1 = s_0$ and $t_{i+1} = s_i t_i s_i$ for $1 \le i < r$. For nonnegative integers a, b with $a + b = r$, a subgroup of W generated by reflections $t_1, \ldots, t_a$ and a subset of $\{s_1, \ldots, s_{a-1}\} \cup \{s_{a+1}, \ldots, s_r\}$ is called a quasi-parabolic subgroup of W. Thus every bicomposition $\lambda = (\lambda^{(1)}, \lambda^{(2)})$ of r (i. e., both $\lambda^{(1)}$ and $\lambda^{(2)}$ are compositions with $|\lambda^{(1)}| + |\lambda^{(2)}| = r$) defines a unique quasi-parabolic subgroup $W_\lambda = B_\lambda \rtimes \mathfrak{S}_\lambda$ with 'top' part $\mathfrak{S}_\lambda$, the Young subgroup corresponding to the composition $\bar{\lambda} = \lambda^{(1)} \vee \lambda^{(2)}$ obtained by juxtaposition, and 'bottom' part $B_\lambda = \langle t_1, \ldots, t_{|\lambda^{(1)}|}\rangle$.

Let $\mathcal{H} = \mathcal{H}(W)$ be the Hecke algebra over $\mathcal{Z}$ associated with W as defined in (1.3), and put $\mathcal{H}' = \mathcal{H} \otimes_{\mathcal{Z}} \mathcal{Z}'$ for a commutative $\mathcal{Z}$-algebra $\mathcal{Z}'$. For simplicity, we put $v_{s_0} = v_0$ and $v_{s_i} = v$ for all $i = 1, \ldots, r-1$. Then $\mathcal{Z} = \mathbb{Z}[v_0, v_0^{-1}, v, v^{-1}]$. Denote the image of v_0 (resp. v) in $\mathcal{Z}'$ by q_0 (resp. q), and let $q_w = q_{w_1} \cdots q_{w_m}$ and $\varepsilon_w = (-1)^{\ell(w)}$, where $w = w_1 \ldots w_m$ with $w_i \in S$ is a reduced expression of w and $\ell(w) = m$.

For a *parabolic* subgroup W_λ of W (thus, $\lambda^{(1)} = (|\lambda^{(1)}|)$), let $\mathcal{H}'_\lambda = \mathcal{H}'(W_\lambda)$ be the subalgebra of $\mathcal{H}'$ generated by τ_w, $w \in W_\lambda$. Then the trivial and sign representations of W_λ (see (1.4)) are generated by $x_\lambda = \sum_{w\in W_\lambda} \tau_w$ and $y_\lambda = \sum_{w\in W_\lambda} \varepsilon_w q_w^{-1} \tau_w$, respectively. To generalize them to quasi-parabolic subgroups W_λ, we define $x_\lambda = \pi_a x_{\bar{\lambda}}$ and $y_\lambda = \pi_a^- y_{\bar{\lambda}}$, where $a = |\lambda^{(1)}|$ and

$$\pi_a = \prod_{i=1}^{a} (q^{i-1} + \tau_{t_i}), \qquad \pi_a^- = \prod_{i=1}^{a} (q_0 q^{i-1} - \tau_{t_i}).$$

Let $\mathcal{T}_\lambda = x_\lambda \mathcal{H}$ and $\tilde{\mathcal{T}}_\lambda = y_\lambda \mathcal{H}$. We also put $\mathcal{T}'_\lambda = \mathcal{T}_{\lambda \mathcal{Z}'}$ and $\tilde{\mathcal{T}}'_\lambda = \tilde{\mathcal{T}}_{\lambda \mathcal{Z}'}$ for any commutative $\mathcal{Z}$-algebra $\mathcal{Z}'$. We shall call $\tilde{\mathcal{T}}'_\lambda = y_\lambda \mathcal{H}'$ a *twisted* (q-)permutation module.

For a positive integer n, let $\Lambda_2(n, r)$ denote the set of all bicompositions $\lambda = (\lambda^{(1)}, \lambda^{(2)})$ of r such that $\lambda^{(1)}$ has $\max(n, r)$ parts and $\lambda^{(2)}$ has n parts (counting zeros). Following [DS2, §3.3], we define the *q-Schur*^{2} algebra as

$$\mathcal{S}_q^2(n, r, \mathcal{Z}') = \mathrm{End}_{\mathcal{H}'}(\mathcal{T}'),$$

where $\mathcal{T}' = \mathcal{T}(n, r)' = \oplus_{\lambda \in \Lambda_2(n,r)} \mathcal{T}'_\lambda$. Similarly, using the twisted permutation modules $\tilde{\mathcal{T}}'_\lambda$, we also define a *twisted* version

$$\tilde{\mathcal{S}}_q^2(n, r, \mathcal{Z}') = \mathrm{End}_{\mathcal{H}'}(\tilde{\mathcal{T}}'),$$

where $\tilde{\mathcal{T}}' = \tilde{\mathcal{T}}(n, r)' = \oplus_{\lambda \in \Lambda_2(n,r)} \tilde{\mathcal{T}}'_\lambda$. Note that the Hecke or twisted Hecke endomorphism algebras introduced in Example 2.5 are centralizer subalgebras of $\mathcal{S}_q^2(n, r, \mathcal{Z}')$ or $\tilde{\mathcal{S}}_q^2(n, r, \mathcal{Z}')$. Thus Lemma 1.1 applies.

The following theorem is the main result of [DS2], giving the shape of decomposition matrices for the q-Schur^2 algebras.

Theorem 3.1. *For any (Noetherian) commutative ring $\mathcal{Z}'$, the q-Schur^2 algebra $\mathcal{S}_q^2(n, r, \mathcal{Z}')$ is quasi-hereditary with a cellular basis.*

We now show that the twisted version $\tilde{\mathcal{S}}_q^2(n, r, \mathcal{Z}')$ is in fact isomorphic to $\mathcal{S}_q^2(n, r, \mathcal{Z}')$. The Hecke algebra $\mathcal{H}'$ admits a ring involution

$$\tilde{\ } : \mathcal{H}' \to \mathcal{H}'; \quad \tilde{q}_s = q_s^{-1}, \tilde{\tau}_w = \varepsilon_w q_w^{-1} \tau_w$$

for all $s \in S$ and $w \in W$, and $\mathcal{Z}'$-algebra involution

$$\sigma : \mathcal{H}' \to \mathcal{H}'; \quad \sigma(\tau_w) = \tau_w^\sigma = \varepsilon_w q_w \tau_{w^{-1}}^{-1}.$$

Clearly, both maps commute with the $\mathcal{Z}'$-algebra anti-automorphism

$$\iota : \mathcal{H}' \to \mathcal{H}'; \quad \iota(\tau_w) = \tau_w^\iota = \tau_{w^{-1}}.$$

The maps $\tilde{\ }$ and σ interchange $x_{\bar{\lambda}}$ and $y_{\bar{\lambda}}$ (up to an invertible scalar). More explicitly, we have $y_{\bar{\lambda}} = \tilde{x}_{\bar{\lambda}} = q_{w_{0,\bar{\lambda}}} x_{\bar{\lambda}}^\sigma$, where $w_{0,\bar{\lambda}}$ is the longest word in $\mathfrak{S}_{\bar{\lambda}}$. Since $\pi_a^- = q^{\binom{a}{2}} \tilde{\pi}_a$, we have $y_\lambda = q^{\binom{a}{2}} \tilde{x}_\lambda$. Thus $\tilde{\mathcal{T}}_\lambda$ is the image of $\mathcal{T}_\lambda$ under the map $\tilde{\ }$. However, the map σ does not interchange π_a and π_a^-.

For any $\mathcal{H}'$-module M and automorphism θ of $\mathcal{H}'$, we define M^θ to be the $\mathcal{H}'$-module with the action twisted by θ, i.e., $m * h = m\theta(h)$ for all $m \in M$ and $h \in \mathcal{H}'$. Thus, if θ is a $\mathcal{Z}'$-*algebra* automorphism and M is a right ideal of $\mathcal{H}'$, then M^θ, as a right $\mathcal{H}'$-module, is isomorphic to the image $\theta(M)$ of M. In particular, for any bicomposition λ such that W_λ is parabolic, we see from above that $\mathcal{T}_\lambda'^\sigma \cong \sigma(\mathcal{T}'_\lambda) = \tilde{\mathcal{T}}'_\lambda$. The following shows that this result holds, too, for quasi-parabolic subgroups.

Proposition 3.2. *Let λ be a bicomposition of r. Then we have $\mathcal{T}_\lambda'^{\sigma} \cong \tilde{\mathcal{T}}_\lambda'$ as right $\mathcal{H}'$-modules. Therefore, we have a $\mathcal{Z}'$-algebra isomorphism*

$$\mathcal{S}_q^2(n,r,\mathcal{Z}') \cong \tilde{\mathcal{S}}_q^2(n,r,\mathcal{Z}').$$

Proof. Certainly we have a right $\mathcal{H}'$-module isomorphism $\mathcal{T}_\lambda'^{\sigma} \cong \sigma(x_\lambda)\mathcal{H}'$. Let w_0 be the longest element of W. Then $w_0 = t_1 \dots t_r$ and the element τ_{w_0} is in the center of $\mathcal{H}'$. Thus the (left) multiplication by τ_{w_0} defines an $\mathcal{H}'$-module isomorphism $f : \mathcal{H}' \to \mathcal{H}'$, and so does its restriction to $\sigma(x_\lambda)\mathcal{H}'$. Since

$$\begin{aligned}\tau_{w_0}\sigma(x_\lambda) &= \tau_{w_0}\sigma(\pi_a)y_{\bar\lambda} = \sigma(\pi_a)y_{\bar\lambda}\tau_{w_0}\\ &= \Big(\prod_{i=1}^{a}(q^{i-1} - q_{t_i}\tau_{t_i}^{-1})\Big)\tau_{t_1\dots t_a}y_{\bar\lambda}\tau_{t_{a+1}\dots t_r}\\ &= \Big(\prod_{i=1}^{a}(q^{i-1}\tau_{t_i} - q_0q^{2(i-1)})\Big)y_{\bar\lambda}\tau_{t_{a+1}\dots t_r}\\ &= (-1)^a q^{\binom{a}{2}} y_\lambda \tau_{t_{a+1}\dots t_r},\end{aligned}$$

where $a = |\lambda^{(1)}|$, f induces an $\mathcal{H}'$-module isomorphism from $\sigma(x_\lambda)\mathcal{H}'$ to $y_\lambda\mathcal{H}' = \tau_{w_0}\sigma(x_\lambda)\mathcal{H}'$, proving the first assertion. The last assertion is clear, since $\mathcal{T}'^{\sigma} \cong \tilde{\mathcal{T}}'$ and $\mathrm{End}_{\mathcal{H}'}(\mathcal{T}') \cong \mathrm{End}_{\mathcal{H}'}(\mathcal{T}'^{\sigma})$. □

From [DS2, (3.2.2)], we immediately have the following.

Corollary 3.3. *Every submodule $y_\lambda\mathcal{H}'$ is a $\mathcal{Z}'$-pure right ideal.*

Remark 3.4. (1) The quasi-heredity of the q-Schur^2 algebras is approached by constructing a cellular basis [GL]. In fact, the existence of certain cellular type bases can be used to obtain a criterion for the quasi-heredity. We shall discuss this in next section. On the other hand, the Hecke algebra of type B itself has a cellular basis which is extended to the cellular basis for the q-Schur^2 algebra. It would be interesting to know in general when a cellular basis can be extended to a cellular basis for an enlarged and quasi-hereditary algebra of which the original algebra is a centralizer subalgebra. Some constructions will be discussed, too, in the next section.

(2) Let G be as in Example 2.4. Then the generalized q-Schur algebra defined there is a centralizer subalgebra of a q-Schur^2 algebra. Thus, by Theorem 1.1, its decomposition matrix is a submatrix of that of the q-Schur^2 algebra, and hence it is lower unitriangular. Note that this result has been obtained by Gruber under the assumption that p is linear. It is likely that these generalized q-Schur algebras could be used to determine the decomposition matrix of unipotent characters of G without assuming that p is linear.

(3) Assume again G as in Example 2.4. For $\mathcal{Z}' \in \{K, \mathcal{O}, k\}$, let $M_{\mathcal{Z}'} = R_T^G(\mathcal{Z}')$ as defined in Example 2.5 with k replaced by $\mathcal{Z}'$. Then, by (3.3) and Lemma 1.2,

$$\mathcal{S}_q^2(n,r,\mathcal{Z}') \cong \mathrm{End}_{\mathcal{Z}'G}(\oplus_{\lambda\in\Lambda_2(n,r)}y_\lambda M_{\mathcal{Z}'}) \cong \mathrm{End}_{\mathcal{Z}'G}(\oplus_{\lambda\in\Lambda_2(n,r)}\sqrt{y_\lambda M_{\mathcal{Z}'}}).$$

It would be very interesting to know if every module $\sqrt{y_\lambda M_{\mathcal{Z}'}}$ satisfies Dipper's Hypothesis [D2] as this would imply that the decomposition matrix of the q-Schur[2] algebra is a submatrix of the decomposition matrix of G (cf. [G, §3]).

4. Standardly based algebras — a basis approach

An important problem in representation theory is the construction of nice bases. For example, Kazhdan and Lusztig constructed in [KL] nice bases for Hecke algebras and their representations, while more recently Lusztig [L4] and Kashiwara [K] constructed the canonical bases both for quantum groups ("globally") and their irreducible modules ("locally"). Graham–Lehrer's cellular bases give a more explicit such global-local relation for certain finite dimensional algebras. On the other hand, it is observed that a q-Schur algebra has two nice 'global' bases, the Kazhdan–Lusztig basis [Du1] and the codeterminant basis [G3], [Gr]. Both determine nice local bases for standard (or Weyl) modules. It is very natural to ask: are there similarly related global and local bases for *any* (split) quasi-hereditary algebra? This question has been answered in [DR1].

Definition 4.1. Let $\mathcal{Z}'$ be a commutative ring with 1, and let A be a $\mathcal{Z}'$-algebra.

(a) For a poset $(\Lambda, \leq)$, A is said to be *standardly based* on Λ if the following conditions hold.

(1) For any $\lambda \in \Lambda$, there are (finite) index sets $I(\lambda)$, $J(\lambda)$ and subsets

$$\mathcal{B}^\lambda = \{a^\lambda_{i,j} \mid (i,j) \in I(\lambda) \times J(\lambda)\}$$

of A such that the union $\mathcal{B} = \cup_{\lambda\in\Lambda}\mathcal{B}^\lambda$ is disjoint and forms a $\mathcal{Z}'$-basis for A.

(2) For any $a \in A$, $a^\lambda_{i,j} \in \mathcal{B}$, we have

$$\begin{cases} a \cdot a^\lambda_{i,j} & \equiv \sum_{i'\in I(\lambda)} f_{i',\lambda}(a,i)a^\lambda_{i',j} \mod A(>\lambda) \\ a^\lambda_{i,j}\cdot a & \equiv \sum_{j'\in J(\lambda)} f_{\lambda,j'}(j,a)a^\lambda_{i,j'} \mod A(>\lambda) \end{cases}$$

where $A(>\lambda)$ is the $\mathcal{Z}'$-submodule of A spanned by $\mathcal{B}^\mu$ with $\mu > \lambda$, and $f_{i',\lambda}(a,i)$, $f_{\lambda,j'}(j,a) \in \mathcal{Z}'$ are independent of j and i, respectively. Such a base $\mathcal{B}$ is called a *standard base* for the algebra A.

(b) For any $\lambda \in \Lambda$, let $f_\lambda : J(\lambda) \times I(\lambda) \to \mathcal{Z}'$ be a function whose value $f_\lambda(j,i')$ at $(j,i') \in J(\lambda) \times I(\lambda)$ is defined by

$$a^\lambda_{ij}a^\lambda_{i'j'} \equiv f_\lambda(j,i')a^\lambda_{ij'} \mod A(>\lambda).$$

The function f_λ induces a bilinear form $\beta_\lambda : A^\lambda \times A^\lambda \to \mathcal{Z}'$ with $\beta_\lambda(a^\lambda_{ij}, a^\lambda_{i'j'}) = f_\lambda(j',i)$, where A^λ is the free $\mathcal{Z}'$-submodule of A spanned by a^λ_{ij} for all $(i,j) \in$

$I(\lambda) \times J(\lambda)$. We say that the standard base is *full* if $\mathrm{im}(\beta_\lambda) = \mathcal{Z}'$ for all $\lambda \in \Lambda$. In this case, A is said to be standardly *full*-based.

The notion of a standardly based algebra is a generalization of Graham–Lehrer's *cellular algebra*. The significance of such generalization is the following result. Part (a) of the theorem for cellular algebras over fields is given in [GL, (3.10)].

Theorem 4.2 ([DR1]). (a) *Let $\mathcal{Z}'$ be a Noetherian commutative ring. If A is standardly full-based, then A is a quasi-hereditary algebra over $\mathcal{Z}'$ in the sense of* [CPS2].

(b) *If $\mathcal{Z}'$ is a local Noetherian commutative ring, then A is split quasi-hereditary if and only if A is standardly full-based.*

Applying Theorem 4.2, we immediately obtain the following example and non-example.

Example 4.3. Borel subalgebras of a q-Schur^m algebra are quasi-hereditary (see [DR2]). Recall that, for $m > 2$, Ariki–Koike algebras associated with the groups $\mathbb{Z}/m\mathbb{Z} \wr \mathfrak{S}_r$ are used in the definition of a q-Schur^m algebra.

The first proof of this result was for Schur algebras, the case when $q = 1 = m$, given by J. A. Green [G2]. However, the method can not be easily "quantized" to the q-Schur^m algebra case (even for the case $m = 1$). However, one checks easily that the defining bases for Borel subalgebras of a q-Schur^m algebra are standard bases with all β_λ (4.1b) surjective. Thus, Theorem 4.2 gives the quasi-heredity.

Non-example 4.4. A finite dimensional (split) symmetric algebra is not quasi-hereditary unless it is semisimple. In particular, the (non-semisimple) group algebra of a finite group over a splitting field is not quasi-hereditary.

When a standardly based algebra is not quasi-hereditary, can we 'enlarge' it into a quasi-hereditary algebra? The question is very plausible for two reasons: one is seen from Theorem 4.2 as a quasi-hereditary algebra is a standardly full-based algebra, and the other is the fact (see [DlR], [CPS2]) that every algebra can be embedded into a quasi-hereditary algebra as a centralizer subalgebra. It is very natural to ask if every standardly based algebra can be embedded into a quasi-hereditary algebra as a centralizer subalgebra such that the standard base *extends* to the standard base for the quasi-hereditary algebra. Examples of algebras with cellular bases such as Hecke algebras of type A and B suggest that we do have such a strong relation. That is, a cellular basis can sometimes be extend to a cellular basis for a quasi-hereditary algebra, and the original algebra is a centralizer subalgebra of the quasi-hereditary algebra. In the rest of this section, we will look at some hypotheses required for such an embedding.

Hypothesis 4.5. Let A be a *quasi-Frobenius* algebra over $\mathcal{Z}'$ (see [CR, §58]) with a standard base $\mathcal{B} = \{a_{ij}^\lambda\}$ in the sense of Definition 4.1 such that $I(\lambda) = J(\lambda)$ for all $\lambda \in \Lambda$. Assume that

(a) for every $\lambda, \mu \in \Lambda$ with $\lambda \geq \mu$, there exists a non-empty finite set $I(\lambda, \mu)$ such that $|I(\lambda, \lambda)| = 1$ and, associated to each $\alpha \in I(\lambda, \mu)$, there is a non-empty subset $I(\lambda)_\alpha$ with $|I(\lambda)_\alpha| = 1$ if $\alpha \in I(\lambda, \lambda)$ and $I(\lambda)_\alpha \cap I(\lambda)_\beta = \emptyset$ if $\alpha \neq \beta$,

and

(b) for each $\lambda \in \Lambda$, there is an element $a_\lambda = a_{ii}^\lambda \in \mathcal{B}^\lambda$ such that $a_\eta = 1$ for some η and the intersection $a_\mu A \cap A a_\nu$ is a free $\mathcal{Z}'$-module with basis of the form

$$\mathcal{X}_{\mu\nu} = \{x_{\alpha\beta}^{\mu\lambda\nu} \mid \text{for all } \lambda, \mu, \nu \in \Lambda, \alpha \in I(\lambda, \mu), \beta \in I(\lambda, \nu)\}$$

(assuming $I(\lambda, \mu) = \emptyset$ if $\lambda \not\geq \mu$), where $x_{\alpha\beta}^{\mu\lambda\nu} = \sum_{i \in I(\lambda)_\alpha, j \in I(\lambda)_\beta} a_{ij}^\lambda$.

Note that, by (4.1a(2)), the element η is the unique minimal element of the poset Λ. When $\mu = \nu = \eta$, condition (b) forces $|I(\lambda, \eta)| = |I(\lambda)|$ for all $\lambda \in \Lambda$. Thus condition (a) implies $|I(\lambda)_\alpha| = 1$ for all $\alpha \in I(\lambda, \eta)$. Consequently, $\mathcal{X}_{\eta\eta}$ must coincide with the standard base $\mathcal{B}$ of A.

Theorem 4.6. *Assume Hypothesis* 4.5 *holds. Then A can be embedded into a quasi-hereditary algebra as a centralizer subalgebra and the standard base of A extends a standard base for the quasi-hereditary algebra.*

Proof. Let $S = \mathrm{End}_A(\oplus_{\mu \in \Lambda} a_\mu A)$. Since A is quasi-Frobenius, it follows that the $\mathcal{Z}'$-module $\mathrm{Hom}_A(a_\nu A, a_\mu A)$ is isomorphic to $a_\mu A \cap A a_\nu$ which is free by the hypothesis. Therefore, $\mathrm{Hom}_A(a_\nu A, a_\mu A)$ is free. Define, for λ, μ, ν and $\alpha \in I(\lambda, \mu)$, $\beta \in I(\lambda, \nu)$, $\phi_{\alpha\beta}^{\mu\lambda\nu} : a_\nu A \to a_\mu A$ by sending $a_\nu a$ to $x_{\alpha\beta}^{\mu\lambda\nu} a$, $a \in A$, and extend them to A-module homomorphisms on $\oplus_{\mu \in \Lambda} a_\mu A$. Then the set

$$\Phi = \{\phi_{\alpha\beta}^{\mu\lambda\nu} \mid \lambda, \mu, \nu \in \Lambda, \alpha \in I(\lambda, \mu), \beta \in I(\lambda, \nu)\}$$

forms a basis for S. We claim that Φ is a standard base for S. Indeed, if we put $I_S(\lambda) = \cup_{\mu \in \Lambda} I(\lambda, \mu)$ for each $\lambda \in \Lambda$, then Φ is a disjoint union of $\Phi^\lambda = \{\phi_{\alpha\beta}^\lambda \mid \alpha, \beta \in I_S(\lambda)\}$ where $\phi_{\alpha\beta}^\lambda = \phi_{\alpha\beta}^{\mu\lambda\nu}$ if $\alpha \in I(\lambda, \mu)$, $\beta \in I(\lambda, \nu)$. On the other hand, we have, for any $\phi = \phi_{\gamma\alpha}^{\varrho\lambda\mu} \in \mathrm{Hom}_A(a_\mu A, a_\varrho A)$,

$$\begin{aligned}\phi\phi_{\alpha\beta}^{\mu\lambda\nu}(a_\nu) &= \phi(x_{\alpha\beta}^{\mu\lambda\nu}) = \phi(a_\mu b) = x_{\gamma\alpha}^{\varrho\lambda\mu} b \\ &= a_\phi x_{\alpha\beta}^{\mu\lambda\nu} (\in a_\varrho A \cap A a_\nu) \\ &\equiv \sum_{\gamma \in I(\lambda, \varrho)} c_\gamma x_{\gamma\beta}^{\varrho\lambda\nu} \bmod A(>\lambda),\end{aligned}$$

where $x_{\alpha\beta}^{\mu\lambda\nu} = a_\mu b$ and $x_{\gamma\alpha}^{\varrho\lambda\mu} = a_\phi a_\mu$ for some $b, a_\phi \in A$. Therefore, $\phi\phi_{\alpha\beta}^\lambda \equiv \sum c_\gamma \phi_{\gamma\nu}^\lambda \bmod S(>\lambda)$. By condition 4.1a(2), we see that c_γ is independent of β. The proof for the second relation is symmetric. So the claim is proved. Finally, since the identity map on $a_\mu A$ is in S, one sees easily that every bilinear form β_λ defined in

Definition 4.1b is surjective. Therefore, by Theorem 4.2, we obtain that S is quasi-hereditary. □

We end this section with the following conjecture.

Conjecture 4.7. *Every standardly based algebra A can be embedded into a quasi-hereditary algebra as a centralizer subalgebra with the property that the standard base of A extends to a standard base for the quasi-hereditary algebra.*

5. Stratified algebras — a homological approach

In [CPS3], Cline, Parshall and Scott introduced a notion of stratified algebras which is a natural generalization of quasi-hereditary algebras.

For an algebra A over a commutative ring $\mathcal{Z}'$, let ${}_A\mathcal{C}$ denote the category of finitely generated left A-modules. An ideal J of A is called a *stratifying ideal* provided that J is a $\mathcal{Z}'$-direct summand of A and, for any $M, N \in \mathrm{Ob}({}_{A/J}\mathcal{C})$, inflation defines an isomorphism

$$\mathrm{Ext}^{\bullet}_{A/J}(M, N) \cong \mathrm{Ext}^{\bullet}_{A}(M, N).$$

A sequence $0 = J_0 \subsetneq J_1 \subsetneq \cdots \subsetneq J_n = A$ of ideals in A such that J_i/J_{i-1} is a stratifying ideal of A/J_{i-1} for every i is called a *stratification of A of length n*. If each J_i/J_{i-1} is a projective (left) A/J_{i-1}-module, the stratification is called *standard*.

Similar to the quasi-heredity for an algebra A, which can be characterized by the existence of a family of so-called standard modules in ${}_A\mathcal{C}$, a stratification for A can be realised through the existence of a *stratifying system* for ${}_A\mathcal{C}$.

Definition 5.1. Let Λ be a finite quasi-poset with associated poset $\bar{\Lambda}$. Suppose for $\lambda \in \Lambda$ there is given $\Delta(\lambda) \in \mathrm{Ob}({}_A\mathcal{C})$ which is projective over $\mathcal{Z}'$. Then $\{\Delta(\lambda)\}_{\lambda\in\Lambda}$ is a *stratifying system* for ${}_A\mathcal{C}$ provided that, for each $\lambda \in \Lambda$, there is given $P(\lambda) \in \mathrm{Ob}(\mathrm{proj}({}_A\mathcal{C}))$ and a surjective morphism $P(\lambda) \twoheadrightarrow \Delta(\lambda)$ satisfying the following three conditions.

(a) For $\lambda, \mu \in \Lambda$, if $\mathrm{Hom}_A(P(\lambda), \Delta(\mu)) \neq 0$, then $\lambda \leq \mu$.

(b) For $\lambda \in \Lambda$, there is given a decreasing filtration $G^{\lambda}_{\bullet}$ of $P(\lambda)$ (of length $t(\lambda)$) with $\mathrm{Gr}_i\, G^{\lambda}_{\bullet} := G^{\lambda}_i / G^{\lambda}_{i+1} \cong \Delta(\nu_{\lambda,i})$ $(0 \leq i < t(\lambda))$ where $\lambda \leq \nu_{\lambda,i} \in \Lambda$ and $\nu_{\lambda,0} = \lambda$.

(c) Any irreducible A-module is a homomorphic image of some $\Delta(\lambda)$.

Suppose that condition (b) is replaced by the stronger condition that,

(b′) for $\lambda \in \Lambda$, there is given a decreasing filtration $G^{\lambda}_{\bullet}$ of $P(\lambda)$ (of length $t(\lambda)$) satisfying $\mathrm{Gr}_i\, G^{\lambda}_{\bullet} \cong \Delta(\nu_{\lambda,i})$ where $\bar{\lambda} < \bar{\nu}_{\lambda,i}$ in $\bar{\Lambda}$ for $0 < i < t(\lambda)$ and $\nu_{\lambda,0} = \lambda$.

Then we say the stratifying system is *strict*.

The following result is due to Cline, Parshall and Scott (see [DPS1]).

Theorem 5.2. *Assume that ${}_A\mathcal{C}$ has a stratifying system $\{\Delta(\lambda)\}_{\lambda\in\Lambda}$. Then the algebra A has a stratification $0 = J_0 \subsetneq J_1 \subsetneq \cdots \subsetneq J_n = A$ of length $n = |\bar{\Lambda}|$. If the stratifying system is strict, then the stratification can be chosen to be standard.*

We are mostly interested in generalized q-Schur algebras. So stratifying an endomorphism algebra is the most interesting case. Cline, Parshall and Scott have developed an abstract Specht/Young module theory which will do the job. In fact, a certain homological property on a 'Specht filtration of a Young module' becomes an important component for the existence of a stratifying system.

Let H be a $\mathcal{Z}'$-algebra which is finitely generated and projective over $\mathcal{Z}'$. Let T be a finitely generated right H-module, projective over $\mathcal{Z}'$, and put $A = \mathrm{End}_H(T)$. Assume $T = \bigoplus_{\lambda\in\Lambda} T_\lambda^{\oplus m_\lambda}$ is a fixed direct sum decomposition, where Λ is a finite quasi-poset. (No assumption that the T_λ are indecomposable or even non-isomorphic is imposed.) Now, we have the following (see [DPS1]).

Theorem 5.3. *For $\lambda \in \Lambda$, assume given an $\mathcal{Z}'$-submodule $S_\lambda \hookrightarrow T_\lambda$ and an increasing filtration $F_\lambda^\bullet$ of T_λ (of length $t(\lambda)$) such that the following three conditions hold.*

(a) *For $\lambda \in \Lambda$, there is a fixed sequence $\nu_{\lambda,0}, \nu_{\lambda,1}, \ldots, \nu_{\lambda,t(\lambda)-1}$ in Λ such that $\nu_{\lambda,0} = \lambda$ and, for $i > 0$, $\nu_{\lambda,i} \geq \lambda$. For $0 < i \leq t(\lambda)$, there is given a fixed isomorphism $\mathrm{Gr}^i\, F_\lambda^\bullet := F_\lambda^i/F_\lambda^{i-1} \cong S_{\nu_{\lambda,i}}$.*

(b) *For $\lambda, \mu \in \Lambda$, $\mathrm{Hom}_H(S_\mu, T_\lambda) \neq 0 \Rightarrow \lambda \leq \mu$.*

(c) (Homological Property) *For all $\lambda \in \Lambda$, we have*

$$\mathrm{Ext}^1_H(T_\lambda/F_\lambda^i, T) = 0, \quad \textit{for all } i.$$

Put $\Delta(\lambda) = \mathrm{Hom}_H(S_\lambda, T)$ and assume that each $\Delta(\lambda)$ is $\mathcal{Z}'$-projective. Then $\{\Delta(\lambda)\}_{\lambda\in\Lambda}$ is a stratifying system for ${}_A\mathcal{C}$ with respect to Λ. If the inequalities in (a) *can be replaced by strict inequalities $\bar{\nu}_{\lambda,i} > \bar{\lambda}$ in $\bar{\Lambda}$ for all $i > 0$, the system is strict.*

A first application of this result is to stratify the Hecke endomorphism algebras defined in Example 2.5. We keep the notation used in 2.5. Thus, the index parameter c_s are defined by $[B : B^s \cap B] = q^{c_s}$, and (W, S) is the finite Coxeter system determined by G. Let $\mathcal{H} = \mathcal{H}(W, S)$ denote the (generic) Hecke algebra over $\mathcal{Z} = \mathbb{Z}[v, v^{-1}]$ defined as in (1.3) with $v_s = v^{c_s}$. Then the specialization $\mathcal{H}_k$ (assuming $\sqrt{q} \in k$) with $v \mapsto \sqrt{q}$ is isomorphic to the Hecke algebra $H(T, k)$ defined in Example 2.5. (We shall call them Hecke algebras of Lie type.)

For $w \in W$ let $\mathcal{R}(w) = \{s \in S \mid ws < w\}$ and $\mathcal{L}(w) = \{s \in S \mid sw < w\}$. If w has a reduced expression $w = s_1 \ldots s_n$, $s_i \in S$, we set $m(w) = \sum_{i=1}^n c_{s_i}$ and $\tilde{\tau}_w = v^{-m(w)}\tau_w$. Then $\{\tilde{\tau}_w\}_{w\in W}$ forms a basis for $\mathcal{H}$. It is shown in [KL] and [L1] that there are canonical elements $\{C_w\}_{w\in W}$, given by[1]

$$C_w = \sum_{y\le w}(-1)^{\ell(w)-\ell(y)}v^{m(w)-m(y)}P_{y,w}(v^{-1})\tilde{\tau}_y \tag{5.4}$$

where $P_{y,w} \in \mathbb{Z}[v]$ with $\deg P_{y,w} < m(w) - m(y)$ if $y < w$ and $P_{w,w} = 1$. These elements $\{C_w\}_{w\in W}$ form a basis which is called the *canonical basis* for $\mathcal{H}$. Furthermore, for $\lambda \subseteq S$, the elements $\{C_w \mid \lambda \subseteq \mathcal{R}(w)\}$ form a basis for $\mathcal{H}y_\lambda$ (see, e.g., [DPS1, (2.3.5)]), and consequently we have

$$\mathcal{H}C_w \subseteq \sum_{\mathcal{R}(w)\subseteq\mathcal{R}(y)} \mathcal{Z}C_y. \tag{5.5}$$

Note that the $\mathcal{Z}$-dual $(\mathcal{H}y_\lambda)^*$ of $\mathcal{H}y_\lambda$ is isomorphic to $y_\lambda\mathcal{H}$.

Let $\Lambda = \mathcal{P}(S)$, the power set of S, and fix an ordering $S = J_1, J_2, \ldots, J_{2^{|S|}} = \emptyset$ on Λ such that $J_i \subseteq J_j \Rightarrow i \ge j$. Then $\lambda \in \Lambda$ determines a subsequence $\lambda^{(1)}, \ldots, \lambda^{(m_\lambda)} = \lambda$ consisting of those $\mu \in \Lambda$ satisfying $\lambda \subseteq \mu$. Let

$$\begin{cases} E_i^{\lambda,\mathcal{R}} = \operatorname{span}\{C_y \mid \mathcal{R}(y) = \lambda^{(j)},\ j \le m_\lambda - i\}, & 0 \le i < m_\lambda, \\ E_{m_\lambda}^{\lambda,\mathcal{R}} = 0, & \\ F_{\lambda,\mathcal{R}}^i = (\mathcal{H}y_\lambda/E_i^{\lambda,\mathcal{R}})^*, & 0 \le i \le m_\lambda. \end{cases} \tag{5.6}$$

By (5.5), we obtain a (decreasing) right-set filtration $E^{\lambda,\mathcal{R}}$ of $\mathcal{H}y_\lambda$ and an (increasing) dual right-set filtration $F_{\lambda,\mathcal{R}}$.

Let

$$\mathcal{T} = \bigoplus_{\lambda\in\Lambda} y_\lambda\mathcal{H}, \quad \mathcal{S} = \operatorname{End}_{\mathcal{H}}(\mathcal{T}), \tag{5.7}$$

where $y_\lambda = \sum_{w\in W_\lambda}(-1)^{\ell(w)}v^{-2m(w)}\tau_w$, and let $S_{\lambda,\mathcal{R}} = F^1_{\lambda,\mathcal{R}}$. As before, we write $M' = M \otimes_{\mathcal{Z}} \mathcal{Z}'$ for a $\mathcal{Z}$-module M.

Theorem 5.8 ([DPS1]). *There exists a quasi-poset structure on Λ with respect to which the data $\mathcal{T}'$, $S'_{\lambda,\mathcal{R}}$, and $F'_{\lambda,\mathcal{R}}$ satisfy the conditions* (a), (b) *and* (c) *of Theorem* 5.3 *and therefore defines a stratifying system for ${}_{\mathcal{S}'}\mathfrak{C}$. In particular, if $|S| > 1$, the algebra $\mathcal{S}'$ has a stratification of length ≥ 3.*

The stratification for $\mathcal{S}'$ defined in Theorem 5.8 is not standard in general. In the next section, we will propose a standard one using a finer filtration than the right-set filtration.

[1] We shall use this basis instead of the C'-basis used in [DPS1], [DPS2] and [DPS3]. From the Hecke algebra point of view, these two bases result in the same generalized q-Schur algebra. However, from the decomposition matrix point of view (cf. Examples 2.3 and 2.4), the C-basis has its advantage in linking with the decomposition matrix of G.

6. Cells and generalized q-Schur algebras

In this section we propose to stratify certain generalized q-Schur algebras with cells. We maintain the notation used in Section 5. In particular, $\mathcal{H}$ is the (generic) Hecke algebra over $\mathcal{Z} = \mathbb{Z}[v, v^{-1}]$ defined with respect to index parameters $\{c_s\}_{s\in S}$. From [KL] and [L1], there is another basis $\{C'_w\}_{w\in W}$, where

$$C'_w = \sum_{y\le w} v^{-m(w)+m(y)} P_{y,w}(v)\tilde{\tau}_y. \tag{6.1}$$

For $y, w \in W$, define polynomials $Q_{y,w} \in \mathcal{Z}[v]$ by

$$\sum_{y\le z\le w} (-1)^{\ell(z)-\ell(y)} P_{y,z} Q_{z,w} = \delta_{y,w},$$

and let

$$D_y = \sum_{y\le w} Q_{y,w}(v^{-1}) v^{m(w)-m(y)}\tilde{\tau}_w. \tag{6.2}$$

Since $y \le w \Leftrightarrow w_0 w \le w_0 y \Leftrightarrow w w_0 \le y w_0$, where w_0 is the longest word in W, we have, by (5.4) and (6.1-2),

$$Q_{y,w} = P_{w_0 w, w_0 y} = P_{w w_0, y w_0} \quad \text{and} \quad D_y = \tilde{\tau}_{w_0} C'_{w_0 y} = C'_{y w_0}\tilde{\tau}_{w_0} \tag{6.3}$$

(see [L2, (5.1.8)] for the equal parameter case).

Let $\mathrm{tr} : \mathcal{H} \to \mathcal{Z}$, $\tau_w \mapsto \delta_{w1}$, be the $\mathcal{Z}$-linear "trace" map. Then $\langle a, b\rangle = \mathrm{tr}(ab)$ defines a non-degenerate, symmetric, associative pairing on $\mathcal{H}$ (see, e.g., [L2, (5.1.9)]). This pairing satisfies $\langle \tilde{\tau}_w, \tilde{\tau}_v\rangle = \delta_{w,v^{-1}}$. Furthermore, we have ([L2], [Br, §3]) $\langle C_w, D_y\rangle = \delta_{w,y^{-1}}$, for all $y, w \in W$. In other words, $\{D_w\}_{w\in W}$ is the basis for $\mathcal{H}$ dual to $\{C_w\}_{w\in W}$.

To define the notion of cells, we define a relation $\le_L$ on W by setting $y \le_L w$ if there exists $x \in W$ such that C_y appears in $C_x C_w$ (with non-zero coefficient). Note that $(W, \le_L)$ is a quasi-poset, i.e., $\le_L$ is a pre-order (which is reflexive and transitive). This is because $y \le_L w$ if and only if there exists sequences $y = y_0, y_1, \dots, y_m = w$ in W and $s_1, \dots, s_m$ in S such that y_{i-1} appears in $C_{s_i} C_{y_i}$ for all $i = 1, \dots, m$.

We define $y \le_R w$ if $y^{-1} \le_L w^{-1}$, and let $\le_{LR}$ be the pre-order generated by $\le_L, \le_R$, and let $\sim_L$, $\sim_R$ and $\sim_{LR}$ be the corresponding equivalence relations (e.g., $x \sim_L y \Leftrightarrow x \le_L y \le_L x$). We shall call the corresponding equivalence classes *left, right* and *two-sided* cells, respectively.

When c_s are all equal, the notion of cells was first introduced by Kazhdan and Lusztig [KL]. We shall call cells in this case the Kazhdan–Lusztig cells. When c_s are unequal, then W is a fixed-point subgroup of a Weyl group $\boldsymbol{W}$ under a graph automorphism σ induced by the Frobenius morphism F. Note that $\boldsymbol{W}$ is the Weyl group of the ambient reductive group $\boldsymbol{G}$, and we assume that all the parameters of the Hecke algebra associated with $\boldsymbol{W}$ are equal. The first two parts of the following result can be found in [Br, 6.8,6.10].

Lemma 6.4. (a) *The left cells of W are the fixed-point sets of (Kazhdan–Lusztig) left cells of $\mathbf{W}$ under σ.*

(b) *For any $x, y \in W$, if $x \leq_L y$ and $x \sim_{LR} y$, then $x \sim_L y$.*

(c) *Let w_0 be the longest word in W. Then*

$$x \leq_L y \Leftrightarrow w_0 y \leq_L w_0 x \Leftrightarrow y w_0 \leq_L x w_0.$$

Proof. Since $\sigma(w_0) = w_0$, (c) follows immediately from (a). □

Let $\boldsymbol{D}$ be the set of distinguished involutions in $\mathbf{W}$ ([L3]). Then $\boldsymbol{D}$ is a cross section of left cells (and right cells as well) in $\mathbf{W}$. By Lemma 6.4, the fixed-point set $\mathcal{D}$ of $\boldsymbol{D}$ under σ is a cross section of left cells in W. In other words, if we denote by Ω the set of left cells, then, for every $\omega \in \Omega$, $\omega \cap \mathcal{D}$ has exactly one element. Call this element d_ω. The pre-order $\leq_R$ (resp. $\leq_{LR}$) induces a partial ordering on Ω: $\omega \leq_L \omega' \Leftrightarrow d_\omega \leq_L d_{\omega'}$ (resp. $\omega \leq_{LR} \omega' \Leftrightarrow d_\omega \leq_{LR} d_{\omega'}$). For any $d \in \mathcal{D}$, we let $\Omega^{\leq_L d} = \{\omega \in \Omega \mid d_\omega \leq_L d\}$, and define

$$\mathcal{H}^{\leq_L d} = \sum_{y \leq_L d} \mathcal{Z} C_y, \qquad \mathcal{H}^{<_L d} = \sum_{y <_L d,\, y \not\sim_L d} \mathcal{Z} C_y.$$

Then their quotient $E_\omega = \mathcal{H}^{\leq_L d} / \mathcal{H}^{<_L d}$ $(d \in \omega)$ is called the *left cell module* associated with the left cell ω. Call the dual $S_\omega = E_\omega^*$ *the dual left cell module*. Note that S_ω is a right $\mathcal{H}$-module.

Lemma 6.5. (a) *Every $\mathcal{H}^{\leq_L d}$ is a left ideal of $\mathcal{H}$ and is a $\mathcal{Z}$-pure submodule of $\mathcal{H}$.*

(b) *For any $\lambda \in \mathcal{P}(S)$, let $d_\lambda = w_{0,\lambda}$ be the longest word in W_λ. Then $d_\lambda \in \mathcal{D}$ and $\mathcal{H}^{\leq_L d_\lambda} = \mathcal{H} y_\lambda$.*

(c) *Let $T_d = (\mathcal{H}^{\leq_L d})^*$. Then*

$$T_d \cong \mathcal{H}/I_d, \quad \text{where } I_d = \sum_{y,\, y \not\leq_L d} \mathcal{Z} D_{y^{-1}}.$$

Proof. The fact that $d_\lambda \in \mathcal{D}$ in the equal parameter case follows from the definition and in the unequal parameter case follows from Lemma 6.4.

Given $h \in \mathcal{H}$, the matrix of h as a right operator on the dual module T_d relative to the basis $\{\zeta_y\}$ dual to that defined by the C_y, $y \leq_L d$, is $(\langle hC_w, D_{y^{-1}} \rangle)_{y,w \leq_L d}$. Since $\langle hC_w, D_{y^{-1}} \rangle = \langle C_w, D_{y^{-1}} h \rangle$, the map $f : T_d \to \mathcal{H}/I_d$, $\zeta_y \mapsto D_{y^{-1}} + I_d$, defines a bijective $\mathcal{H}$-module homomorphism. □

Fix a listing $\omega_1, \ldots, \omega_m$ of Ω such that $\omega_i \leq_L \omega_j \Rightarrow i < j$. This listing induces for each $d \in \mathcal{D}$ a similar listing $\omega_{d,1}, \ldots, \omega_{d,n_d}$ of $\Omega^{\leq_L d}$. Then $d \in \omega_{d,n_d}$, and we have a (decreasing) filtration E^d for $\mathcal{H}^{\leq_L d}$ and an (increasing) filtration F_d for T_d

defined by putting

$$\begin{cases} E_i^d = \mathrm{span}\{C_y \mid y \in \omega_{d,j}, j \le n_d - i\}, & 0 \le i < n_d \\ E_{n_d}^d = 0, & \\ F_d^i = (\mathcal{H}^{\le_L d}/E_i^d)^*, & 0 \le i \le n_d. \end{cases} \tag{6.6}$$

The filtration E^d (resp. F_d) is a left cell filtration of $\mathcal{H}^{\le_L d}$ (resp. a dual left cell filtration of T_d), that is, we have $E_{i-1}^d/E_i^d \cong E_{\omega_{d,n_d-i+1}}$ (resp. $F_d^i/F_d^{i-1} \cong E^*_{\omega_{d,n_d-i+1}}$) for all $i = 1, \ldots, n_d$.

We now introduce generalized q-Schur algebras associated to subsets of $\mathcal{D}$.

Definition 6.7. Let Λ be a subset of $\mathcal{D}$ and define, for any commutative $\mathcal{Z}$-algebra $\mathcal{Z}'$, $\mathcal{S}(\Lambda) = \mathrm{End}_{\mathcal{H}}(\oplus_{d\in\Lambda} T_d)$ and $\mathcal{S}_{\mathcal{Z}'}(\Lambda) = \mathrm{End}_{\mathcal{H}_{\mathcal{Z}'}}(\oplus_{d\in\Lambda} T_{d,\mathcal{Z}'})$.

Note from Lemma 6.5b that, for a suitable Λ, $\mathcal{S}_{\mathcal{Z}'}(\Lambda)$ can be Morita equivalent to the algebra $\mathcal{S}_{\mathcal{Z}'}$ defined in (5.7), or to the algebra introduced in Example 2.4. In particular, $\mathcal{S}_{\mathcal{Z}'}(\mathcal{D})$ contains the algebra $\mathcal{S}_{\mathcal{Z}'}$ defined in (5.7) as a centralizer subalgebra.

We end this section with a conjecture on the structure of some $\mathcal{S}(\Lambda)$.

Conjecture 6.8. *There exists a subset Λ of $\mathcal{D}$ with the following property: putting $\mathcal{T}^+ = \oplus_{d\in\Lambda} T_d$, $\mathcal{S}^+_{\mathcal{Z}'} = \mathcal{S}_{\mathcal{Z}'}(\Lambda)$ and $\Delta^+(\omega)_{\mathcal{Z}'} = \mathrm{Hom}_{\mathcal{H}}(S_\omega, \mathcal{T}^+)_{\mathcal{Z}'}$, for $\omega \in \Omega$, then $\{\Delta^+(\omega)_{\mathcal{Z}'}\}_{\omega\in\Lambda}$ is a strict stratifying system relative to the quasi-poset $(\Omega, \le_{LR}^{op})$ for the module category ${}_{\mathcal{S}^+_{\mathcal{Z}'}}\mathcal{C}$, where $\le_{LR}^{op}$ is the partial ordering opposite to $\le_{LR}$.*

Remark 6.9. (a) If we introduce an equivalence relation on Ω:

$$\omega \sim \omega' \Leftrightarrow E_{\omega K} \cong E_{\omega' K}$$

(and thus, $\Leftrightarrow E_\omega \cong E_{\omega'}$, by [Gy, Theorem A]), where $K = \mathbb{Q}(v)$, then Λ can be any cross section of the equivalence classes. Of course, a "nicer" one should be chosen like the one in the type A case, where Λ consists only of longest elements of Young subgroups.

(b) This conjecture is in fact Conjecture (2.5.2) in [DPS1]. However, here we have presented a candidate for $\mathcal{S}^+$.

By the construction for $\mathcal{S}^+_{\mathcal{Z}'}$, one sees easily that the conditions (a) and (b) in Theorem 5.3 with $(\Lambda, \le) = (\Omega, \le_{LR}^{\mathrm{op}})$ hold automatically. Now, Lemma 6.4b implies that, if the system $\{\Delta^+(\omega)_{\mathcal{Z}'}\}_\omega$ were a stratifying system, it would be strict. So the proof of the conjecture can be reduced to show the following homological property.

Homological Property 6.10. For any $d, d' \in \mathcal{D}$, we have

$$\mathrm{Ext}^1_{\mathcal{H}}(T_d/F_d^i, T_{d'}) = 0.$$

By Lemma 6.5b, when $d = d_\lambda$ and $d' = d_\mu$ for some $\lambda, \mu \in \mathcal{P}(S)$, the homological property holds ([DPS1, (2.3.9)]) where the proof is reduced to prove a base change property [DPS1, (1.2.13),(2.3.8)]. So we finally have to prove the following.

Base-change Property 6.11. Keep the notation introduced in (6.4). For any $d, d' \in \mathcal{D}$, and $\mathcal{Z}$-algebra $\mathcal{Z}'$, base change defines an isomorphism

$$\mathrm{Hom}_{\mathcal{H}}(\mathcal{H}^{\leq_L d'}, E_i^d)_{\mathcal{Z}'} \cong \mathrm{Hom}_{\mathcal{H}_{\mathcal{Z}'}}(\mathcal{H}_{\mathcal{Z}'}^{\leq_L d'}, E_{i\mathcal{Z}'}^d).$$

The truth of Conjecture 6.5 would provide a nice "Specht module theory" for the algebras $\mathcal{H}$ and $\mathcal{S}^+$ with "Specht modules" being dual left cell modules and would turn the theory of cell representations into the representation theory of a homologically nicer algebra $\mathcal{S}^+$.

Remark 6.12. (a) Let $\mathcal{H}$ be defined by G and Λ a subset of $\mathcal{D}$. For $\mathcal{Z}' \in \{K, \mathcal{O}, k\}$, let $T_d R_T^G(\mathcal{Z}')$ denote the quotient $R_T^G(\mathcal{Z}')/\sqrt{I_d R_T^G(\mathcal{Z}')}$. It would be interesting to know if the following isomorphism holds:

$$\mathcal{S}_{\mathcal{Z}'}(\Lambda) \cong \mathrm{End}_{\mathcal{Z}'G}(\oplus_{d\in\Lambda} T_d R_T^G(\mathcal{Z}')).$$

(b) The algebra $\mathcal{S}^+$, or some tensor product of $\mathcal{S}^+$'s, should play a similar rôle as the q-Schur algebras do in the type A case (compare Example 2.3). This can be seen from the rôle that cell representations played in the representations of finite groups of Lie type. For simplicity, we briefly look at the case when G is split. Here σ acts on $\boldsymbol{W}$ trivially and $W = \boldsymbol{W}$. Let $\bar{\Omega}$ be the set of two-sided cells in W. It is known that the set $\mathrm{Irr}(W)$ of (ordinary) irreducible characters of W are divided into the families $\cup_{c\in\bar{\Omega}}\mathcal{F}_c$, where $\mathcal{F}_c$ consists of irreducible constituents of the specialization at $v = 1$ of the Kazhdan–Lusztig two-sided cell module defined by c ([BV, (2.29)]). For $c \in \bar{\Omega}$, define $\mathcal{E}(G, 1)_c$ to be the set of irreducible constituents appearing in the (virtual) characters $R_\phi = \frac{1}{|W|}\sum_{w\in W}\phi(w)R^G_{T_w,1}$ where $\phi \in \mathcal{F}_c$. Then the set $\mathcal{E}(G, 1)$ of unipotent characters is divided into families $\cup_{c\in\bar{\Omega}}\mathcal{E}(G, 1)_c$ indexed by the set of two-sided cells.

If W is of type A, then each family contains exactly one element, and $\mathcal{S}^+$ is Morita equivalent to the q-Schur algebra $\mathcal{S}_q(n)$. Note that, in this case, the decomposition matrix of a specialized $\mathcal{S}^+$ is that of the unipotent blocks $\mathcal{E}_p(G, 1)$. (A p-block containing a unipotent character is called *unipotent*. Notice that, by an unpublished result of Geck (see [H2, 1.4]), a p-block is unipotent if and only if it is contained in $\mathcal{E}_p(G, 1)$.) It would be very natural to expect that similar result would be true in general.

7. A quantum group approach

We end this section by briefly having a review of the quantum group approach to generalized q-Schur algebras.

Let l be a positive integer and p a prime. Let ϕ denote the cyclotomic polynomial ϕ_l if $l \neq 1$, and put $\phi = \phi_p$ if $l = 1$. Let $\mathfrak{m}$ be a maximal ideal of $\mathcal{Z} = \mathbb{Z}[v, v^{-1}]$

containing p and ϕ, and such that q becomes a primitive l^{th}-root of unity in the residue field $k = \mathcal{Z}/\mathfrak{m}$. The quotient field of $\mathcal{Z}$ is denoted K.

We let U_K be a quantized enveloping algebra over K associated with some finite root system Σ, and let $U_{\mathcal{Z}}$ denote the integral subalgebra of U_K over $\mathcal{Z}$, i.e., the q-analogue of the Kostant $\mathbb{Z}$-form.

Let Γ be any saturated set of dominant weights, and let $V(\gamma)_K$ denote the irreducible type 1 U_K-module with highest weight γ, for $\gamma \in \Gamma$. Put $V(\Gamma)_K = \bigoplus_{\gamma\in\Gamma} V(\gamma)_K$, and let $\mathcal{S}_{\mathcal{Z}}[\Gamma]$ be the image of $U_{\mathcal{Z}}$ under the natural map

$$U_{\mathcal{Z}} \subseteq U_K \to \mathrm{End}_{\mathcal{Z}}(V(\Gamma)_K).$$

We define $\mathcal{S}_{\mathcal{Z}}[\Gamma]$ to be the *generalized q-Schur algebra* over $\mathcal{Z}$ associated with Γ. We have the following result (see [DS2, §5]).

Theorem 7.1. *Let $\mathcal{O} = \mathcal{Z}_{\mathfrak{m}}$. Assume that l is either divisible by 2 or 3, or is an integer smaller than the Coxeter number of the root system Σ.*

(a) *The generalized q-Schur algebra $\mathcal{S}[\Gamma] = \mathcal{S}_{\mathcal{Z}}[\Gamma] \otimes_{\mathcal{Z}} \mathcal{O}$ is $\mathcal{O}$-quasihereditary and split. The category $\mathcal{C}[\Gamma]$ of $\mathcal{O}$-finite type 1 modules of $U = U_{\mathcal{Z}} \otimes_{\mathcal{Z}} \mathcal{O}$ whose composition factors all have highest weights in Γ coincides precisely with the category of $\mathcal{O}$-finite $\mathcal{S}[\Gamma]$-modules. In particular, $\mathcal{C}[\Gamma]$ is an $\mathcal{O}$-finite highest weight category.*

(b) *Let Γ be the 'lowest lp-alcove'. Suppose $\gamma \in \Gamma$ is a weight such that the Weyl module $V(\gamma)_k$ for $U_k = U_{\mathcal{Z}} \otimes k$ is multiplicity-free. Let $\mathcal{Z}(\phi)$ denote the residue field of the localization $\mathcal{Z}_{(\phi)}$. Then the irreducible quotients of $V(\gamma)_k$ and of the Weyl module $V(\gamma)_{\mathcal{Z}(\phi)}$ for $U_{\mathcal{Z}(\phi)}$ have the same dimension.*

It would be interesting to know any connection between $\mathcal{S}[\Gamma]$ and the generalized q-Schur algebras defined in previous sections when the corresponding group is of type other than A.

References

[BV] Barbasch, D., Vogan, D., Primitive ideals and orbital integrals in complex exceptional groups, J. Algebra 80 (1983), 350–382.

[Br] Bremke, K., On generalized cells in affine Weyl groups, J. Algebra 191 (1997), 149–173.

[BM] Broué M., Michel, J., Blocs et séries de Lusztig dans un groupe réductif fini, J. Reine Angew. Math. 395 (1989), 56–67.

[C] Cabanes, M., Algèbres de Hecke comme algèbres symétriques et théorème de Dipper, C. R. Acad. Sci. Paris 327 (1998), 531–536.

[Cr1] Carter, R., Finite groups of Lie type, John Wiley & Sons, New York 1985.

[Cr2] Carter, R., Deligne-Lusztig theory and block theory, Proceedings of the Hyderabad Conference on Algebraic Groups (Hyderabad, 1989), Manoj Prakashan, Madras 1991, 59–80.

[CPS1] Cline, E., Parshall, B., Scott, L., Finite dimensional algebras and highest weight categories, J. Reine Angew. Math. 391 (1988), 85–99.

[CPS2] Cline, E., Parshall, B., Scott, L., Integral and graded quasi-hereditary algebras, I, J. Algebra 131 (1990), 126–160.

[CPS3] Cline, E., Parshall, B., Scott, L., Stratifying Endomorphism Algebras, Mem. Amer. Math. Soc. 591, Amer. Math. Soc., Providence, RI, 1996.

[CR] Curtis, C., Reiner, I., Representation theory of finite groups and associative algebras, Wiley, New York 1962.

[DlR] Dlab, V., Ringel, C., Every semi-primary ring is the endomorphism ring of a projective module over a quasi-hereditary ring, Proc. Amer. Math. Soc. 107 (1989), 1–5.

[D1] Dipper, R., Polynomial representations of finite general linear groups in non-describing characteristic, Progr. Math. 96 Birkhäuser, Basel 1991, 343–370.

[D2] Dipper, R., On quotients of Hom-functors and representations of finite general linear groups, I, J. Algebra 130 (1990), 235–259.

[DD] Dipper, R., Du, J., Harish-Chandra vertices and Steinberg's tensor product theorems for finite general linear groups, Proc. London. Math. Soc. 75 (1997), 559–599.

[DJ] Dipper, R., James, G., The q-Schur algebra, Proc. London Math. Soc. 59 (1989), 23–50.

[DJM] Dipper, R., James, G., Mathas, A., The (Q, q)-Schur algebras, Proc. Lond. Math. Soc. 77 (1998), 327–361.

[Du1] Du, J., Kahzdan-Lusztig bases and isomorphism theorems for q-Schur algebras, Contemp. Math. 139, Birkhäuser, Boston 1992, 121–140.

[Du2] Du, J., IC bases and quantum linear groups, Proc. Sympos. Pure Math. 56, Vol. 2, Amer. Math. Soc., Providence, RI, 1994, 135–148.

[Du3] Du, J., A note on quantized Weyl reciprocity at roots of unity, Algebra Colloq. 2 (1995), 363–372.

[DPS1] Du, J., Parshall, B., Scott, L., Stratifying endomorphism algebras associated to Hecke algebras, J. Algebra 203 (1998), 169–210.

[DPS2] Du, J., Parshall, B., Scott, L., Cells and q-Schur algebras, Transform. Groups 3 (1998), 33–49.

[DPS3] Du, J., Parshall, B., Scott, L., Quantum Weyl reciprocity and tilting modules, Comm. Math. Physics 195 (1998), 321–352.

[DR1] Du, J., Rui, H., Based algebras and standard bases for quasi-hereditary algebras, Trans. Amer. Math. Soc. 350 (1998), 3207–3235.

[DR2] Du, J., Rui, H., Borel type subalgebras for the q-Schur^m algebra, J. Algebra 213 (1999), 567–595.

[DS1] Du, J., Scott, L. L., Lusztig conjectures, old and new, I, J. Reine Angew. Math. 455 (1994), 141–182.

[DS2] Du, J., Scott, L. L., The q-Schur^2 algebra, Trans. Amer. Math. Soc. 352 (2000), 4325–4353.

[G] Geck, M., Modular Harish-Chandra series, Hecke algebras and (generalized) q-Schur algebras, in: Modular Representation Theory of Finite Groups (Michael J. Collins, Brian J. Parshall, Leonard L. Scott, eds.) Walter de Gruyter, Berlin 2001, 1–66.

[GH] Geck, M., Hiss, G., Modular representations of finite groups of Lie type in non-defining characteristic, in: Finite Reductive groups: Related structures and representations (M. Cabanes, ed.), Birkhäuser, Basel 1997, 195–249.

[GHM] Geck, M., Hiss, G., Malle, G., Towards a classification of the irreducible representations in non-describing characteristic of finite group of Lie type, Math. Z. 222 (1996), 353–386.

[GL] Graham, J., Lehrer, G., Cellular algebras, Invent. Math. 123 (1996) 1–34.

[G1] Green, J. A., The characters of finite general linear groups, Trans. Amer. Math. Soc. 80 (1955), 402–447.

[G2] Green, J. A., On certain subalgebras of the Schur algebra, J. Algebra 131 (1990), 265–285.

[G3] Green, J. A., Combinatorics and the Schur algebra, J. Pure Appl. Algebra 88 (1993) 89–106.

[Gr] Green, R., A Straightening formula for the quantized codeterminants, Comm. Algebra 24 (1996), 2887–2913.

[GrH] Gruber, J., Hiss, G., Decomposition numbers of finite classical groups for linear primes, J. Reine Angew. Math. 485 (1997), 55–91.

[Gy] Gyoja, A., Cells and modular representations of Hecke algebras, Osaka J. Math. 33 (1996), 307–341.

[H1] Hiss, G., Harish-Chandra series of Brauer characters in a finite group with split BN-pair, J. London Math. Soc. 48 (1991), 219–228.

[H2] Hiss, G., Decomposition numbers of finite groups of Lie type in non-defining characteristic, in: Representation Theory of Finite Groups and Finite-Dimensional Algebras (Michler, G. O., Ringel, C. M., eds.), Progr. Math. 95, Birkhäuser, Basel 1991, 405–418.

[HL1] Howlett, R., Lehrer, G., Induced cuspidal representations and generalized Hecke rings, Invent. Math. 58 (1980), 37–64.

[HL2] Howlett, R., Lehrer, G., On the Harish-Chandra induction and restriction for the modules of Levi subgroups, J. Algebra 165 (1994), 172–183.

[K] Kashiwara, M., Crystal bases of the q-analogue of universal enveloping algebras, Duke Math. J. 63 (1991), 465–516.

[KL] Kazhdan, D., Lusztig, G., Representations of Coxeter groups and Hecke algebras, Invent. Math. 53 (1979), 165–184.

[L1] Lusztig, G., Left cells in Weyl groups, Lecture Notes in Math. 1024, Springer-Verlag, Berlin 1983, 99–111.

[L2] Lusztig, G., Characters of Reductive Groups over a Finite Field, Princeton University Press, New Jersey 1984.

[L3] Lusztig, G., Cells in the affine Weyl groups, II, J. Algebra 109 (1987) 539–548.

[L4] Lusztig, G., Canonical bases arising from quantized enveloping algebras, J. Amer. Math. Soc. 3 (1990), 447–498.

School of Mathematics, University of New South Wales, Sydney 2052, Australia

E-mail: j.du@unsw.edu.au

Filtrations on projective modules for Iwahori–Hecke algebras

Meinolf Geck and Raphaël Rouquier

Abstract. We consider a generic Iwahori–Hecke algebra $H_{\mathcal{O}}$ associated with a finite Weyl group, defined over a suitable discrete valuation ring $\mathcal{O}$. We define filtrations on $H_{\mathcal{O}}$-modules in terms of Lusztig's a-function. For a projective module, we show that the quotients of this filtration are direct sums of irreducible lattices. As an application, we prove refinements of the results on decomposition numbers obtained by the first named author in [4].

1. Introduction

Let H be the generic Iwahori–Hecke algebra associated with a finite Weyl group W, defined over the ring $A = \mathbb{Z}[v, v^{-1}]$ where v is an indeterminate (see, for example, [1, §68]). Let K be the field of fractions of A and $\mathcal{O} \subseteq K$ be a discrete valuation ring with residue field k such that $A \subseteq \mathcal{O}$ and the image of v in k has finite order. We have a corresponding decomposition map d_k^H between the Grothendieck groups of H_K and H_k (see (2.1) below). One of the main problems in the representation theory of Iwahori–Hecke algebras is the determination of such decomposition maps. For a survey of known results and their applications to the representation theory of finite reductive groups, see [3].

This paper is a continuation of the work begun in [4]. We assume that the characteristic of k is either 0 or a good prime for W. Using Lusztig's a-function, it is possible to attach a-values to the simple modules of H_K and H_k. Then it is shown in [4] that d_k^H is compatible with these a-values in the following sense. Let $\mathrm{Irr}(H_K)$ and $\mathrm{Irr}(H_k)$ be complete sets of representatives of the isomorphism classes of simple modules for H_K and H_k, respectively. Then, in the appropriate Grothendieck group of finite-dimensional modules over H_K and H_k, respectively, we have equations

$$d_k^H([V]) = \sum_{M \in \mathrm{Irr}(H_k)} d_{V,M}\,[M] \quad \text{where } V \in \mathrm{Irr}(H_K).$$

The compatibility result mentioned above can now be stated as follows; see [4, Theorem 3.3]. For any $V \in \mathrm{Irr}(H_K)$ and $M \in \mathrm{Irr}(H_k)$ we have

$$d_{V,M} \neq 0 \implies a_M \leq a_V. \tag{1.1}$$

Moreover, there exists a subset $B \subseteq \mathrm{Irr}(H_K)$ and a bijection $B \leftrightarrow \mathrm{Irr}(H_k)$, $V \leftrightarrow \overline{V}$, such that the following holds (see (2.5) below):

$$d_{V,\overline{V}} = 1 \quad \text{and} \quad a_V = a_{\overline{V}} \quad \text{for all } V \in B. \tag{1.2}$$

The above two conditions together imply that the decomposition matrix of d_k^H has a lower triangular shape with 1 along the diagonal. There is in fact an explicit construction of the subset $B \subseteq \mathrm{Irr}(H_K)$ which in turn is based on a construction due to Lusztig, in terms of the "asymptotic" algebra J.

The aim of this paper is to prove a number of refinements of the above results, which have been announced in [4, Remark 3.7]. These refinements essentially assert that certain decomposition numbers $d_{V,M}$ must be 0. For example, we show in Corollary 4.3 that for any $V \in \mathrm{Irr}(H_K)$ and $M \in \mathrm{Irr}(H_k)$ we have the following implication:

$$d_{V,M} \neq 0 \quad \text{and} \quad a_V = a_M \quad \Longrightarrow \quad V \in B \quad \text{and} \quad M = \overline{V}. \tag{1.3}$$

Combining this with (1.1) shows that there is a unique subset $B \subseteq \mathrm{Irr}(H_K)$ and a unique bijection $B \leftrightarrow \mathrm{Irr}(H_k)$ such that (1.2) holds.

These results follow from the study of certain filtrations of $H_{\mathcal{O}}$-modules defined in terms of the a-function; see Definition 3.2. For a projective module, we obtain strong assertions about the quotients of such a filtration in Theorems 3.4 and 4.2. This is partly based on the results in [4], which will be recalled in Section 2. The applications, especially to decomposition numbers, are discussed in Section 4.

2. Decomposition maps and the asymptotic algebra

Throughout this paper, W is a finite Weyl group with set of simple reflections S. Let $A = \mathbb{Z}[v, v^{-1}]$ be the ring of Laurent polynomials in an indeterminate v over $\mathbb{Z}$. Let H be the corresponding generic Iwahori–Hecke algebra over A with parameter v^2. Then H has a free A-basis $\{T_w\}_{w \in W}$ with the well-known multiplication rules; i.e., we have $(T_s - v^2)(T_s + 1) = 0$ for $s \in S$ and $T_w = T_{s_1} \dots T_{s_m}$ whenever $w = s_1 \dots s_m$ with $s_i \in S$ is a reduced expression.

If $A \to R$ is any homomorphism into a commutative ring R and Y is any A-module, we regard R as an A-module and set $Y_R := R \otimes_A Y$. This applies in particular to H itself. If R is a field we denote by $R_0(H_R)$ the Grothendieck group of finite-dimensional H_R-modules. (All modules are left modules, unless explicitly stated otherwise.)

(2.1) Decomposition maps are defined in the following setting (see [5] and [3, §2]). Let K be the field of fractions of A. By extension of scalars, we obtain a K-algebra H_K. Let $\mathfrak{p} \subset A$ be a non-zero prime ideal. We consider the corresponding localized ring $A_{\mathfrak{p}}$, and denote its residue field by $k_{\mathfrak{p}}$. We assume that the image of v in $k_{\mathfrak{p}}$ has finite order. Then, by [4, Corollary 3.6], the algebra $H_{k_{\mathfrak{p}}}$ is split.

There exists a discrete valuation ring $\mathcal{O} \subseteq K$ with maximal ideal $J(\mathcal{O})$ such that $A \subseteq \mathcal{O}$ and $J(\mathcal{O}) \cap A = \mathfrak{p}$. (This follows from the fact that A is a regular ring, cf. [3, §2]). Let k be the residue field of $\mathcal{O}$; we may regard k as an extension of $k_{\mathfrak{p}}$. Since $H_{k_{\mathfrak{p}}}$ is split, the scalar extension from $k_{\mathfrak{p}}$ to k defines an isomorphism $R_0(H_{k_{\mathfrak{p}}}) \cong R_0(H_k)$ which preserves the classes of simple modules. We will henceforth identify

$$R_0(H_{k_{\mathfrak{p}}}) = R_0(H_k).$$

Then we have a corresponding decomposition map

$$d_k^H : R_0(H_K) \to R_0(H_k),$$

defined in the usual way by choosing $\mathcal{O}$-forms for the simple H_K-modules and reducing them modulo the maximal ideal of $\mathcal{O}$ (see [2, Section I.17]). Note that d_k^H only depends on $\mathfrak{p}$ but not on the choice of $\mathcal{O}$ since A is integrally closed in K and $H_{k_{\mathfrak{p}}}$ is split (see [3, Proposition 2.3]).

(**2.2**) It is shown in [4] that the decomposition map d_k^H has an alternative interpretation as follows. Let J be Lusztig's "asymptotic" algebra and $\phi\colon H \to J_A$ be the corresponding A-algebra homomorphism; see [7]. (By definition, J is a free $\mathbb{Z}$-algebra with a basis labelled by the elements of W, and we set $J_A = A \otimes_{\mathbb{Z}} J$.) If $A \to R$ is any homomorphism into a commutative ring R, we obtain a corresponding homomorphism of R-algebras $\phi_R\colon H_R \to J_R$.

Consider the k-algebra homomorphism $\phi_k\colon H_k \to J_k$, which in turn determines a group homomorphism

$$(\phi_k)_*\colon R_0(J_k) \to R_0(H_k).$$

Assume from now on that the characteristic of k is either 0 or a good prime for W. Then it is shown in [4, Remark 2.5] that $R_0(J_k)$ and $R_0(H_K)$ can be naturally identified and that, under this identification, $(\phi_k)_*$ coincides with d_k^H. The identification of $R_0(J_k)$ and $R_0(H_K)$ is based on the following two facts.

(1) The K-algebra homomorphism $\phi_K\colon H_K \to J_K$ is an isomorphism; see [7, Theorem 2.8].

(2) The algebras J_K and J_k are both semisimple, and the decomposition map $d_k^J\colon R_0(J_K) \to R_0(J_k)$ (defined in a similar way as d_k^H above) is an isomorphism preserving the classes of simple modules; see [4, Remark 2.5].

Thus the map $(\phi_K)_* \circ (d_k^J)^{-1}\colon R_0(J_k) \to R_0(H_K)$ is also an isomorphism which preserves the classes of simple modules. Given a simple J_k-module E, we denote by E_* a simple H_K-module corresponding to E under that map. With this notation, we have the following identity in $R_0(H_k)$ (see [4, Theorem 3.3]):

$$d_k^H([E_*]) = (\phi_k)_*([E]) \quad \text{for all simple } J_k\text{-modules } E.$$

This means that the problem of determining d_k^H is equivalent to that of determining $(\phi_k)_*$.

(**2.3**) It is possible to attach non-negative integers ("a-values") to the simple modules of H_k, J_k, and H_K, as follows.

First, let $\{C_w\}_{w\in W}$ be the Kazhdan–Lusztig basis of H and $a\colon W \to \mathbb{N}_0$ be Lusztig's a-function. Then, following [8, Lemma 1.9], to any simple H_k-module M we attach a non-negative integer a_M by the requirement that

$$\begin{aligned} C_w M &= 0 \quad \text{for all } w \in W \text{ with } a(w) > a_M, \\ C_w M &\neq 0 \quad \text{for some } w \in W \text{ with } a(w) = a_M. \end{aligned}$$

Next, let $\{t_w\}_{w\in W}$ be the natural basis of J. To any simple J_k-module E we can also attach a non-negative integer a_E by the requirement that $t_w E \neq 0$ for some $w \in W$ with $a(w) = a_E$. This is well-defined since $J_k = \bigoplus_i J_k^i$ is a decomposition into 2-sided ideals where J_k^i is the subspace generated by all t_w with $a(w) = i$ (see [8, (1.3)(d)]).

Finally, we can also attach an a-value a_V to any $V \in \mathrm{Irr}(H_K)$, in an analogous was as this was done above for the simple H_k-modules. By [9, (3.3), (3.4)], that a-value can also be determined in terms of the generic degree associated to V. With the notation in (2.2), we have

$$a_{E_*} = a_E \quad \text{for every simple } J_k\text{-module } E.$$

The basic relation between the a-values for the modules of H_k and J_k is as follows.

Theorem 2.4 ([8, Lemma 1.9]). *For any simple H_k-module M, there exists a J_k-module $\tilde{M}_J$ such that, in $R_0(H_k)$, we have $(\phi_k)_*([\tilde{M}_J)] = [M] +$ sum of terms $[M']$ where M' are simple H_k-modules with $a_{M'} < a_M$. Moreover, for all composition factors E of $\tilde{M}_J$, we have $a_E = a_M$.*

(Lusztig gives an explicit construction for $\tilde{M}_J$; we will come back to this point in Proposition 4.5 below.)

(**2.5**) We can now describe the subset $B \subseteq \mathrm{Irr}(H_K)$ and the bijection $B \leftrightarrow \mathrm{Irr}(H_k)$, $V \leftrightarrow \overline{V}$, satisfying condition (1.2). Let $M \in \mathrm{Irr}(H_k)$ and $\tilde{M}_J$ be as in Theorem 2.4. Then $\tilde{M}_J$ has a unique composition factor $E(M)$ such that $[M]$ occurs in the decomposition of $(\phi_k)_*([E(M)])$. In $R_0(H_k)$, we have $(\phi_k)_*([E(M)] = [M] +$ sum of terms $[M']$ where M' are simple H_k-modules with $a_{M'} < a_M$. It is easily seen that non-isomorphic modules M, M' yield non-isomorphic modules $E(M)$, $E(M')$. So we can set

$$B = \{E(M)_* \mid M \in \mathrm{Irr}(H_k)\}$$

and define a bijection $B \leftrightarrow \mathrm{Irr}(H_k)$ by $E(M)_* \leftrightarrow M$. Note that we have

$$a_{E(M)_*} = a_{E(M)} = a_M.$$

For details of the proof, see [4, Section 3]. In Proposition 4.5 below, we will show that $\tilde{M}_J$ is simple and so $\tilde{M}_J = E(M)$.

3. Filtrations on modules

Let $\mathcal{O} \subseteq K$ be a discrete valuation ring as in (2.1). Then $(K, \mathcal{O}, k)$ is a "modular system" for the $\mathcal{O}$-algebra $H_\mathcal{O}$, as in the usual setting for modular representation theory (e.g., in [2, Section I.17]).

We will mainly be dealing with $H_\mathcal{O}$-modules which are finitely generated and free over $\mathcal{O}$ (such modules will be called *lattices*). Note that, since H_K is split semisimple (see [6]), a result of Heller shows that the Krull–Schmidt–Azuyama Theorem holds for $H_\mathcal{O}$-lattices (see [1, Theorem 30.18]). Moreover, idempotents can be lifted from H_k to $H_\mathcal{O}$ (see [1, Ex. 6.16]). Thus the results in [2, Section I.17] are valid without passing to the completion of $\mathcal{O}$.

Recall that all modules are left modules, unless explicitly stated otherwise. For any lattice V, we shall write $V_K = K \otimes_\mathcal{O} V$ and $V_k = k \otimes_\mathcal{O} V$.

(**3.1**) For the following constructions, see [8, Section 1]. For any $i \geq 0$, let

$$H^{\geqslant i} = A\text{-submodule of } H \text{ generated by all } C_w \text{ with } a(w) \geq i.$$

Note that $H^{\geqslant 0} = H$ and $H^{\geqslant N} = \{0\}$ where N is the length of the longest element of W (see [7, Proposition 1.2]). We obtain a filtration

$$\{0\} = H^{\geqslant N} \subseteq H^{\geqslant N-1} \subseteq \cdots \subseteq H^{\geqslant 1} \subseteq H^{\geqslant 0} = H \tag{1}$$

where each term $H^{\geqslant i}$ is a 2-sided ideal of H which is finitely generated and free over A. Let $H^i := H^{\geqslant i}/H^{\geqslant i+1}$; this is an (H, H)-bimodule which is finitely generated and free over A. (A basis is given by the images of the C_w with $a(w) = i$.) There is also a natural (J_A, J_A)-bimodule structure on H^i which is compatible with the action of H itself; denoting the action of J_A by $\circ$, we have for $f \in H^i$, $h \in H$ and $j \in J_A$,

$$hf = \phi(h) \circ f, \quad j \circ (fh) = (j \circ f)h, \quad (hf) \circ j = h(f \circ j), \tag{2}$$

where $\phi \colon H \to J_A$ is Lusztig's homomorphism.

If $A \to R$ is any homomorphism into a commutative ring R, we can apply extension of scalars and obtain 2-sided ideals $H_R^{\geqslant i}$ in H_R with similar properties as above. In particular, each quotient H_R^i admits left and right actions by elements of H_R and of J_R, and these satisfy similar compatibility properties as above; i.e., we have for $f \in H_R^i$, $h \in H_R$ and $j \in J_R$,

$$hf = \phi_R(h) \circ f, \quad j \circ (fh) = (j \circ f)h, \quad (hf) \circ j = h(f \circ j). \tag{2'}$$

In particular, this means that H_R^i is a (J_R, H_R)-bimodule.

Definition 3.2. Consider the filtration of $H_\mathcal{O}$ obtained by extending scalars from A to $\mathcal{O}$ in (3.1)(1):

$$\{0\} = H_\mathcal{O}^{\geqslant N} \subseteq H_\mathcal{O}^{\geqslant N-1} \subseteq \cdots \subseteq H_\mathcal{O}^{\geqslant 1} \subseteq H_\mathcal{O}^{\geqslant 0} = H_\mathcal{O}.$$

Recall that each term $H_{\mathcal{O}}^{\geqslant i}$ is a 2-sided ideal in $H_{\mathcal{O}}$; moreover, $H_{\mathcal{O}}^{\geqslant i}$ and $H_{\mathcal{O}}^{i}$ are finitely generated and free over $\mathcal{O}$. Let V be any $H_{\mathcal{O}}$-module. Then we have a filtration

$$\{0\} = V^{\geqslant N} \subseteq V^{\geqslant N-1} \subseteq \cdots \subseteq V^{\geqslant 1} \subseteq V^{\geqslant 0} = V, \quad \text{where} \quad V^{\geqslant i} := H_{\mathcal{O}}^{\geqslant i} V.$$

We also set $V^i := V^{\geqslant i}/V^{\geqslant i+1}$ for any i.

Note that, *a priori*, each V^i is an $H_{\mathcal{O}}$-module but it is not clear that it is free over $\mathcal{O}$, even if V is an $H_{\mathcal{O}}$-lattice.

(3.3) Let P be a projective $H_{\mathcal{O}}$-module which is of the form $P = H_{\mathcal{O}}e$ where $e \in H_{\mathcal{O}}$ is an idempotent. We consider the filtration of P as given by Definition 3.2. Let $i \geq 0$. Then we have

$$P^{\geqslant i} = H_{\mathcal{O}}^{\geqslant i} P = H_{\mathcal{O}}^{\geqslant i} e \quad \text{and} \quad P^{\geqslant i+1} = H_{\mathcal{O}}^{\geqslant i+1} P = H_{\mathcal{O}}^{\geqslant i+1} e.$$

Now consider the exact sequence

$$\{0\} \to H_{\mathcal{O}}^{\geqslant i+1} \to H_{\mathcal{O}}^{\geqslant i} \to H_{\mathcal{O}}^{i} \to \{0\}.$$

Multiplying with e yields a sequence

$$\{0\} \to H_{\mathcal{O}}^{\geqslant i+1} e \to H_{\mathcal{O}}^{\geqslant i} e \to H_{\mathcal{O}}^{i} e \to \{0\}$$

which is also exact. (To see this, also work with the idempotent $1 - e$.) It follows that we have a canonical isomorphism of $H_{\mathcal{O}}$-modules

$$P^i \cong H_{\mathcal{O}}^i e.$$

Hence P^i is isomorphic to a direct summand of $H_{\mathcal{O}}^i$. This shows, in particular, that P^i is an $H_{\mathcal{O}}$-lattice.

Theorem 3.4. *Recall that the characteristic of the residue field of $\mathcal{O}$ is either* 0 *or a prime which is good for W. Let P be a projective $H_{\mathcal{O}}$-module as in* (3.3) *and $i \geq 0$ be such that $P^i \neq \{0\}$. We consider a decomposition $P^i = \bigoplus_\lambda V_\lambda$, where each term V_λ is an indecomposable $H_{\mathcal{O}}$-lattice. Then $(V_\lambda)_K$ is simple for all λ and the a-value of $(V_\lambda)_K$ is i.*

Proof. By (3.3), we can assume that $P^i = H_{\mathcal{O}}^i e$. The crux of the proof consists of using the compatible left actions of $J_{\mathcal{O}}$ and of $H_{\mathcal{O}}$ on $H_{\mathcal{O}}^i$; see (3.1). Since P^i is a submodule of $H_{\mathcal{O}}^i$ obtained by right multiplication with some element of $H_{\mathcal{O}}$, it is clear that P^i is also a $J_{\mathcal{O}}$-submodule of $H_{\mathcal{O}}^i$. Let us now regard P^i as a $J_{\mathcal{O}}$-module and write $P^i = \bigoplus_\lambda E_\lambda$ where each E_λ is an indecomposable $J_{\mathcal{O}}$-lattice. By Lemma 3.6 below, $(E_\lambda)_K$ is a simple J_K-module for all λ.

Now we return to the $H_{\mathcal{O}}$-module structure on P^i. Formula (2′) in (3.1) shows that the natural left action of $H_{\mathcal{O}}$ on $H_{\mathcal{O}}^i$ actually factors through the action of $J_{\mathcal{O}}$ via Lusztig's homomorphism $\phi_{\mathcal{O}}$. It follows that $P^i = \bigoplus_\lambda E_\lambda$ is also a decomposition of P^i into $H_{\mathcal{O}}$-lattices. Extending scalars from $\mathcal{O}$ to K yields a decomposition $P_K^i = \bigoplus_\lambda (E_\lambda)_K$, where the H_K-action on each $(E_\lambda)_K$ is pulled back from the J_K-action via

Lusztig's homomorphism ϕ_K. But ϕ_K is an isomorphism; see (2.2)(1). Hence each $(E_\lambda)_K$ is also a simple H_K-module.

Thus we have shown that P^i is direct sum of $H_{\mathcal{O}}$-lattices which are simple after tensoring with K. Then a similar statement holds for any decomposition of P^i as a direct sum of indecomposable $H_{\mathcal{O}}$-lattices since the Krull–Schmidt–Azumaya Theorem holds (see the remarks at the beginning of this section).

It remains to prove the statement about the a-values. Since P^i is a direct summand of $H^i_{\mathcal{O}}$ and V_λ is a direct summand of P^i, we conclude that $(V_\lambda)_K$ is a direct summand of H^i_K. Now, by definition, the latter space is isomorphic to $\bigoplus_\Gamma [\Gamma]_K$, where Γ runs over the two-sided cells of W on which the a-function is constantly equal to i and $[\Gamma]_K$ is the corresponding H_K-module. Then [9, (3.4)] shows that every simple H_K-module occurring in $[\Gamma]_K$, where Γ is a two-sided cell as above, has a-value i. □

Remark 3.5. The above proof shows that each $H_{\mathcal{O}}$-lattice V_λ occurring in the decomposition of P^i also has a structure as $J_{\mathcal{O}}$-module in such a way that the original $H_{\mathcal{O}}$-action is pulled back from that of $J_{\mathcal{O}}$ via Lusztig's homomorphism $\phi_{\mathcal{O}}\colon H_{\mathcal{O}} \to J_{\mathcal{O}}$.

The following result, which is used in the proof of Theorem 3.4, shows that $J_{\mathcal{O}}$ is a hereditary order in J_K (see [1, §26B]).

Lemma 3.6. *Recall the assumptions on the residue field of $\mathcal{O}$ as in Theorem* 3.4. *Let V be a $J_{\mathcal{O}}$-lattice. Then V is projective. If V is indecomposable, then V_K is a simple J_K-module.*

Proof. This easily follows from the fact that J_k is semisimple (see [4, Remark 2.5]). For the convenience of the reader, we give the details here. Since J_k is semisimple, the module V_k is semisimple and hence projective. By [1, Theorem 30.11], this implies that V itself is projective. Now assume that V is such that V_K is not simple. We must show that then V is not indecomposable. By [1, Proposition 23.7], there exists a $J_{\mathcal{O}}$-sublattice $V' \subseteq V$ such that $\{0\} \neq V' \neq V$ and V/V' is also a $J_{\mathcal{O}}$-lattice. Then, as we have already seen, V/V' is projective and so the sequence $\{0\} \to V' \to V \to V/V' \to \{0\}$ splits. Thus we have shown that V is not indecomposable, as desired. □

4. Applications

We now discuss some applications of Theorem 3.4. We keep the set-up and the notation of Section 3.

(4.1) Let $\mathrm{Irr}(H_K)$ and $\mathrm{Irr}(H_k)$ be as in Section 1. Then, in $R_0(H_k)$, we have equations

$$d_k^H([V]) = \sum_{M \in \mathrm{Irr}(H_k)} d_{V,M}\,[M] \quad \text{for } V \in \mathrm{Irr}(H_K), \tag{1}$$

where $d_{V,M}$ are the decomposition numbers. By property (1.1), we know that the sum need only be extended over those M for which $a_M \leq a_V$.

The numbers $d_{V,M}$ also have an alternative interpretation in terms of projective modules as follows. It is well-known (see [2, Chapter 1]) that for each simple H_k-module M there exists a projective indecomposable $H_{\mathcal{O}}$-module $P(M)$, unique up to isomorphism, such that $P(M)_k$ has M as a simple quotient. Then, by Brauer reciprocity (see [2, Theorem I.17.8]), we have the following identity in $R_0(H_K)$:

$$[P(M)_K] = \sum_{V \in \mathrm{Irr}(H_K)} d_{V,M}\,[V] \quad \text{for all } M \in \mathrm{Irr}(H_k). \tag{2}$$

Now let $E(M)_*$ be as in (2.5). Using (1.1) and the equality $a_{E(M)_*} = a_M$, we can also write

$$[P(M)_K] = [E(M)_*] + \sum_{\substack{V \in \mathrm{Irr}(H_K) \\ a_V \geq a_M}} d_{V,M}\,[V] \quad \text{for } M \in \mathrm{Irr}(H_k). \tag{3}$$

The following result shows that, in equation (3), the sum need only be extended over those V for which $a_V > a_M$.

Theorem 4.2. *Let M be a simple H_k-module and $a = a_M$. Then P_K^a is simple and isomorphic to $E(M)_*$. In $R_0(H_K)$, we have $[P(M)_K] = [P_K^a] +$ sum of terms $[V]$ where V are simple H_K-modules with $a_V > a$.*

Proof. Let $P = P(M)$. We can assume that $P = H_{\mathcal{O}}e$ where $e \in H_{\mathcal{O}}$ is a primitive idempotent, so that we are in the set-up of (3.3).

Now let $i \geq 0$ be maximal such that $P = P^{\geqslant i}$. Then we have a surjective homomorphism of $H_{\mathcal{O}}$-modules $P \to P^i$. The fact that P has a unique simple quotient implies that P^i must be indecomposable. Combining this with Theorem 3.4 shows that P_K^i is simple. Moreover, all composition factors of P_K other than P_K^i have a-values $\geq i+1$. Thus equation (4.1)(2) takes the form

$$[P_K] = [P_K^i] + \sum_{\substack{V \in \mathrm{Irr}(H_K) \\ a_V > i}} d_{V,M}\,[V].$$

Comparison with equation (4.1)(3) shows that $i = a$ and $P_K^i \cong E(M)_*$, as desired. □

The following result was announced in [4, Remark 3.7(1)].

Corollary 4.3. *For each $M \in \mathrm{Irr}(H_k)$ there exists a unique $V \in \mathrm{Irr}(H_K)$ such that $a_M = a_V$ and $d_{V,M} \neq 0$. Consequently, property* (1.3) *holds. We have $d_{V,M} = 1$.*

Proof. Let $M \in \mathrm{Irr}(H_k)$ and $V \in \mathrm{Irr}(H_K)$ be such that $a_V = a_M$ and $d_{V,M} \neq 0$. Then $[V]$ occurs in the sum on the right hand side of equation (4.1)(3). But, by Theorem 4.2, there is only one such term with a-value equal to a_M, and this is $E(M)_*$. It follows that $V \cong E(M)_*$ and $d_{V,M} = 1$, and the proof is complete. □

The above results have the following application to Brauer trees, as announced in [4, Remark 4.5].

(**4.4**) Assume that k has characteristic 0, and consider a block of "defect 1" as in [4, (4.3)]. Assume that the block contains $e+1 \geq 2$ simple H_K-modules $V_1, \ldots, V_{e+1}$ such that the corresponding Brauer tree is given by

where the labelling is such that $a_{V_1} \leq a_{V_{e+1}}$. We claim that then we must have

$$a_{V_1} < a_{V_2} < \cdots < a_{V_{e+1}}.$$

For the proof, let us fix some $i \in \{1, \ldots, e\}$. By [4, Theorem 4.4], we already know that $a_{V_i} \leq a_{V_{i+1}}$. Hence it is enough to prove that $a_{V_i} \neq a_{V_{i+1}}$. Now the edge on the tree connecting V_i and V_{i+1} corresponds to a simple H_k-module M and, by Brauer reciprocity, we have $[P(M)_K] = [V_i] + [V_{i+1}]$. Theorem 4.2 shows that V_i and V_{i+1} must have different a-values, as required.

The following result was announced in [4, Remark 3.7(2)].

Proposition 4.5. *Let M be a simple H_k-module and set $a = a_M$. Consider the (J_k, H_k)-bimodule H_k^a and let $\tilde{M}_J := H_k^a \otimes_{H_k} M$. Then $\tilde{M}_J$ is a simple J_k-module isomorphic to $E(M)$ (see (2.5)).*

Proof. Note that the above definition of $\tilde{M}_J$ is the one used in Theorem 2.4. Hence, by definition, $E(M)$ is a composition factor of $\tilde{M}_J$. So it remains to show that $\tilde{M}_J$ is a simple J_k-module. Again, as in the proof of Theorem 4.2, we use the compatible left actions of J_k and H_k on H_k^a.

Let $e \in H_{\mathcal{O}}$ be a primitive idempotent such that $P(M) \cong H_{\mathcal{O}}e$. As in (3.3), we have $P(M)^a \cong H_{\mathcal{O}}^a e$. We claim that, if we set

$$Q := H_{\mathcal{O}}^a \otimes_{H_{\mathcal{O}}} H_{\mathcal{O}}e,$$

then Q_k is simple when regarded as a (left) J_k-module. This can be seen as follows.

We consider the natural $H_{\mathcal{O}}$-module structure on Q given by left multiplication. Now observe that the canonical map $Q \to H_{\mathcal{O}}^a e$ given by multiplication is an isomorphism. (To see this, also work with the idempotent $1-e$.) Hence we actually have $Q \cong P(M)^a$. Using Theorem 4.2, we conclude that Q_K is a simple H_K-module. Now the left action of $H_{\mathcal{O}}$ on $H_{\mathcal{O}}^a$ (and hence also on Q), factors through the action of $J_{\mathcal{O}}$ via Lusztig's homomorphism $\phi_{\mathcal{O}}$ (see (3.1)(2′)). Since Φ_K is an isomorphism (see (2.2)(1)), we deduce that Q_K is also simple when regarded as a J_K-module. Using (2.2)(2), it follows that Q_k is a simple J_k-module, as claimed.

Now, by definition, $P(M)_k$ is a projective indecomposable H_k-module with M as a simple quotient, and we have

$$Q_k \cong (H_{\mathcal{O}}^a \otimes_{H_{\mathcal{O}}} P(M))_k \cong H_k^a \otimes_{H_k} P(M)_k.$$

The quotient map $P(M)_k \to M$ induces a surjective homomorphism of J_k-modules $Q_k \to H_k^a \otimes_{H_k} M = \tilde{M}_J$. Since Q_k is simple, this map must be an isomorphism. So $\tilde{M}_J$ is also simple. □

Remark 4.6. We now have two constructions of the set $B \subseteq \mathrm{Irr}(H_K)$, one in terms of the decomposition numbers $d_{V,M}$ (see (4.3)) and one in terms of the modules $\tilde{M}_J$ constructed above (see (2.5)). It would be interesting to find a more direct characterization of the simple H_K-modules which belong to B. For this purpose, it seems to be necessary to obtain more precise statements about the action of C_w on $V \in \mathrm{Irr}(H_K)$, where $w \in W$ is such that $C_w V \neq 0$ and $a(w) = a_V$. We hope to come back to this point elsewhere.

(**4.7**) Finally, we remark that the above results also yield results about projective modules and decomposition numbers of finite Weyl groups. For this purpose, we have to consider a prime ideal $\mathfrak{p} \subset A$ such that the image of v in k is 1. Then H_k is nothing but the group algebra $k[W]$.

If k has characteristic $p > 0$, then d_k^H coincides with the usual p-modular decomposition map for W; see [3, Example 3.2].

The results in Section 3 yield the following. Let P be a projective $k[W]$-module. Then P has a filtration

$$\{0\} = P^{\geqslant N} \subseteq P^{\geqslant N-1} \subseteq \cdots \subseteq P^{\geqslant 1} \subseteq P^{\geqslant 0} = P, \quad \text{where } P^{\geqslant i} := k[W]_k^{\geqslant i} P.$$

Theorem 4.2 shows that each quotient $P^i = P^{\geqslant i}/P^{\geqslant i+1}$ is a direct sum of $k[W]$-modules which are p-modular reductions of simple modules of W over a field of characteristic 0 whose a-value is i.

The existence of such filtrations on projective modules seems to be a new result, even in the case where W is of type A_n, i.e., a symmetric group.

References

[1] C. W. Curtis and I. Reiner, Methods of representation theory Vol. I and II, Wiley, New York 1981 and 1987.

[2] W. Feit, The representation theory of finite groups, North–Holland Publishing Company, Amsterdam, New York, Oxford 1982.

[3] M. Geck, Representations of Hecke algebras at roots of unity, Séminaire Bourbaki, 50ème année, 1997–98, Astérisque No. 252 (1998), Exp. 836, 33–55.

[4] M. Geck, Kazhdan–Lusztig cells and decomposition numbers, Represent. Theory 2 (1998), 264–277 (electronic).

[5] M. Geck and R. Rouquier, Centers and simple modules for Iwahori–Hecke algebras, in: Finite reductive groups, related structures and representations (M. Cabanes, ed.), Progr. Math. 141, Birkhäuser, Boston 1997, 251–272.

[6] G. Lusztig, On a theorem of Benson and Curtis, J. Algebra 71 (1981), 490–498.

[7] —, Cells in affine Weyl groups II, J. Algebra 109 (1987), 536–548.

[8] —, Cells in affine Weyl groups III, J. Fac. Sci. Univ. Tokyo 34 (1987), 223–243.

[9] —, Leading coefficients of character values of Hecke algebra, Proc. Symp. Pure Math. 47, Amer. Math. Soc., Providence, RI, 1987, 235–262.

UFR de Mathématiques and UMR 7586, Université Paris 7, 2 Place Jussieu, F–75251 Paris, France

Current address: Institut Girard Desargues, Université Lyon 1, 21 av. Claude Bernard, 69622 Villeurbanne cedex, France

E-mail: geck@desargues.univ-lyon1.fr

UFR de Mathématiques and UMR 7586, Université Paris 7, 2 Place Jussieu, 75251 Paris, France

E-mail: rouquier@math.jussieu.fr

Projective representations for some exceptional finite groups of Lie type

*Corneliu Hoffman**

Abstract. Let $G(q)$ an exceptional finite group of Lie type, k an algebraically closed field of characteristic not dividing q and V an irreducible projective representation of G over k. In this paper we obtain lower bounds for the dimension of V in the case G is of type $E_6(q)$, $E_7(q)$, $E_8(q)$, improving the results of [LS]. The new bounds are very close to the dimension of the minimal representations in characteristic zero.

1. Introduction

Let $G(q)$ be a finite group of Lie type defined over $\mathbb{F}_q$. Estimates for the degrees of the minimal projective representations of G over fields of characteristic not dividing q have been obtained in [LS] and [SZ] and have found many applications especially in determining the subgroup structure of the finite simple groups.

With the development of the Deligne–Lusztig theory, the representations in characteristic zero have been studied and consequently the bounds have been improved for some of the classical groups in [TZ], [GPPS], [H].

Recently in [L] the small irreducible complex characters of exceptional groups of Lie type have been determined. The degrees of these characters although a polynomial in q of the same degree as the bounds given in [LS] are in fact much larger that these estimates. The aim of this paper is to improve the bounds of [LS] for some of these groups for all cross characteristics. Note that in the case of characteristic zero the results of [L] settle the problem. Therefore we can assume that the characteristic is positive. Nevertheless we will need at times to consider modules in characteristic zero.

To fix the notation, we consider $G = G(q)$ an exceptional group of type $E_i(q)$, $i = 6, 7, 8$, and V a faithful irreducible projective representation of G over an algebraically closed field of characteristic not dividing q. By abuse of notation we will use V for the representation as well as for the module associated to it.

*I need to thank my former advisor Robert Guralnick for all his support and advice over the years. I also thank Pham Huu Tiep and Michael Collins for reading the preliminary version of this work and providing useful comments.

We will prove the following (here Φ_i denotes the i-th cyclotomic polynomial):

Theorem 1. *Let $G = G(q)$ be a finite simple group of Lie type, and let V be a faithful irreducible projective representation of G over an algebraically closed field of characteristic not dividing q.*

(1) *If $G \cong E_6(q)$, then* $\dim V \geq q\Phi_8(q)\Phi_9(q) - 1$.

(2) *If $G \cong E_7(q)$ then* $\dim V \geq q\Phi_7(q)\Phi_{12}(q)\Phi_{14}(q) - 2$.

(3) *If $G \cong E_8(q)$ then* $\dim V \geq q\Phi_4^2(q)\Phi_8(q)\Phi_{12}(q)\Phi_{20}(q)\Phi_{24}(q) - 3$.

If V is a projective representation of G then V can be viewed as a representation of a central extension $\bar{G}$ of G. We will adopt the following notation: if H is a subgroup of G we denote by $\bar{H}$ the subgroup of $\bar{G}$ containing $Z(\bar{G})$ such that $\bar{H}/Z(\bar{G}) = H$.

Let $P = LQ$ be a parabolic subgroup of G such that the unipotent radical Q is a special group of order q^{2n+1} and L acts transitively on $Z(Q)^*$. If V is an irreducible representation for a central extension $\bar{G}$ of G, note that $\bar{Q} = Q_0 \times Z(\bar{G})$ where Q and Q_0 have the same structure as L-modules (this is true for groups of type $E_i(q)$ because the order of the Schur multiplier is relatively prime to q). We will use the following lemma.

Lemma 1. *Let $G = G(q)$, $P = LQ$ be as above. Let V be an irreducible module for a central extension $\bar{G}$ of G of dimension smaller than the bounds in the theorem. Then $V = [V, Z(Q_0)] \oplus [N, Q_0] \oplus C_V(Q_0)$ where $\bar{Q} = Q_0 \times Z(\bar{G})$ is as above and $N = C_V(Z(Q_0))$. Moreover $[N, Q_0] \neq 0$ and* $\dim[V, Z(Q_0)] = (q-1)q^n$.

Proof. The action of $Z(Q_0)$ on V and then the action of Q_0 on $C_V(Z(Q_0))$ yield a decomposition

$$V = [V, Z(Q_0)] \oplus [N, Q_0] \oplus C_V(Q_0)$$

which is a direct sum of $\bar{P}$-modules.

One can find a root subgroup $U_\alpha \leq Q_0$ such that $U_\alpha \cap Z(Q_0) = 1$ and such that U_α and $Z(Q_0)$ are conjugate in G ([LS]). If $M = [V, Z(Q_0)]$ then the Brauer character $t_M(x) = 0$ for all $x \notin Z(Q_0)$ ([LS] Lemma 2.3). Let $x \in U_\alpha$, $y \in Z(Q)$, with x, y conjugate in G, $N = C_V(Z(Q_0))$, and note that $t_N(y) = \dim C_V(Z(Q_0))$, $t_M(x) \neq t_M(y)$ and $t_V(x) = t_V(y)$ so $t_N(x) \neq \dim C_V(Z(Q_0))$. This means that x does not act trivially on $C_V(Z(Q_0))$ and hence $[N, Q_0] \neq 0$.

Moreover all irreducible Q_0-submodules of $M = [V, Z(Q_0)]$ have dimension q^n and the corresponding Brauer characters are determined by the action of $Z(Q_0)$; therefore these submodules are permuted transitively by $\bar{L}$ (see [LS]). It follows that $\dim[V, Z(Q_0)] = m(q-1)q^n$ where m is the multiplicity of each of these modules. If $m \geq 2$ then this already will surpass the bounds in the theorem so we can assume $m = 1$. □

Now the decomposition $V = [V, Z(Q_0)] \oplus [N, Q_0] \oplus C_V(Q_0)$ holds for any projective module V; also, since V actually affords a (normal) representation for Q, we will write this decomposition with Q in place of Q_0 from here on. We will use this decomposition to obtain lower bounds for dim V; Lemma 1 applies to the first term, and the length of the smallest nontrivial orbit of L acting on $Q/Z(Q)$ gives a lower bound for $\dim[N, Q]$. (This is true because $Q/Z(Q)$ is abelian so the action of L on it is dual to the action of L on its characters.)

Moreover, one can regard $C_V(Q)$ as a $\bar{L}$-module and prove that $\bar{L}'$ will not act trivially on $C_V(Q)$. To do this will need the following result.

Lemma 2. *Let G be a finite group of Lie type, $P = P_J$ a parabolic subgroup and U an irreducible G-module in characteristic zero. Consider W the Weyl group of G and W_J its subgroup corresponding with P_J. Then the number of constituents of U_{P_J} of degree one is bounded above by the number of irreducible constituents of the permutation representation of W on W/W_J.*

Proof. If M is a constituent of U_{P_J} of degree one and λ is the corresponding character of P, then $U \leq M_P^G$ (this is because U is irreducible). Moreover, if λ, μ are two (not necessary distinct) linear characters of P, then $(\lambda_P^G, \mu_P^G)_G \neq 0$ (here $(\alpha, \beta)_H$ is the usual scalar product of complex characters). By Frobenius reciprocity we have that $(\lambda_P^G, \mu_P^G)_G = (\lambda, (\mu_P^G)_P)_P$.

We know by Mackey's theorem that $(M_P^G)_P)_P = \sum_{x \in P\backslash G/P} ({}^x M_{P\cap {}^x P})^P$. Note also that, by the Bruhat decomposition, two linear characters that agree on $P \cap {}^x P$ agree on the maximal torus of P and so they must agree on P. We obtain that

$$(\lambda_P^G, \mu_P^G) = (\lambda, (\mu_P^G)_P)_P = \sum_{x \in P\backslash G/P} (\lambda, ({}^x\mu_{P\cap {}^x P})^P) = \sum_{x \in P\backslash G/P} (\lambda_{P\cap {}^x P}, {}^x\mu_{P\cap {}^x P}).$$

The conclusion is that there are at most $|P\backslash G/P|$ possible representations of P of degree one that can occur in U_P, these being the conjugates of M under the representatives of the double cosets. Furthermore $|\ P\backslash G/P\ |=|\ W_J\backslash W/W_J\ |= (1_{W_J}^W, 1_{W_J}^W)$ and this is the number of the irreducible factors of the permutation representation. □

2. The case $G = E_6(q)$

In this case we will take as in [LS] the parabolic $P = LQ$ where L, the Levi complement of P, is of type A_5. Then Q is a special group of order q^{20+1} and $Q/Z(Q)$ as an L-module has the form $\Lambda^3 N$ where N is the natural 6-dimensional module.

Fortunately, using the results in [CH] or [R] we can compute the sizes of the orbits of L on $\Lambda^3 N$. There are five orbits. Their corresponding representatives and lengths are given by (where $\mathcal{O}_x$ denotes the orbit of x):

$$|\mathcal{O}_{e_1 \wedge e_2 \wedge e_3}| = (q^5 - 1)(q^2 + 1)(q^3 + 1)$$

$$|\mathcal{O}_{e_1\wedge e_2\wedge e_3+e_4\wedge e_5\wedge e_6}| = \frac{1}{2}q^9(q-1)(q^2+1)(q^3+1)(q^5-1)$$

$$|\mathcal{O}_{f_\lambda}| = \frac{1}{2}q^9(q-1)(q^2-1)(q^3-1)(q^5-1)$$

$$|\mathcal{O}_{e_1\wedge e_2\wedge e_3+e_1\wedge e_3\wedge e_4+e_5\wedge e_6\wedge e_1}| = q^4(q^4-1)(q^5-1)(q^6-1)$$

$$|\mathcal{O}_{e_1\wedge e_2\wedge e_3+e_1\wedge e_4\wedge e_5}| = q^2(q^3-1)(q^5-1)(q^6-1)/(q-1).$$

To define f_λ, we pick $\lambda \in F_q$ such that the polynomial

$$P_\lambda(X) = \begin{cases} X^2 - \lambda & \text{if } q \text{ odd} \\ X^2 + \lambda X + 1 & \text{if } q \text{ even} \end{cases}$$

is irreducible. Then we define

$$f_\lambda = \begin{cases} e_1 \wedge e_2 \wedge e_3 + \lambda(e_1 \wedge e_2 \wedge e_3 + e_3 \wedge e_4 \wedge e_5 + e_4 \wedge e_2 \wedge e_6) & \text{if } q \text{ odd} \\ e_1 \wedge e_2 \wedge e_6 + e_1 \wedge e_5 \wedge e_3 + e_2 \wedge e_3 \wedge e_4 & \\ \quad + \lambda(e_1 \wedge e_2 \wedge e_3 + e_3 \wedge e_4 \wedge e_5 + e_4 \wedge e_2 \wedge e_6) & \\ \qquad + (\lambda^2+1) e_4 \wedge e_5 \wedge e_6 & \text{if } q \text{ even.} \end{cases}$$

Note that f_λ and f_μ are always in the same orbit by Theorem 2.2 of [CH].

In particular the smallest orbit has length $(q^5-1)(q^2+1)(q^3+1)$ so that $\dim[N, Q] \geq (q^5-1)(q^2+1)(q^3+1)$. Also by Lemma 1 we get that $\dim[V, Z(Q_0)] = (q-1)q^{10}$, hence $\dim[V, Q] \geq q^{10}(q-1) + (q^5-1)(q^2+1)(q^3+1)$. Note that this already improves [LS].

We do know from [L] that in characteristic zero there is a $\bar{G}$-module U of dimension $q\Phi_8\Phi_9$ and this is the smallest possible dimension solving the problem in this case. If we repeat the argument above for this module we obtain that $\dim C_U(Q_0) = (q^6-1)/(q-1)$.

For the rest of the section assume that V is the initial module in characteristic $p > 0$ not dividing q and that U is the minimal module in characteristic 0.

Using the tables in [ATL] we know that the Weyl group of $E_6(q)$ is $\mathrm{Sp}_4(3) \cdot 2$ and that the decomposition in irreducible factors of the permutation representation of $\mathrm{Sp}_4(3)$ on $\mathrm{Sp}_4(3)/S_6$ is (in the notation of [ATL]) $1a + 15b + 20a$.

Using Lemma 2 there are at most 6 linear characters of P in U. In particular L does not act linearly on $C_U(Q)$ so using the results in [TZ] there is a nontrivial L-factor of $C_U(Q)$ of dimension at least $(q^6-1)/(q-1) - 2$.

Consider $x \in L$ a long root element and $y \in Q$ conjugate to x. If $M = [V, Q]$ we note that $t_M(x)$, $t_M(y)$ will not depend on the structure of M but only on its dimension (t_M is the Brauer character of M). This is because x just permutes the irreducible Q-factors of M and the structure of M as a y-module is already prescribed by the dimensional constraints. The same statement holds for the values of $\xi_{M'}$, the ordinary character of $M' = [U, Q]$.

In particular $t_M(x) = \xi_{M'}(x)$, $t_M(y) = \xi_{M'}(y)$. We also have that $t_V(x) = t_V(y)$ and y acts trivially on $C_V(Q)$. Hence x acts trivially on $C_V(Q)$ iff $t_M(x) = t_M(y)$.

This cannot be possible because $\xi_{M'}(x) \neq \xi_{M'}(y)$. Therefore x does not act trivially on $C_V(Q)$ and we can use the results of [GPPS] to obtain that $\dim C_V(Q) \geq (q^6-1)/(q-1)-2$; hence $\dim V \geq q\Phi_8(q)\Phi_9(q)-2$.

Note that in particular $\dim V \geq q\Phi_8(q)\Phi_9(q)-1$ unless $p|(q^6-1)/(q-1)$.

Next assume that $\dim V = q\Phi_8(q)\Phi_9(q)-2$ so that $p|(q^6-1)/(q-1)$. Consider $P_1 = L_1Q_1$ a parabolic of type D_5. In this case Q_1 is elementary abelian of order q^{16} and L_1 acts on it via the spin representation. Also $V = [Q_1, V] \oplus C_V(Q_1)$ as a P_1-module.

Therefore $\dim[Q_1, V]$ is at least as large as the smallest orbit of L_1 on the nontrivial characters of Q_1.

Using [I] we get that $Q_1 \setminus \{0\}$ splits into two orbits under the action of L_1. The two orbit lengths are $(q^3+1)(q^8-1)$ and $q^3(q^5-1)(q^8-1)$.

In particular the smallest orbit has length $(q^3+1)(q^8-1)$ and so this has to be the dimension of $[Q_1, V]$. In particular this means $[Q_1, V]$ is a permutation module for L_1 since the $(q^3+1)(q^8-1)$ eigenvectors for Q_1 form a basis that is permuted by L_1. Let M denote the L_1-module $C_V(Q_1)$. It follows that $\dim M = (q^4+1)(q^3+q+1)-2$.

Since $[Q_1, V]$ is a permutation module for L_1, $\dim C_{[Q_1,V]}(L_1) = 1$ and so $C_M(L_1)$ is a hyperplane in $C_V(L_1)$. If we repeat the argument for the opposite parabolic $\tilde{P}_1 = L_1\tilde{Q}_1$, we get that either $\dim C_M(L_1) \leq 1$ or $C_M(L_1) \cap C_{\tilde{M}}(L_1) \neq 0$ and this leads to a contradiction.

In particular L_1 does not act trivially on M; hence using [H] we get that the dimension of a nontrivial factor of L_1 is at least $(q^5-1)(q^4+q)/(q^2-1)-1$ (this is because $p|(q^6-1)/(q-1)$ so p cannot divide $(q^5-1)/(q-1)$). This is larger than $\dim M$ hence a contradiction. In conclusion $\dim V \geq q\Phi_8(q)\Phi_9(q)-1$.

3. The case $G = E_7(q)$

In this case we consider $P = LQ$ of type D_6. It will follow that the unipotent radical Q is a special group of order q^{32+1} and that $Q/Z(Q)$ is isomorphic to the spin representation of L'. Again [I] describes the orbit lengths in this case.

If q is odd, then $\Omega_{12}(q)$ will have $q+2$ orbits on the spin representation.

- $\frac{q-1}{2}$ orbits of length $q^{15}(q^3-1)(q^5-1)(q^8-1)$.
- $\frac{q-1}{2}$ orbits of length $q^{15}(q^3+1)(q^5+1)(q^8-1)$.
- One orbit of length $(q^3+1)(q^5+1)(q^8-1)$.
- One orbit of length $q^3(q^2+1)(q^4+1)(q^6-1)(q^{10}-1)$.
- One orbit of length $q^7(q^8-1)(q^6-1)(q^{10}-1)$.

If q is even, then $\Omega_{12}(q)$ will have $2q+1$ orbits on the spin module.

- $q-1$ orbits of length $\frac{1}{2}q^{15}(q^3+1)(q^5+1)(q^8-1)$.
- $q-1$ orbits of length $\frac{1}{2}q^{15}(q^3-1)(q^5-1)(q^8-1)$.
- One orbit of length $(q^3+1)(q^5+1)(q^8-1)$.
- One orbit of length $q^3(q^2+1)(q^4+1)(q^6-1)(q^{10}-1)$.
- One orbit of length $q^7(q^8-1)(q^6-1)(q^{10}-1)$.

In particular, for all q, the shortest orbit has length $(q^3+1)(q^8-1)(q^5+1)$ so Lemma 1 gives that if V is an irreducible representation of G then

$$\dim[V,Q] \geq q^{16}(q-1)+(q^3+1)(q^8-1)(q^5+1).$$

Furthermore, by [L], there is a complex representation U of $E_7(q)$ of dimension

$$q\Phi_7\Phi_{12}\Phi_{14} = q(q^6+q^5+q^4+q^3+q^2+q+1)(q^6-q^5+q^4-q^3+q^2-q+1)(q^4-q^2+1).$$

Also note that

$$q\Phi_7\Phi_{12}\Phi_{14} - q^{16}(q-1)-(q^3+1)(q^8-1)(q^5+1) = \frac{(q^6-1)(q^5+q)}{q^2-1}+1.$$

By Lemma 2, the number of linear characters of P occurring in V cannot be larger that the number of irreducible factors in the permutation representation of W on cosets of W_J. The Weyl group of $E_7(q)$ is $2\times \mathrm{Sp}_6(2)$ and the subgroup corresponding to W_J is in fact $2^5\cdot S_6$. The decomposition of the permutation representation of $\mathrm{Sp}_6(2)$ on $\mathrm{Sp}_6(2)/(2^5\cdot S_6)$ is $1a+27a+35b$ (in [ATL] notation) so there are at most 6 linear characters of P inside V. Since we have that $(q^6-1)(q^5+q)/(q^2-1)>6$, if we repeat the argument above for U instead of V we see that L' will not act trivially on $C_U(Q)$. We can use a Brauer characters argument as in Section 2 to get that L' will not act trivially on $C_V(Q)$; therefore, by [H], $\dim C_V(Q)\geq (q^6-1)(q^5+q)/(q^2-1)-2$. Adding the two estimates we get $\dim V\geq q\Phi_7\Phi_{12}\Phi_{14}-3$.

Assume that $\dim V = q\Phi_7\Phi_{12}\Phi_{14}-3$. Consider $P_1 = L_1Q_1$, the parabolic of type $E_6(q)$. In this case Q is elementary abelian of order q^{27} and L_1 act on it irreducibly.

Let $V=[Q_1,V]\oplus C_V(Q_1)$; again we need to find the lengths of the orbits of L_1 on Q_1. Using Lemma 5.4 of [LiS] we get that in the action of $E_6(q)$ on the space of lines of Q_1 there are three orbits. Their respective lengths are

$$(q^9-1)(q^8+q^4+1)$$

$$q^{12}(q^5-1)(q^9-1)$$

$$q^4(q^5-1)(q^9-1)(q^8+q^4+1)/(q-1).$$

Also the smallest orbit is in fact an affine orbit. In particular it follows that $\dim C_V(Q_1) = (q^9-1)(q^8+q^4+1)$ and so

$$\dim C_V(Q_1) = (q^6+1)(q^9+q^6-q-1)/(q^4-1) - 3 = q\Phi_8(q)\Phi_9(q) - 2.$$

Also by a similar argument as in Section 2, it follows that L_1 does not act trivially on $C_V(Q_1)$ so that the smallest possible nontrivial composition factor is of dimension $q\Phi_8(q)\Phi_9(q) - 1$, a contradiction.

In conclusion, $\dim V \geq q\Phi_7\Phi_{12}\Phi_{14} - 2$, proving the theorem.

4. The case $G = E_8(q)$

For this case we consider the parabolic $P = LQ$ of type $E_7(q)$. The unipotent radical Q is a special group of order q^{56+1} and $Q/Z(Q)$ is isomorphic to the 56-dimensional $E_7(q)$-module of highest weight λ_7.

Note that $H = R(L)$, the radical of the Levi complement, is a 1-dimensional torus and that it is centralized by the Levi complement L. Assume that a subgroup $H' \leq H$ fixes all the root subgroups $U_\alpha \subset Q$ that are not in $Z(Q)$. Since $Z(Q) = [Q, Q]$, it follows that H' fixes the whole of Q and so it fixes the entire Borel subgroup. This is impossible since $Z(G) = 1$. Therefore the torus $H = R(L)$ acts faithfully on $Q/Z(Q)$; in particular, since the action is linear, in order to find the lengths of the orbits of L it is enough to compute the orbits of $E_7(q)$ on the 1-spaces of the 56-dimensional module $Q/Z(Q)$.

The stabilizers of 1-spaces for this module have been computed in [LiS] and [C] so we know that there are five orbits each having stabilizers ([LiS] Lemma 4.3).

- A parabolic subgroup of type $E_6(q)$.
- $E_6(q) \cdot 2$ (graph automorphism).
- ${}^2E_6(q) \cdot 2$ (field automorphism).
- A subgroup of type $q^{1+32} \cdot B_5(q) \cdot (q-1)$.
- A subgroup of type $q^{26} \cdot F_4(q) \cdot (q-1)$.

In particular the smallest orbit is in fact an affine orbit, corresponds to a parabolic subgroup of type $E_6(q)$ and has length $(q^5+1)(q^{14}-1)(q^9+1)$. Let V be an irreducible G-module. We have $\dim V \geq q^{28}(q-1) + (q^5+1)(q^{14}-1)(q^9+1)$.

From [L] we know that $E_8(q)$ has an irreducible ordinary representation U of dimension

$$q\Phi_4^2\Phi_8\Phi_{12}\Phi_{20}\Phi_{24} = q(q^2-1)^2(q^4+1)(q^4-q^2+1)(q^8-q^6+q^4-q^2+1)(q^8-q^4+1)$$

and

$$q\Phi_4^2\Phi_8\Phi_{12}\Phi_{20}\Phi_{24} - q^{28}(q-1) - (q^5+1)(q^{14}-1)(q^9+1) = q\Phi_7\Phi_{12}\Phi_{14} + 1.$$

The Weyl groups are $W = 2.\mathrm{O}_8^+(2).2$ and $W_J = \mathrm{Sp}_6(2) \times 2$ and, by [ATL], the permutation character of $\mathrm{O}_8^+(2).2$ on $\mathrm{O}_8^+(2).2/\mathrm{Sp}_6(2) \times 2$ is $1a + 35a + 84a$; so by Lemma 2 there are at most six linear characters of L on U and, repeating the argument above for U instead of V, we get that L cannot act trivially on $C_U(Q)$. Hence we can use a similar argument as in Section 2 to show that L does not act trivially on $C_V(Q)$. Theorem 1 follows from the bound for $E_7(q)$ obtained in the previous section and Lemma 1.

References

[ATL] J. H. Conway, R. T. Curtis, S. P. Norton, R. A. Parker and R. A. Wilson, An ATLAS of Finite Groups, Oxford University Press, Oxford 1985.

[CH] A. M. Cohen and A. G. Helminck, Trilinear alternating forms on a vector space of dimension 7, Comm. Algebra 16 (1988), 1–25.

[C] B. N. Cooperstein, The fifty-six-dimensional module for E_7, I. A four form for E_7, J. Algebra 173 (1995), 361–389.

[GPPS] R. Guralnick, T. Pentilla, C. Praeger and J. Saxl, Linear groups with orders having certain primitive prime divisors, Proc. London Math. Soc. 78 (1999), 167–214.

[H] C. Hoffman, Cross characteristic projective representations for some classical groups, J. Algebra 229 (2000), 666–677.

[I] J.-I. Igusa, A classification of spinors up to dimension twelve, Amer. J. Math. 88 (1985), 337–344.

[LS] V. Landázuri and G. M. Seitz, On the minimal degrees of projective representations of the finite Chevalley groups, J. Algebra 32 (1974), 418–443.

[LiS] M. W. Liebeck and J. Saxl, On the orders of maximal subgroups of the finite exceptional groups of Lie type, Proc. London Math. Soc. 55 (1987), 299–330.

[L] F. Lübeck, Smallest degrees of complex characters of exceptional groups of Lie type, Comm. Algebra, to appear.

[R] G. E. Röhrle, On extraspecial parabolic subgroups, in: Linear Algebraic Groups and Their Representations (R. S. Elman et al., eds.), Angeles, CA, 1992), Contemp. Math. 153, Amer. Math. Soc., Providence, RI, 1993, 143–155.

[SZ] G. M. Seitz and A. E. Zalesskii, On the minimal degrees of projective representations of the finite Chevalley groups, II, J. Algebra 158 (1993), 233–243.

[TZ] P. H. Tiep and A. Zalesskii, Minimal characters of the finite classical groups, Comm. Algebra 24 (1996), 2093–2167.

Department of Mathematics and Statistics, Bowling Green State University, Bowling Green, OH 43403, U.S.A.

E-mail: hoffman@bgnet.bgsu.edu

Splitting fields and twisted group rings for the finite general linear groups

Nicholas J. Kuhn

Abstract. Let the Galois group $\mathrm{Gal}(\mathbb{F}_q\mathbb{F}_p)$ act diagonally on the group ring $\mathbb{F}_q[\mathrm{GL}_n(\mathbb{F}_q)]$, and let $\mathbb{F}_p[\mathrm{GL}_n(\mathbb{F}_q)]_\phi$ denote the fixed point algebra. This is a twisted version of the group ring $\mathbb{F}_p[\mathrm{GL}_n(\mathbb{F}_q)]$: these two $\mathbb{F}_p$-algebras become isomorphic after extending the scalars to $\mathbb{F}_q$. Although $\mathbb{F}_q$ is the minimal field splitting $\mathbb{F}_p[\mathrm{GL}_n(\mathbb{F}_q)]$, the prime field $\mathbb{F}_p$ splits $\mathbb{F}_p[\mathrm{GL}_n(\mathbb{F}_q)]_\phi$. The analogous result also holds for the semigroup ring $\mathbb{F}_p[M_n(\mathbb{F}_q)]$.

1. Introduction

Let p be a prime. If G is a finite group, or more generally, a finite semigroup, we say that a characteristic p field $\mathbb{F}$ is the *p-splitting field of* G if $\mathbb{F}$ is the minimal field splitting the semigroup ring $\mathbb{F}_p[G]$. Such an $\mathbb{F}$ will always be a finite field.

Now let $q = p^s$, and let $\mathbb{F}_q$ be the finite field with q elements. Suppose that the Galois group $\mathrm{Gal}(\mathbb{F}_q : \mathbb{F}_p)$ acts on G, i.e. we have a homomorphism

$$\phi : \mathrm{Gal}(\mathbb{F}_q : \mathbb{F}_p) \to \mathrm{Aut}(G).$$

We use ϕ to define an action of $\mathrm{Gal}(\mathbb{F}_q : \mathbb{F}_p)$ on $\mathbb{F}_q[G]$ as follows: given $\alpha \in \mathrm{Gal}(\mathbb{F}_q : \mathbb{F}_p)$, $\lambda \in \mathbb{F}_q$, and $g \in G$, we let $\alpha \cdot (\lambda[g]) = \alpha(\lambda)[\phi(\alpha)(g)]$. Define $\mathbb{F}_p[G]_\phi$ to be the fixed point algebra $\mathbb{F}_q[G]^{\mathrm{Gal}(\mathbb{F}_q:\mathbb{F}_p)}$.

Basic Galois theory, as presented in [10, Lemma 3.4], implies that the canonical map

$$\mathbb{F}_q \otimes_{\mathbb{F}_p} \mathbb{F}_p[G]_\phi \to \mathbb{F}_q[G]$$

is an isomorphism of $\mathbb{F}_q$-algebras. Thus the $\mathbb{F}_p$-algebra $\mathbb{F}_p[G]_\phi$ can be viewed as a twisted semigroup ring: the two $\mathbb{F}_p$-algebras $\mathbb{F}_p[G]$ and $\mathbb{F}_p[G]_\phi$, though not necessarily isomorphic, become isomorphic after extension of scalars.

Given a pair (G, ϕ) as above, we say that $\mathbb{F}$ is the *p-splitting field of* (G, ϕ) if $\mathbb{F}$ is the minimal field splitting the semigroup ring $\mathbb{F}_p[G]_\phi$.

The point of this note is to explain the following family of examples, in which a natural twisting yields a *smaller* splitting field. Thus, in some sense, it is the twisted algebra that could be viewed as the more fundamental object.

Theorem 1.1. *Let G be either the group $\mathrm{GL}_n(\mathbb{F}_q)$ or the (multiplicative) semigroup $M_n(\mathbb{F}_q)$. Let ϕ be the 'canonical' action of $\mathrm{Gal}(\mathbb{F}_q : \mathbb{F}_p)$ on these: $\alpha \cdot (\lambda_{ij}) = (\alpha(\lambda_{ij}))$ for all $\alpha \in \mathrm{Gal}(\mathbb{F}_q : \mathbb{F}_p)$. Then*

(1) *$\mathbb{F}_q$ is the p-splitting field for G, but*

(2) *$\mathbb{F}_p$ is the p-splitting field for (G, ϕ).*

The theorem is proved in §3. Our proof procceds by making use of a 'twisted' generic representation category, which is related to both the symmetric groups and our twisted semigroup rings. The well known fact that $\mathbb{F}_p$ splits the symmetric groups is then 'lifted' to an analogous fact about the twisted generic representation category, and then 'projected' down to the twisted semigroup algebras, using some properties of splitting fields for abelian categories presented in §2.

It is likely that group theorists more versed in algebraic groups techniques than the author will be able to construct their own proofs, using weights of representations and the like. We remark that the key input to our theorem, Theorem 3.1, is roughly a theorem at the level of the Steinberg Tensor Product Theorem, and is proved in [8] from first principles.

In §4, we reveal how we came across the theorem: it is useful when one wishes to study modules for the finite general linear groups via the associated Schur algebras. This allows many arguments in the literature[1] that assume one is working over an algebraically closed field to work equally well over the prime field.

2. Splitting fields and recollement

With $\mathbb{F}$ a field, let $\mathcal{A}$ be an $\mathbb{F}$-linear abelian category. We say that $\mathbb{F}$ *splits* $\mathcal{A}$ if $\mathrm{End}_{\mathcal{A}}(A) = \mathbb{F}$ for all simple objects $A \in \mathcal{A}$. In this section we record a couple of situations in which knowing that $\mathbb{F}$ splits various categories implies that $\mathbb{F}$ splits a related category.

The first involves recollment diagrams (see e.g. [6]). One has a diagram of $\mathbb{F}$-linear abelian categories

$$\begin{array}{ccccc} & \xleftarrow{q} & & \xleftarrow{l} & \\ \mathcal{A} & \xrightarrow{i} & \mathcal{B} & \xrightarrow{e} & \mathcal{C} \\ & \xleftarrow{p} & & \xleftarrow{r} & \end{array}$$

satisfying the following conditions:

[1] For example, results in [1, 3, 7].

(a) l is left adjoint to e, which is left adjoint to r. Thus e is exact.

(b) q is left adjoint to i, which is left adjoint to p. Thus i is exact.

(c) i is a full embedding, and $eB \simeq 0$ if and only if $B \simeq i(A)$ for some A in $\mathcal{A}$.

(d) The adjunctions $\epsilon_C : erC \to C$ and $\eta_C : C \to elC$ are isomorphisms for all C in $\mathcal{C}$.

Lemma 2.1. *In this situation, $\mathbb{F}$ splits $\mathcal{B}$ if and only if $\mathbb{F}$ splits $\mathcal{A}$ and $\mathcal{C}$.*

Proof. As noted in [6, Proposition 4.7], there is a bijection between sets of isomorphism classes of simple objects:

$$\{\text{simples in } \mathcal{A}\} \amalg \{\text{simples in } \mathcal{C}\} \longleftrightarrow \{\text{simples in } \mathcal{B}\}$$

given by sending a simple $A \in \mathcal{A}$ to $i(A)$ and a simple $C \in \mathcal{C}$ to $c(C)$. Here $c : \mathcal{C} \to \mathcal{B}$ is the 'intermediate extension' defined by $c(C) = \operatorname{Im}\{l(C) \to r(C)\}$, and satisfies $ec(C) \simeq C$, naturally, for all $C \in \mathcal{C}$. The lemma follows: both i and c will induce isomorphisms of endomorphism rings of the simple objects. □

Now let R and S be finite dimensional $\mathbb{F}$-algebras. One says that $\mathbb{F}$ *splits* R if $\mathbb{F}$ splits the category of R-modules. The following is well known (see e.g. [2, Theorem 10.38]):

Lemma 2.2. *If $\mathbb{F}$ splits R and S, then $\mathbb{F}$ splits $R \otimes_{\mathbb{F}} S$.*

3. Proof of Theorem 1.1

As in [5], let $\mathcal{F}(q)$ be the $\mathbb{F}_q$-linear abelian category with objects the functors

$$F : \text{finite dimensional } \mathbb{F}_q\text{-vector spaces} \longrightarrow \mathbb{F}_q\text{-vector spaces},$$

and with morphisms the natural transformations. We like to view an object $F \in \mathcal{F}(q)$ as a 'generic representation' of the general linear groups over $\mathbb{F}_q$, as $F(\mathbb{F}_q^n)$ becomes an $\mathbb{F}_q[\mathrm{GL}_n(\mathbb{F}_q)]$-module (and an $\mathbb{F}_q[M_n(\mathbb{F}_q)]$-module) for all n.

Let $\mathbb{F}_q[M_{m,n}(\mathbb{F}_q)]$ denote the $\mathbb{F}_q$-vector space with basis $M_{m,n}(\mathbb{F}_q)$, the set of $m \times n$ matrices over $\mathbb{F}_q$. Note that the structure of $F \in \mathcal{F}(q)$ is encoded in suitably compatible $\mathbb{F}_q$-linear maps

$$\mathbb{F}_q[M_{m,n}(\mathbb{F}_q)] \otimes F(\mathbb{F}_q^n) \to F(\mathbb{F}_q^m).$$

Otherwise said, $\mathcal{F}(q)$ is equivalent to the category of $\mathbb{F}_q$-linear representations of the $\mathbb{F}_q$-linear category $\mathcal{L}(q)$ with

- objects $0, 1, 2, \ldots$.

- morphisms from n to m equal to $\mathbb{F}_q[M_{m,n}(\mathbb{F}_q)]$.
- composition of morphisms is induced by matrix multiplication.

We need to define an $\mathbb{F}_p$-linear variant of $\mathcal{F}(q)$, $\mathcal{F}(p,s)$, which is to $\mathcal{F}(q)$ as $\mathbb{F}_p[\mathrm{GL}_n(\mathbb{F}_q)]_\phi$ is to $\mathbb{F}_q[\mathrm{GL}_n(\mathbb{F}_q)]$.

First note that the Galois group $\mathrm{Gal}(\mathbb{F}_q : \mathbb{F}_p)$ acts on $\mathbb{F}_q[M_{m,n}(\mathbb{F}_q)]$ as in the introduction, and this is compatible with matrix multiplication. Let $\mathbb{F}_p[M_{m,n}(\mathbb{F}_q)]_\phi$ denote the fixed set. Define $\mathcal{L}(p,s)$ to be the $\mathbb{F}_p$-linear category with

- objects $0, 1, 2, \ldots$.
- morphisms from n to m equal to $\mathbb{F}_p[M_{m,n}(\mathbb{F}_q)]_\phi$.
- composition of morphisms is induced by matrix multiplication.

Then we define $\mathcal{F}(p,s)$ to be the category of $\mathbb{F}_p$-linear representations of $\mathcal{L}(p,s)$, i.e. the category of additive functors from $\mathcal{L}(p,s)$ to $\mathbb{F}_p$-vector spaces.

We now describe how the category $\mathcal{F}(p,s)$ is related, on one hand, to the categories of modules for the symmetric groups, and, on the other hand, to modules over the twisted semigroups rings of $\mathrm{GL}_n(\mathbb{F}_q)$ and $M_n(\mathbb{F}_q)$.

From [8], we learn the next theorem.

Theorem 3.1. *There exist a chain of thick subcategories of $\mathcal{F}(p,s)$,*

$$\mathcal{F}_0 \subset \mathcal{F}_1 \subset \mathcal{F}_2 \subset \mathcal{F}_3 \subset \cdots,$$

such that

(a) *every simple object of $\mathcal{F}(p,s)$ is in some $\mathcal{F}_d$, and*

(b) *for each d, there are positive integers $r_1, \ldots, r_s$, and a recollement diagram*

$$\mathcal{F}_{d-1} \underset{\longleftarrow}{\overset{\longleftarrow}{\xrightarrow{i}}} \mathcal{F}_d \underset{\longleftarrow}{\overset{\longleftarrow}{\xrightarrow{e}}} \mathbb{F}_p[\Sigma_{r_1} \times \cdots \times \Sigma_{r_s}]\text{-modules.}$$

(The functor labelled e is a generalized Schur functor.)

The methods of [6, §2] easily yield the next two results.

Theorem 3.2. *There is a recollement diagram*

$$\mathit{Kernel}(e) \underset{\longleftarrow}{\overset{\longleftarrow}{\xrightarrow{i}}} \mathcal{F}(p,s) \underset{\longleftarrow}{\overset{\longleftarrow}{\xrightarrow{e}}} \mathbb{F}_p[M_n(\mathbb{F}_q)]_\phi\text{-modules.}$$

(The functor labelled e is just evaluation on the object $n \in \mathcal{L}(p,s)$.)

Theorem 3.3. *There is a recollement diagram*

$$\mathbb{F}_p[\mathrm{GL}_n(\mathbb{F}_q)]_\phi\text{-modules} \underset{\longleftarrow}{\overset{\longleftarrow}{\xrightarrow{i}}} \mathbb{F}_p[M_n(\mathbb{F}_q)]_\phi\text{-modules} \underset{\longleftarrow}{\overset{\longleftarrow}{\xrightarrow{e}}} \mathbb{F}_p[M_{n-1}(\mathbb{F}_q)]_\phi\text{-modules}.$$

(The functor labelled i is induced by the epic map of algebras $\mathbb{F}_p[M_n(\mathbb{F}_q)] \to \mathbb{F}_p[\mathrm{GL}_n(\mathbb{F}_q)]$ sending noninvertible matrices to 0.)

Armed with these three theorems, the main theorem follows easily, starting from the following well known lemma.

Lemma 3.4. *$\mathbb{F}_p$ is the p-splitting field for Σ_r, for all r.*

This follows from general representation theory [9, p. 103], since the symmetric groups have the property that if two elements generate the same subgroup, they are conjugate elements.

Proof of Theorem 1.1. The last lemma and Lemma 2.2 combine to show that $\mathbb{F}_p$ splits the category of $\mathbb{F}_p[\Sigma_{r_1} \times \cdots \times \Sigma_{r_s}]$-modules. Using induction on d and Lemma 2.1, we conclude that $\mathbb{F}_p$ splits each of the categories $\mathcal{F}_d$ occurring in Theorem 3.1, and thus splits $\mathcal{F}(p,s)$. Then Theorem 3.2 implies that $\mathbb{F}_p$ is a splitting field for $\mathbb{F}_p[M_n(\mathbb{F}_q)]_\phi$, and Theorem 3.3 implies that $\mathbb{F}_p$ is a splitting field for $\mathbb{F}_p[\mathrm{GL}_n(\mathbb{F}_q)]_\phi$. This proves statement (2) of the theorem.

Statement (1) of the theorem is well known, but we quickly can deduce it from statement (2). From statement (2), it follows that $\mathbb{F}_q$ splits both $\mathbb{F}_p[M_n(\mathbb{F}_q)]$ and $\mathbb{F}_p[\mathrm{GL}_n(\mathbb{F}_q)]$. To see that this is the minimal splitting field for these algebras, one just needs to observe that $\mathrm{End}_{\mathrm{GL}_n(\mathbb{F}_q)}(\mathbb{F}_q^n) = \mathbb{F}_q$. □

4. Twisted semigroup rings and Schur algebras

The classic Schur algebras arise as one studies the polynomial representations of the algebraic group GL_n [4]. It is natural to compare a GL_n-module to the family of $\mathrm{GL}_n(\mathbb{F}_q)$-modules, defined, for all q, by restriction. Traditionally, these representations are vector spaces over the algebraically closed field $\bar{\mathbb{F}}_p$. The point of this short section is to observe that by using the twisted algebras we can work over the prime field $\mathbb{F}_p$ to the same effect.

Let's first recall the definition of the Schur algebras. Let $S(n,r)$ be the $\mathbb{F}_p$-algebra $(M_n(\mathbb{F}_p)^{\otimes r})^{\Sigma_r}$, with the multiplication induced by matrix multiplication[2]. The category of $S(n,r)$ modules is equivalent to the category of homogeneous polynomial representations of GL_n of degree r.

[2]Equivalently, $S(n,r) = \mathrm{End}_{\Sigma_r}((\mathbb{F}_p^n)^{\otimes r})$.

Now note that $\mathbb{F}_q \otimes_{\mathbb{F}_p} S(n,r) = (M_n(\mathbb{F}_q)^{\otimes_{\mathbb{F}_q} r})^{\Sigma_r}$. There are maps of $\mathbb{F}_q$-algebras

$$\Theta : \mathbb{F}_q[M_n(\mathbb{F}_q)] \to \mathbb{F}_q \otimes_{\mathbb{F}_p} S(n,r)$$

defined by sending a matrix A to $A^{\otimes r}$. Inspection of the Galois group action reveals that Θ is equivariant if the semigroup ring has precisely the diagonal action of the introduction (and the Galois group acts on $\mathbb{F}_q \otimes_{\mathbb{F}_p} S(n,r)$ in the obvious way). Taking fixed points, we get maps of $\mathbb{F}_p$-algebras

$$\Theta : \mathbb{F}_p[M_n(\mathbb{F}_q)]_\phi \to S(n,r)$$

for all n, q, r.

The maps Θ are compatible as q varies. An inclusion of fields $\mathbb{F}_{p^s} \subset \mathbb{F}_{p^t}$ induces an inclusion of algebras $\mathbb{F}_p[M_n(\mathbb{F}_{p^s})]_\phi \subset \mathbb{F}_p[M_n(\mathbb{F}_{p^t})]_\phi$, and we have the following lemma.

Lemma 4.1. *There are commutative diagrams*

$$\begin{array}{ccc} \mathbb{F}_p[M_n(\mathbb{F}_{p^s})]_\phi & \longrightarrow & \mathbb{F}_p[M_n(\mathbb{F}_{p^t})]_\phi \\ \downarrow{\scriptstyle\Theta} & & \downarrow{\scriptstyle\Theta} \\ S(n,r) & = & S(n,r)\,. \end{array}$$

Corresponding to the Frobenius twist of GL_n-modules are algebra maps

$$\Phi : S(n,pr) \to S(n,r)$$

induced by sending $A^{\otimes p}$ to A. Meanwhile, the $\mathbb{F}_p$-algebra maps

$$\Phi : \mathbb{F}_q[M_n(\mathbb{F}_q)] \to \mathbb{F}_q[M_n(\mathbb{F}_q)]$$

which send a matrix (a_{ij}) to (a_{ij}^p) are equivariant isomorphisms, and so induce automorphisms of $\mathbb{F}_p$-algebras

$$\Phi : \mathbb{F}_p[M_n(\mathbb{F}_q)]_\phi \to \mathbb{F}_p[M_n(\mathbb{F}_q)]_\phi.$$

These are compatible; we have the following lemma.

Lemma 4.2. *There are commutative diagrams*

$$\begin{array}{ccc} \mathbb{F}_p[M_n(\mathbb{F}_q)]_\phi & \xrightarrow{\Phi} & \mathbb{F}_p[M_n(\mathbb{F}_q)]_\phi \\ \downarrow{\scriptstyle\Theta} & & \downarrow{\scriptstyle\Theta} \\ S(n,pr) & \xrightarrow{\Phi} & S(n,r)\,. \end{array}$$

Acknowledgement. This research was partially supported by the National Science Foundation.

References

[1] E. Cline, B. Parshall, L. Scott, and W. van der Kallen, Rational and generic cohomology, Invent. Math. 39 (1977), 143–163.

[2] C. W. Curtis and I. Reiner, Methods of Representation Theory: Vol. I, John Wiley and Sons, New York 1981.

[3] V. Franjou, E. M. Friedlander, A. Skorichenko, and A. Suslin, General linear and functor cohomology over finite fields, Ann. of Math. 150 (1999), 663–728.

[4] J. A. Green, Polynomial Representations of GL_n, Lecture Notes in Math. 830, Springer-Verlag, Berlin 1980.

[5] N. J. Kuhn, Generic representation theory of the finite general linear groups and the Steenrod algebra: I, Amer. J. Math. 116 (1994), 327–360.

[6] N. J. Kuhn, Generic representation theory of the finite general linear groups and the Steenrod algebra: II, K-Theory 8 (1994), 395–428.

[7] N. J. Kuhn, Rational cohomology and cohomological stability in generic representation theory, Amer. J. Math. 120 (1998), 1317–1341.

[8] N. J. Kuhn, A stratification of the generic representation category and generalized Schur algebras, preprint.

[9] J.-P. Serre, Linear Representations of Finite Groups, Grad. Texts in Math. 42, Springer-Verlag, New York 1977.

[10] H. N. Ward, Representations of symplectic groups, J. Algebra 20 (1972), 182–195.

Department of Mathematics, University of Virginia, Charlottesville, VA 22903, U.S.A.

E-mail: njk4x@virginia.edu

Irreducible tensor products of representations of finite quasi-simple groups of Lie type

*Kay Magaard and Pham Huu Tiep**

Abstract. Let G be a finite quasi-simple group of Lie type defined over a field of characteristic p. We determine whether G can possess irreducible Brauer characters α, β (of degree > 1) in characteristic other than p such that $\alpha \otimes \beta$ is irreducible.

1. Introduction

Let G be a finite group and r be a fixed prime (or $r = 0$). Consider the following statement:

$\mathcal{P}(r)$: $\alpha \otimes \beta$ is not irreducible for any nonlinear Brauer characters $\alpha, \beta \in \mathrm{IBr}_r(G)$

(by $\mathrm{IBr}_r(G)$ with $r = 0$, we mean $\mathrm{Irr}(G)$).

One motivation for studying the property $\mathcal{P}(r)$ comes from the desire to classify the overgroups of the irreducible subgroups of the finite classical groups $G(q)$. According to Aschbacher's Theorem [A], if M is a maximal subgroup of $G(q)$, then either M belongs to one of 8 collections $\mathcal{C}_i$, $1 \leq i \leq 8$, of "natural" subgroups of $G(q)$, or $M \in \mathcal{S}$, a collection of quasi-simple groups which act irreducibly on the natural module V of $G(q)$. For instance, $\mathcal{C}_4$ consists of stabilizers of tensor decompositions of V. Conversely, if $M \in \bigcup_{i=1}^{8} \mathcal{C}_i$, then the maximality of M is determined in [KL]. Now suppose $M \in \mathcal{S}$. Could it happen that $M < N < G$ for some $N \in \mathcal{C}_4$ – in other words, could the (irreducible) representation of M on V be tensor-decomposable?

This problem of tensor-decomposable irreducible modules has been investigated by Isaacs [Is] for solvable groups G. In particular, he conjectures that if the solvable group G has two faithful complex characters α and β such that $\alpha\beta$ is irreducible, then G is cyclic.

---

*Part of this work was done while the authors were participating in the special semester on Representations of Algebraic Groups and Related Finite Groups at the Isaac Newton Institute (Cambridge). It is a pleasure to thank the Institute for its hospitality and support.

The authors are grateful to M. Geck, P. Sin and A. Turull for helpful discussion, and to M. Collins for a number of comments that greatly improved the exposition of the paper. It is the first author's pleasure to thank the SFB 343 at the Universität Bielefeld for its hospitality and support.

The goal of this paper is to find all finite quasi-simple groups with property $\mathcal{P}(r)$. It has been shown by Bessenrodt and Kleshchev [BK1] that the symmetric group $\mathbb{S}_n$ ($n \geq 5$) satisfies $\mathcal{P}(0)$. They also classified all complex irreducible characters α, β of degree > 1 of the alternating group $\mathbb{A}_n$ ($n \geq 5$) such that $\alpha \otimes \beta$ is irreducible: it happens if and only if $n = a^2$ for some $a \geq 3$, one of the characters α, β is the nontrivial irreducible component of the natural permutation character, and the other one corresponds to the partition (a^a). More recently, Bessenrodt and Kleshchev have shown in [BK2] that $\mathbb{S}_n$ satisfies $\mathcal{P}(r)$ for any odd prime r, and that $\mathbb{S}_n$ satisfies $\mathcal{P}(2)$ if n is odd. The modular case for $\mathbb{A}_n$ is considered in [BK3]. The cases of complex characters of $2\mathbb{S}_n$ and $2\mathbb{A}_n$ are being studied. We note that there are examples of irreducible tensor products of complex spin representations of $\mathbb{A}_9$, $\mathbb{A}_{10}$. Also, some examples of irreducible tensor products of representations of $\mathbb{S}_n$ in characteristic 2 have been found by Gow and Kleshchev [GK], see also [JJ].

We investigated the 26 sporadic finite simple groups (and their covers) using the character tables that we found in GAP. We found that M_{11}, J_1 and $2HS$ satisfy $\mathcal{P}(r)$ for all primes r, and that HN, Ly, He, Fi_{23} and J_4 satisfy $\mathcal{P}(0)$ whereas all other sporadic groups do not even satisfy $\mathcal{P}(0)$. The list of tensor-decomposable irreducible representations that we found can be reproduced in GAP with minimal effort and is hence not included in this paper. Instead, we list a couple of examples of irreducible tensor products $\Phi \otimes \Psi$ for $G = BM, 2 \cdot BM$:

$$\begin{aligned} &\Phi(1) = 4371, && \Psi(1) = 53936390144, \\ &\Phi(1) = 96256, && \Psi(1) = 90807234375, \end{aligned}$$

and for $G = M$:

$$\begin{aligned} &\Phi(1) = 196883, && \Psi(1) = 8980616927734375, \\ &\Phi(1) = 19360062527, && \Psi(1) = 8980616927734375, \\ &\Phi(1) = 21296876, && \Psi(1) = 3503434660075044981. \end{aligned}$$

We focus our attention on the finite quasi-simple groups G of Lie type defined over a field $\mathbb{F}_q$ of characteristic p, $q = p^f$, and $p \neq r$.

Observe that G cannot satisfy $\mathcal{P}(p)$. Indeed, let us view G as the fixed point subgroup $\mathcal{G}^F$, where $\mathcal{G}$ is a simple algebraic group defined over $\overline{\mathbb{F}}_p$ and F is the corresponding Frobenius endomorphism. Then the Steinberg tensor product theorem asserts that any irreducible $\overline{\mathbb{F}}_pG$-module can be written as the tensor product $V_1 \otimes V_2^{(p)} \otimes \cdots \otimes V_f^{(p^{f-1})}$ for some restricted irreducible $\overline{\mathbb{F}}_pG$-modules $V_1, \ldots, V_f$; conversely, any such tensor product is irreducible. This remarkable theorem gives us many examples of modules whose tensor product is irreducible. Even more, in combination with a result of Seitz, this theorem allows us to describe all pairs (U, V) of $\overline{\mathbb{F}}_pG$-modules such that $U \otimes V$ is irreducible. For, let

$$V = V_1 \otimes V_2^{(p)} \otimes \cdots \otimes V_f^{(p^{f-1})}, \quad U = U_1 \otimes U_2^{(p)} \otimes \cdots \otimes U_f^{(p^{f-1})},$$

where each U_i, V_i is restricted and irreducible. Then

$$U \otimes V \simeq (U_1 \otimes V_1) \otimes (U_2 \otimes V_2)^{(p)} \otimes \cdots \otimes (U_f \otimes V_f)^{(p^{f-1})}.$$

Suppose $U \otimes V$ is irreducible. Then, for each i, the G-module $W_i := U_i \otimes V_i$ is irreducible. In particular, the $\mathcal{G}$-module W_i is irreducible. Now [S1, (1.6)] states that W_i is restricted, and it also gives us the list of all pairs (U_i, V_i) of restricted modules such that the $\mathcal{G}$-module W_i is irreducible. Conversely, suppose that we are given irreducible restricted modules U_i, V_i, $1 \leq i \leq f$, such that each W_i is irreducible. Since W_i is restricted in this case, the G-module W_i is irreducible. Hence the G-module $U \otimes V$ is irreducible by the Steinberg tensor product theorem.

The main result of the paper is the following.

Theorem 1.1. *Let G be a finite quasi-simple group of Lie type defined over a field $\mathbb{F}_q$. Then either G has property $\mathcal{P}(r)$ for every r not dividing q, or one of the following holds.*

(i) $q \leq 3$.

(ii) $G = \mathrm{Sp}_{2n}(5)$.

(iii) $G = \mathrm{Sp}_{2n}(q)$, *q even.*

(iv) $G/Z(G) = \mathrm{L}_3(4)$.

(v) $G = F_4(q)$, *q even.*

(vi) $G = {}^2F_4(q)$, *q even.*

Throughout the paper, "r not dividing q" means that either $r = 0$, or r is coprime to q. We prove more detailed results in the body of the paper. In particular we give examples where property $\mathcal{P}(r)$ does not hold. For example, it is shown in Proposition 5.4 that when G is $\mathrm{Sp}_{2n}(3)$ or $\mathrm{Sp}_{2n}(5)$, then $\mathcal{P}(0)$ does not hold. Also, $\mathcal{P}(0)$ fails for $\mathrm{SU}_n(2)$ with $n \geq 3$, cf. [LST], for $\mathrm{Sp}_{2n}(2)$ with $n \geq 3$, cf. [MMT], and for $4\mathrm{L}_3(4)$ (direct computation using GAP). Thus the cases (i)–(iv) in Theorem 1.1 actually occur. We conjecture, but are not yet able to show, that neither of the cases (v) or (vi) can happen. In particular, the cases (v) and (vi) cannot happen if $r = 0$; see Proposition 8.3. Furthermore we conjecture that $\mathrm{Sp}_{2n}(q)$, with $2|q$ and $q \neq 2$, does have property $\mathcal{P}(r)$.

For the reader's convenience, we formulate the main result in the complex case.

Theorem 1.2. *Let G be a finite quasi-simple group of Lie type defined over a field $\mathbb{F}_q$. Then either G has property $\mathcal{P}(0)$, or one of the following holds.*

(i) $q \leq 3$.

(ii) $G = \mathrm{Sp}_{2n}(5)$.

(iii) $G = \mathrm{Sp}_{2n}(q)$, *q even.*

(iv) $G/Z(G) = \mathrm{L}_3(4)$.

The dual problem of determining whether the product of two conjugacy classes of a finite group could form a single conjugacy class has been investigated by Arad [Ar]. Also, property $\mathcal{P}(0)$ is closely related to factorizations of quasi-primitive characters intensively investigated by Ferguson and Turull [FT].

2. Preliminaries

We begin with the following trivial statements.

Lemma 2.1. *Let $G = H * C$ be a central product of a subgroup H and an abelian subgroup C. Then any irreducible $\mathbb{F}H$-module V extends to G.*

Proof. Let W be a simple submodule of V^G. Then $W = A \otimes B$, where A is an irreducible H-module and B an irreducible C-module. Since C is abelian, $\dim(B) = 1$. Next, $0 \neq \mathrm{Hom}(W, V^G) \simeq \mathrm{Hom}(W|_H, V)$; hence $W|_H = V$. □

Lemma 2.2. *Let G be a finite group and H a normal subgroup of G with G/H abelian. Let V be a G-module and let U be a simple submodule, resp. a simple quotient, of $V|_H$. Suppose U extends to G. Then the G-module V has a submodule, resp. a quotient, $\tilde{U}$, which is an extension of U.*

Proof. 1) We claim that, if W is an arbitrary composition factor of the G-module U^G, then $W|_H \simeq U$. Indeed, since U is G-invariant, $U^G|_H$ is the sum of $|G/H|$ copies of U. Hence, $W|_H$ is a direct sum of copies of U. By [Hup, (V.17.5)], $W = W_1 \otimes W_2$, where W_1 affords a projective representation of G and W_2 a projective representation of G/H. We can choose W_1 to be an extension $\hat{U}$ of U to G. Since the factor sets in W_1 and W_2 are inverse sets, W_2 also affords a linear representation of G/H. But G/H is abelian; hence $\dim W_2 = 1$ and W is an extension of U.

2) Consider the case where U is a submodule of $V|_H$. Then $0 \neq \mathrm{Hom}(U, V|_H) \simeq \mathrm{Hom}(U^G, V)$. Thus V has a submodule V_1 which is a nonzero quotient of U^G. By 1), V_1 has a submodule which is an extension of U.

3) Consider the case where U is a quotient of $V|_H$. Then $0 \neq \mathrm{Hom}(V|_H, U) \simeq \mathrm{Hom}(V, U^G)$. Thus V has a quotient V_1 which is a nonzero submodule of U^G. By 1), V_1 has a quotient which is an extension of U. □

Observe that one cannot drop the assumption G/H is abelian in Lemma 2.2. For example, one can have examples where V is the tensor product of an extension of U and an irreducible G/H-module of dimension > 1, and yet V is irreducible.

We are going to consider representations of a finite group G over an algebraically closed field $\mathbb{F}$ of characteristic r. Usually, G contains a p-subgroup of *special type* for a prime $p \neq r$, that is, $[Q, Q] = Z(Q) = \Phi(Q)$ has exponent p and $[x, Q] = Z(W)$ for all $x \in Q \setminus Z(Q)$. More concretely, if p is odd then Q is the set

$$\{[X, Y, z] | X, Y \in \mathbb{F}_q^m, z \in \mathbb{F}_q\}$$

(for a natural integer m and a power $q = p^f$) endowed with the following group operation

$$[X, Y, z] \circ [X', Y', z'] = [X + X', Y + Y', z + z' + \sum_{i=1}^{m}(x_i y_i' - x_i' y_i)],$$

if

$$X = (x_1, \ldots, x_m), \quad Y = (y_1, \ldots, y_m),$$
$$X' = (x_1', \ldots, x_m'), \quad Y' = (y_1', \ldots, y_m').$$

Clearly, $[Q, Q] = Z(Q) = \Phi(Q)$ coincides with the subgroup $\{t_c := [0, 0, c] \mid c \in \mathbb{F}_q\}$. Moreover, $[x, Q] = Z(Q)$ for any $x \in Q \setminus Z(Q)$; hence Q is of special type. If G is a finite group of Lie type defined over a field of characteristic p, then we can select Q such that $P := N_G(Q)$ is a "big" parabolic subgroup of G.

Let Q be a p-group of special type. For any $\mathbb{F}Q$-module V, let

$$\Omega(V) = \{\lambda \in \mathrm{Irr}(Z(Q)) \setminus \{1\} \mid \lambda \text{ occurs in } V|_{Z(Q)}\}.$$

In what follows, whenever we talk about $\Omega(V)$, we will have a fixed subgroup Q, specified in advance, of a finite group X, an $\mathbb{F}X$-module V, and, abusing the notation, we will denote $\Omega(V|_Q)$ by $\Omega(V)$. Furthermore, when we consider a finite group G (say $\mathrm{GL}_n(q)$) and a normal subgroup S (say $\mathrm{SL}_n(q)$), it can easily be checked in all cases that the subgroup Q in question is the same for G and for S. For any $\lambda \in \Omega(V)$, let V_λ be the λ-eigenspace for $Z(Q)$ on V, and let

$$d_\lambda(V) = \dim(V_\lambda)/\sqrt{|Q/Z(Q)|}.$$

Lemma 2.3. *Let G be a finite group with a p-subgroup Q of special type and* $\mathrm{char}(\mathbb{F}) \neq p$. *Let $C \leq N_G(Q) \cap C_G(Z(Q))$. Suppose that V and W are $\mathbb{F}G$-modules such that there is a $\lambda \in \mathrm{IBr}_r(Z(Q)) \setminus \{1\}$ with $\lambda \in \Omega(V)$ and $\lambda^{-1} \in \Omega(W)$. Then $V \otimes W$ has a C-submodule F of dimension $d_\lambda(V)d_{\lambda^{-1}}(W)$, on which Q acts trivially.*

Proof. Since $\mathrm{char}(\mathbb{F}) \neq p$, we can write

$$V = \Big(\bigoplus_{\mu \in \Omega(V)} V_\mu\Big) \oplus V_1, \quad W = \Big(\bigoplus_{\mu \in \Omega(W)} W_\mu\Big) \oplus W_1.$$

But C centralizes $Z(Q)$; hence C preserves each of the summands in these decompositions. In particular, $U := V_\lambda \otimes W_{\lambda^{-1}}$ is a direct summand of the C-module $V \otimes W$. Let α be the unique irreducible character (over $\mathbb{F}$) of Q whose restriction to $Z(Q)$ contains λ. Then the character of Q on V_λ, respectively on $W_{\lambda^{-1}}$, equals $d_\lambda(V)\alpha$, respectively $d_{\lambda^{-1}}(W)\bar{\alpha}$. It is easy to see that $\alpha\bar{\alpha}$ is the regular character of $Q/Z(Q)$. Hence Q acts on U with character $d_\lambda(V)d_{\lambda^{-1}}(W)\sum_{\beta \in \mathrm{IBr}_r(Q/Z(Q))} \beta$. Let F be the fixed point subspace of Q on U. Then clearly $\dim(F) = d_\lambda(V)d_{\lambda^{-1}}(W)$, F is C-invariant, and Q acts trivially on F. □

A key ingredient of our proof is the following.

Proposition 2.4. *Let G be a finite group with a p-subgroup Q of special type and* $\mathrm{char}(\mathbb{F}) \neq p$. *Let $Q \leq C = N_G(Q) \cap C_G(Z(Q))$. Suppose that V and W are $\mathbb{F}G$-modules such that there is $\lambda \in \mathrm{IBr}_r(Z(Q)) \setminus \{1\}$ with $\lambda \in \Omega(V)$ and $\lambda^{-1} \in \Omega(W)$. Suppose in addition that G has a subgroup $N \leq N_G(Q)$ such that $N/Q = C/Q * B$ for some abelian subgroup B. Then $V \otimes W$ has an N-submodule T of dimension at most $d_\lambda(V)d_{\lambda^{-1}}(W)$. In particular, if the G-module $V \otimes W$ is irreducible, then*

$$\dim(V) \cdot \dim(W) \leq (G : N)d_\lambda(V)d_{\lambda^{-1}}(W).$$

Proof. By Lemma 2.3, the C-module $V \otimes W$ has a submodule F of dimension $d_\lambda(V)d_{\lambda^{-1}}(W)$, on which Q acts trivially. Let F' be a simple C-submodule of F. By Lemma 2.1, F' extends to N. By Lemma 2.2, the N-module $V \otimes W$ has a submodule T which is an extension of F'. If the G-module $V \otimes W$ is irreducible, then $0 \neq \mathrm{Hom}_N(T, (V \otimes W)|_N) \simeq \mathrm{Hom}_G(T^G, V \otimes W)$, whence $V \otimes W$ is a quotient of T^G, and so

$$\dim(V) \cdot \dim(W) \leq \dim(T^G) \leq (G : N)d_\lambda(V)d_{\lambda^{-1}}(W). \qquad \square$$

Extending the proof of [S2, (1.4)] and [Hup, (V.17.5)], one can prove the following supplement to Lemma 2.3, which we will use in the proofs of Propositions 3.3 and 3.4.

Lemma 2.5. *Keep the notation and assumptions of Lemma* 2.3. *In addition, suppose that*

(i) *C is a semi-direct product of Q by $A \times B$;*

(ii) *A acts faithfully on Q and A is perfect;*

(iii) *B has trivial Schur multiplier;*

(iv) $(|Z(Q)|, |Q/Z(Q)|) \neq (3, 9), (9, 81)$.

Then the C-module T mentioned in Lemma 2.3 *is actually the tensor product of two C/Q-modules, of dimensions $d_\lambda(V)$ and $d_{\lambda^{-1}}(W)$, respectively.*

Lemma 2.6. *Let G be a finite group with property $\mathcal{P}(r)$. Let H be a normal subgroup of G of prime index t, and let $\gamma, \delta \in \mathrm{IBr}_r(H)$ be characters of degree* > 1 *such that*

(i) *γ extends to G;*

(ii) *for a fixed $h \in G \setminus H$, if $\delta^h \neq \delta$, then $\gamma \otimes \delta^h \neq \gamma \otimes \delta$.*

Then the H-character $\gamma \otimes \delta$ is not irreducible.

Proof. Consider the inertia group I of δ. Then $H \leq I \leq G$. If $I = G$, then δ extends to G by [F, 3.2.14], and so $\gamma \otimes \delta$ is reducible by the condition $\mathcal{P}(r)$ for G. Suppose $I < G$. Then $I = H$, and δ has exactly t distinct conjugates δ_i, $i = 1, \ldots, t$, and $\delta^G|_H = \sum_{i=1}^t \delta_i$. Let γ' denote an extension of γ to G. Suppose $\gamma \otimes \delta$ is irreducible. Then $\gamma \otimes \delta_i$, $i = 1, \ldots, t$, are all irreducible and distinct; hence the G-character $\gamma' \otimes \delta^G$ is irreducible, again contradicting the condition $\mathcal{P}(r)$ for G. $\square$

3. Linear groups

Let $G = \mathrm{GL}_n(q)$, $S = \mathrm{SL}_n(q)$ with $q = p^f$. It is easy to check that all covers of $\mathrm{L}_2(q)$ with $q > 3$ and $q \neq 9$ have property $\mathcal{P}(r)$ for all r. On the other hand, the universal cover of $\mathrm{L}_2(9) \simeq \mathrm{Sp}_4(2)'$ does not have property $\mathcal{P}(r)$ for any r. In what follows we assume that $n \geq 3$, $(n, q) \neq (3, 2), (3, 4), (4, 2), (4, 3)$. Let $\mathbb{F}_q^n$ be the natural module for G, with a basis $(e_1, \dots, e_n)$. We consider the parabolic subgroup $P = St_S(\langle e_2\rangle, \langle e_2, \dots, e_n\rangle_{\mathbb{F}_q})$ of S. Then $Q = O_p(P)$ is a group of order q^{2n-3} of special type. The center $Z(Q)$ has order q and contains $q - 1$ transvections t_c, where $c \in \mathbb{F}_q^\bullet$ and t_c has matrix $\mathrm{diag}\left(\begin{pmatrix} 1 & 0 \\ c & 1 \end{pmatrix}, I_{n-2}\right)$ with respect to the given basis of V. It is easy to see that Q has a complement

$$\{\mathrm{diag}(a, b, X) \mid a, b \in \mathbb{F}_q^\bullet, X \in \mathrm{GL}_{n-2}(q), ab \cdot \det(X) = 1\},$$

and all the transvections t_c are conjugate in P. Our assumption on (n, q) guarantees that S is the universal cover of $\mathrm{PSL}_n(q)$. It is shown in [GT] that S has a few nontrivial irreducible representations in cross-characteristic of degree $(q^n - 1)/(q - 1) - a$ for $a = 2, 1$, or 0; those representations will be referred to as *Weil representations*.

Let V be any irreducible $\mathbb{F}S$-module of dimension > 1. Then the subgroup $\{\mathrm{diag}(a^{-1}, 1, a, I_{n-3}) \mid a \in \mathbb{F}_q^\bullet\}$ acts transitively on the set $\mathrm{IBr}_r(Z(Q)) \setminus \{1\}$; hence $\Omega(V) = \mathrm{IBr}_r(Z(Q)) \setminus \{1\}$, and all $d_\lambda(V)$ are the same. We begin with the following estimate.

Lemma 3.1. *In the above notation,*

$$d_\lambda(V) \leq \frac{\dim(V)}{q^{n-1} - 1}.$$

If $d_\lambda(V) = 1$ and $n \geq 5$, then V is one of the Weil modules.

Proof. Let $R = St_S(e_2)$ and let $A = O_p(R)$. Then A is an elementary abelian group of order q^{n-1}. Let $V(\mu)$ be the homogeneous component of $V|_A$ corresponding to $\mu \in \mathrm{IBr}_r(A)$. Fix a nontrivial $\mu \in \mathrm{IBr}_r(A)$. Then $\dim(V) = \dim(V(1)) + (q^{n-1} - 1)\dim(V(\mu))$. In particular, $\dim(V(\mu)) \leq \dim(V)/(q^{n-1} - 1)$. Let χ be the Brauer character of V. Then $\chi(t_1) = \dim(V(1)) - \dim(V(\mu))$, and so $\dim(V(\mu)) = (\dim(V) - \chi(t_1)/q^{n-1}$. In the meantime,

$$\dim(V) = \dim(V_1) + q^{n-2}(q - 1)d_\lambda(V), \quad \chi(t_1) = \dim(V_0) - q^{n-2}d_\lambda(V),$$

whence

$$d_\lambda(V) = \frac{\dim(V) - \chi(t_1)}{q^{n-1}} = \dim(V(\mu)) \leq \frac{\dim(V)}{q^{n-1} - 1}.$$

Next suppose that $d_\lambda(V) = 1$. Then $\dim V(\mu) = 1$ and $\dim(V) = \dim(V(1)) + (q^{n-1} - 1)$. Observe that $V(1) = C_V(A)$. Now we can look at the opposite parabolic subgroup R' and its radical A' of order q^{n-1}. Since $S = \langle A, A'\rangle$ (cf. [GM]) and V is irreducible and non-linear, $C_V(A) \cap C_V(A') = 0$. The above computation shows that $\dim(C_V(A)) = (\dim(V) + (q^{n-1} - 1)\chi(t_1))/q^{n-1}$. Similarly, if $t' \in A'$ is a non-identity element, then $\dim(C_V(A')) = (\dim(V) + (q^{n-1} - 1)\chi(t'))/q^{n-1}$. Observe that t_1 and t' are S-conjugate as $n > 2$. Hence $\dim(C_V(A)) = \dim(C_V(A'))$ and so $\dim(V) \geq 2\dim(C_V(A))$. We conclude that $\dim(V) \leq 2(q^{n-1} - 1)$. It remains to apply the main result of [GT]. □

Observe that Lemma 3.1 remains true for an $\mathbb{F}G$-module V.

Proposition 3.2. *Suppose $n \geq 2$. Then the groups $\mathrm{SL}_n(q)$ with $q \geq 4$, and $\mathrm{GL}_n(q)$ with $q \geq 3$, have property $\mathcal{P}(r)$ for any r not dividing q.*

Proof. The cases (3, 4) and (4, 3) can be checked using GAP. Thus we may assume that $n \geq 3$, and $(n, q) \neq (3, 4), (4, 3)$. First we assume that $G = \mathrm{GL}_n(q)$, $q \geq 3$, has irreducible $\mathbb{F}G$-modules V, W of dimension > 1 such that $V \otimes W$ is irreducible. Consider the parabolic subgroup $\hat{P} = St_G(\langle e_2\rangle, \langle e_2, \dots, e_n\rangle_{\mathbb{F}_p})$ of G. Then $Q = O_p(\hat{P})$, and Q has a complement

$$D := \{\mathrm{diag}(a, b, X) \mid a, b \in \mathbb{F}_q^\bullet, X \in \mathrm{GL}_{n-2}(q)\}$$

in $\hat{P}$. Consider the following subgroups

$$A := \{\mathrm{diag}(a, a, X) \mid a \in \mathbb{F}_q^\bullet, X \in \mathrm{GL}_{n-2}(q)\}, \quad B := \{\mathrm{diag}(1, a, I_{n-2}) \mid a \in \mathbb{F}_q^\bullet\}$$

of D. Clearly, A normalizes Q. Set $C = QA$. Then C centralizes $Z(Q)$. Moreover, $\hat{P} = CB$, and B centralizes C/Q. Pick any $\lambda \in \Omega(V)$. Applying Proposition 2.4, we get

$$\dim(V) \cdot \dim(W) \leq (G : \hat{P}) d_\lambda(V) d_{\lambda^{-1}}(W).$$

Combined with Lemma 3.1, this inequality yields

$$1 \leq \frac{(q^n - 1)(q^{n-1} - 1)}{(q-1)^2 \cdot (q^{n-1} - 1)^2},$$

a contradiction since $q \geq 3$.

Next we assume that $S = \mathrm{SL}_n(q)$, $q \geq 4$, has irreducible $\mathbb{F}S$-modules V, W of dimension > 1 such that $V \otimes W$ is irreducible. Consider the complement

$$D := \{\mathrm{diag}(a, b, X) \mid a, b \in \mathbb{F}_q^\bullet, X \in \mathrm{GL}_{n-2}(q), ab \cdot \det(X) = 1\}$$

of Q in P. Consider the following subgroups

$$A := \{\mathrm{diag}(a, a, X) \mid a \in \mathbb{F}_q^\bullet, \det(X) = a^{-2}\}, \quad B := \{\mathrm{diag}(a, a^{-1}, I_{n-2}) \mid a \in \mathbb{F}_q^\bullet\}$$

of D. Again, A normalizes Q. Setting $C = QA$, one sees that C centralizes $Z(Q)$ and B centralizes C/Q. By Proposition 2.4,

$$\dim(V) \cdot \dim(W) \leq (S : BC) d_\lambda(V) d_{\lambda^{-1}}(W).$$

Combined with Lemma 3.1, this inequality yields

$$1 \le \frac{(q^n-1)(q^{n-1}-1)\cdot \gcd(2,q-1)}{(q-1)^2\cdot(q^{n-1}-1)^2},$$

a contradiction since $q \ge 4$. □

Proposition 3.3. *The groups* $\mathrm{SL}_n(2)$, $n \ge 5$, *have property* $\mathcal{P}(r)$ *for any* $r \ne 2$.

Proof. We proceed by induction on $n \ge 5$. The cases $n = 5, 6$ can be checked directly using GAP.

Consider the case $n \ge 7$ and assume the contrary: $S = \mathrm{SL}_n(2)$ has irreducible $\mathbb{F}S$-modules V, W of dimension > 1 such that $V \otimes W$ is irreducible. We pick any $\lambda \in \Omega(V)$ and distinguish two subcases.

1) $d_\lambda(V), d_{\lambda^{-1}}(W) > 1$.

Notice that, since $q = 2$, P is exactly the subgroup $C = QA$ defined in the first part of the proof of Proposition 3.2. By Lemma 2.5 applied to P in place of C, P has a submodule T inside $V \otimes W$, where T is acted on trivially by Q and it is the tensor product of two P/Q-modules of dimension $d_\lambda(V)$ and $d_{\lambda^{-1}}(W)$. Since $P/Q \simeq \mathrm{SL}_{n-2}(2)$, by the induction hypothesis T is not irreducible. Hence we can choose a composition factor U of the P-module T such that $\dim(U) \le \dim(T)/2$. Moreover, we can choose U as a submodule or a quotient of T. In the former case,

$$0 \ne \mathrm{Hom}(U, (V \otimes W)|_P) \simeq \mathrm{Hom}(U^S, V \otimes W);$$

hence $V \otimes W$ is a quotient of U^S. In the latter case,

$$0 \ne \mathrm{Hom}((V \otimes W)|_P, U) \simeq \mathrm{Hom}(V \otimes W, U^S);$$

hence $V \otimes W$ is a submodule of U^S.

Let L be a Levi subgroup of P. Then $L \simeq \mathrm{SL}_{n-2}(2)$, and U is actually an L-module. Observe that U^S is the Harish Chandra induction $R_L^S(U)$ (where we use the parabolic subgroup P containing L). We introduce two more parabolic subgroups. The first one, $P_1 = St_S(\langle e_1\rangle, \langle e_1, e_2, \dots, e_{n-1}\rangle)$, is obviously conjugate to P. We can view U as a module U_1 for a Levi subgroup L_1 of P_1, and then $U^S = R_L^S(U) \simeq R_{L_1}^S(U_1)$. The second one is $P_2 = St_S(\langle e_1\rangle, \langle e_1, e_2\rangle)$ having $L_2 := L$ as a Levi subgroup. Let $\mathcal{S} = \{(1,2), (2,3), \dots, (n-1,n)\}$ be the standard set of generators for the Weyl group of $S = \mathrm{SL}_n(2)$. Since

$$P_1 = \left\{ \left[\begin{array}{c|c|c} 1 & * & * \\ \hline 0 & X & * \\ \hline 0 & 0 & 1 \end{array}\right] \;\middle|\; X \in \mathrm{SL}_{n-2}(2) \right\}, \quad P_2 = \left\{ \left[\begin{array}{cc|c} 1 & * & * \\ 0 & 1 & * \\ \hline 0 & 0 & X \end{array}\right] \;\middle|\; X \in \mathrm{SL}_{n-2}(2) \right\},$$

P_1 is the standard parabolic subgroup corresponding to the subset

$$\mathcal{I}_1 = \{(2,3), (3,4), \dots, (n-2, n-1)\},$$

and P_2 is the standard parabolic subgroup corresponding to the subset

$$\mathcal{I}_2 = \{(3,4), (4,5), \ldots, (n-1, n)\}$$

of $\mathcal{S}$. Clearly, $\mathcal{I}_1$ and $\mathcal{I}_2$ are conjugate via the cycle $w = (2, 3, 4, \ldots, n-1, n)$ of the Weyl group of S. Now the independence property of the Harish-Chandra induction (cf. [HL], [DD]) implies that $R^S_{L_1}(U_1) \simeq R^S_{L_2}(U_2)$, where $U_2 = {}^wU_1$ (the w-conjugate of U_1). Thus U^S is isomorphic to the Harish-Chandra induction $(1 \circ 1 \circ U_2) \uparrow S$ from P_2 to S of the $P_2/O_p(P_2)$-module U_2 in the notation of [GT].

Next we can write

$$(1 \circ 1 \circ U_2) \uparrow S \simeq (((1 \circ 1) \uparrow \mathrm{SL}_2(2)) \circ U_2) \uparrow S.$$

But $(1 \circ 1) \uparrow \mathrm{SL}_2(2)$ is a reducible 3-dimensional module. Therefore, any composition factor of U^S has dimension at most $(2/3)\dim(U^S)$.

The above arguments show that $\dim(V \otimes W) \leq (1/3)(S : P)\dim(T)$. Combined with Lemma 3.1, this yields

$$1 \leq \frac{1}{3} \cdot \frac{(2^n - 1)(2^{n-1} - 1)}{(2^{n-1} - 1)^2},$$

a contradiction.

2) Either $d_\lambda(V)$ or $d_{\lambda^{-1}}(W) = 1$.

For definiteness let us suppose $d_\lambda(V) = 1$. The above argument with Harish-Chandra induction shows that $\dim(V \otimes W) \leq (2/3) \cdot (S : P) d_\lambda(V) d_{\lambda^{-1}}(W)$. Recall that $d_\lambda(V) = 1$, and $d_{\lambda^{-1}}(W) \leq \dim(W)/(2^{n-1} - 1)$ by Lemma 3.1. By [GT], $\dim(V) \geq 2^n - 3$. We come to the inequality

$$2^n - 3 \leq \frac{2}{3} \cdot \frac{(2^n - 1)(2^{n-1} - 1)}{2^{n-1} - 1},$$

again a contradiction. □

Proposition 3.4. *Let $n \geq 3$ and let V, W be two nontrivial irreducible modules of* $\mathrm{SL}_n(3)$ *in characteristic $r \neq 3$ such that $V \otimes W$ is absolutely irreducible. Then $r > 0$, n is even, and at least one of the modules V, W is a Weil module.*

Proof. We proceed by induction on $n \geq 3$. The cases $n = 3, 4$ can be checked directly using GAP. So we assume that $n \geq 5$. Pick any $\lambda \in \Omega(V)$ and consider two cases.

1) $d_\lambda(V), d_{\lambda^{-1}}(W) > 1$.

Let P' be the (unique) normal subgroup of type $Q \cdot \mathrm{SL}_{n-2}(3)$ in P. By Lemma 2.5 applied to P' in place of C, P' has a submodule T inside $V \otimes W$, where T is acted on trivially by Q and it is the tensor product of two P'/Q-modules of dimension $d_\lambda(V)$ and $d_{\lambda^{-1}}(W)$. Since $P'/Q \simeq \mathrm{SL}_{n-2}(3)$, by the induction hypothesis T is not irreducible. Hence we can choose a composition factor L of the P'-module T such that $\dim(L) \leq \dim(T)/2$. Moreover, we can choose L as a submodule or a quotient of T. Arguing as in the proof of Proposition 3.3, we see that $V \otimes W$ is either a quotient or a submodule of L^S.

Consider the subgroup $D := \{\mathrm{diag}(a, a, I_{n-2}) \mid a \in \mathbb{F}_q^\bullet\}$ in P. Since D centralizes P'/Q, L extends to a $P'D$-module of the same dimension, whence $L^{P'D}$ consists of two constituents of the same dimension. Hence we arrive at the bound $\dim(V \otimes W) \leq (1/2)(S : P'D)\dim(T)$. Combined with Lemma 3.1, this yields

$$1 \leq \frac{1}{2} \cdot \frac{(3^n - 1)(3^{n-1} - 1)}{2(3^{n-1} - 1)^2},$$

a contradiction.

2) Either $d_\lambda(V)$ or $d_{\lambda^{-1}}(W) = 1$. Clearly, the case n is odd follows from Proposition 3.2, since $\mathrm{GL}_n(3) \simeq \mathrm{SL}_n(3) \times \mathbb{Z}_2$ when n is odd.

Now we assume that n is even. Let α, resp. β, be the character afforded by V, resp. W. For definiteness we assume that $d_\lambda(V) = 1$, whence α is one of the two Weil characters of S by Lemma 3.1. By the virtue of Proposition 3.2, we may also assume that β is not extendible to G. In the rest of the proof we assume that $r = 0$.

First we consider the case $\alpha(1) = (3^n - 1)/3$. For any $g \in S$, let $d(g)$ denote the dimension of the fixed point subspace of g on the natural module of S. Assume g is unipotent. Then $d(g) \geq 1$. Following the proof of [TZ, Lemma 4.1], one can show that $\alpha(g) = (3^{d(g)} - 1)/2 \geq 1$. Also, it is easy to show that α extends to G. By our assumption, β^G is an irreducible complex character of G, and $\beta^G|_S = \beta + \beta'$ for some G-conjugate β' of β. Let T be a Sylow 3-subgroup of S. By a result of Zelevinsky [Z], there is a character λ of T such that the multiplicity of λ in $\beta^G|_T$ is one. From this it follows that λ enters exactly one of the two characters $\beta|_T$ and $\beta'|_T$, whence $\beta|_T \neq \beta'|_T$. Thus there is a unipotent element $g \in S$ such that $\beta(g) \neq \beta'(g)$. Since $\alpha(g) \geq 1$, $\alpha \otimes \beta \neq \alpha \otimes \beta'$. We have seen that the pair (α, β) satisfies the assumption of Lemma 2.6 (with $H = S$). By Lemma 2.6 and Proposition 3.2, $\alpha \otimes \beta$ cannot be irreducible, a contradiction.

Next we consider the case $\alpha(1) = (3^n - 3)/2$. Let $R = AL$ denote the first maximal parabolic subgroup of S with $A := \mathbb{Z}_3^{n-1}$ and $L := \mathrm{GL}_{n-1}(3)$. Then $\alpha = (1_R)^S - 1_S$, whence $(\alpha|_R, 1_R)_R = 1$. We claim that $\beta|_R$ is irreducible. For, suppose that $\beta|_R$ is reducible: $\beta|_R = \gamma + \gamma'$ for some R-characters γ and γ' with $\gamma(1) \leq \beta(1)/2$. Then $(\alpha \otimes \beta, \gamma^S)_S = (\alpha|_R \otimes \beta|_R, \gamma)_R \geq 1$, but $\alpha \otimes \beta$ is irreducible by assumption; hence

$$\frac{3^n - 3}{2}\beta(1) = \alpha(1)\beta(1) \leq \gamma^S(1) = (S : R)\gamma(1) \leq (S : R)\beta(1)/2 = \frac{3^n - 1}{4}\beta(1),$$

a contradiction. Next, the transitivity of L on $\mathrm{Irr}(A) \backslash \{1_A\}$ and the irreducibility of $\beta|_R$ imply that $(3^{n-1} - 1) | \beta(1)$. Let ℓ be a primitive prime divisor (cf. [Zs]) of $3^{n-1} - 1$, and let ℓ^a be the largest power of ℓ dividing $3^{n-1} - 1$. Then ℓ^{2a} divides $(\alpha \otimes \beta)(1)$ and does not divide $|S|$. Therefore $\alpha \otimes \beta$ cannot be irreducible, again a contradiction. □

Thus we obtain a complete result in the complex case:

Theorem 3.5. *Suppose $n \geq 2$. Then the groups $\mathrm{SL}_n(q)$ and $\mathrm{GL}_n(q)$ have property $\mathcal{P}(0)$.*

4. Unitary groups

Let $G = \mathrm{GU}_n(q)$, $S = \mathrm{SU}_n(q)$ with $q = p^f$, $n \geq 3$, $(n,q) \neq (3,2), (4,2), (4,3), (6,2)$. Let $\mathbb{F}_{q^2}^n$ be the natural module for G, endowed with a Hermitian form with Gram matrix $\mathrm{diag}\left(\begin{pmatrix} 0 & 1 \\ 1 & 0 \end{pmatrix}, I_{n-2}\right)$ with respect to a basis $(e_1, \dots, e_n)$. We consider the parabolic subgroup $P = St_S(\langle e_1 \rangle_{\mathbb{F}_{q^2}})$ of S. Then $Q = O_p(P)$ is a group of order q^{2n-3} of special type. The center $Z(Q)$ has order q and contains $q-1$ transvections t_c, where $c \in \mathbb{F}_{q^2}^\bullet$, $c + c^q = 0$. It is easy to see that Q has a complement

$$\left\{\mathrm{diag}(a, a^{-q}, X) \mid a \in \mathbb{F}_{q^2}^\bullet, X \in \mathrm{GU}_{n-2}(q), \det(X) = a^{q-1}\right\},$$

and all the transvections t_c are conjugate in P. Our assumption on (n,q) guarantees that S is the universal cover of $\mathrm{PSU}_n(q)$.

Let V be any irreducible $\mathbb{F}S$-module of dimension > 1. Since the subgroup $\left\{\mathrm{diag}(a, a^{-q}, a^{q-1}, I_{n-3}) \mid a \in \mathbb{F}_{q^2}^\bullet\right\}$ acts transitively on $\mathrm{IBr}_r(Z(Q)) \setminus \{1\}$, $\Omega(V) = \mathrm{IBr}_r(Z(Q)) \setminus \{1\}$ and all $d_\lambda(V)$ are the same. In particular,

$$d_\lambda(V) \leq \dim(V)/q^{n-2}(q-1).$$

Proposition 4.1. *Suppose $n \geq 3$. Then the groups $\mathrm{SU}_n(q)$ with $q \geq 4$, and $\mathrm{GU}_n(q)$ with $q \geq 3$, have property $\mathcal{P}(r)$ for any r not dividing q.*

Proof. The case $(4,3)$ can be checked directly using GAP; hence we may assume that $(n,q) \neq (4,3)$. First we assume that $G = \mathrm{GU}_n(q)$, $q \geq 3$, has irreducible $\mathbb{F}G$-modules V, W of dimension > 1 such that $V \otimes W$ is irreducible. Pick any $\lambda \in \Omega(V)$ and consider the parabolic subgroup $\hat{P} = St_G(\langle e_1 \rangle_{\mathbb{F}_{q^2}})$ of G. Then $Q = O_p(\hat{P})$, and Q has a complement

$$D := \left\{\mathrm{diag}(a, a^{-q}, X) \mid a \in \mathbb{F}_{q^2}^\bullet, X \in \mathrm{GU}_{n-2}(q)\right\}$$

in $\hat{P}$. Consider the following subgroups

$$A := \left\{\mathrm{diag}(1, 1, X) \mid X \in \mathrm{GU}_{n-2}(q)\right\}, \quad B := \left\{\mathrm{diag}(a, a^{-q}, I_{n-2}) \mid a \in \mathbb{F}_{q^2}^\bullet\right\}$$

of D. Then A normalizes Q, and $C := QA$ centralizes $Z(Q)$. Moreover, $\hat{P} = CB$, and B centralizes C/Q. Hence

$$\dim(V) \cdot \dim(W) \leq (G : \hat{P}) d_\lambda(V) d_{\lambda^{-1}}(W)$$

by Proposition 2.4. Thus

$$q^{2n-4}(q-1)^2 \leq \frac{\dim(V)\cdot\dim(W)}{d_\lambda(V)\cdot d_{\lambda^{-1}}(W)} \leq \frac{(q^n-(-1)^n)(q^{n-1}-(-1)^{n-1})}{q^2-1},$$

a contradiction since $q \geq 3$ and $n \geq 3$.

Next we assume that $S = \mathrm{SU}_n(q)$, $q \geq 4$, has irreducible $\mathbb{F}S$-modules V, W of dimension > 1 such that $V \otimes W$ is irreducible. Consider the complement

$$D := \{\mathrm{diag}(a, a^{-q}, X) \mid a \in \mathbb{F}_{q^2}^\bullet, X \in \mathrm{GU}_{n-2}(q), \det(X) = a^{q-1}\}$$

of Q in P. Consider the following subgroups

$$A := \{\mathrm{diag}(a, a, X) \in D \mid a^{q+1} = 1\},\quad B := \{\mathrm{diag}(a, a^{-1}, I_{n-2}) \mid a^{q-1} = 1\}$$

of D. Then A normalizes Q and $C := QA$ centralizes $Z(Q)$, and B centralizes C/Q. Hence $\dim(V)\cdot\dim(W) \leq (S : BC)d_\lambda(V)d_{\lambda^{-1}}(W)$ by Proposition 2.4. Thus

$$q^{2n-4}(q-1)^2 \leq \frac{\dim(V)\cdot\dim(W)}{d_\lambda(V)\cdot d_{\lambda^{-1}}(W)} \leq \frac{(q^n-(-1)^n)(q^{n-1}-(-1)^{n-1})}{q^2-1}\cdot\gcd(2, q-1),$$

a contradiction since $q \geq 4$ and $n \geq 3$. □

5. Symplectic groups in odd characteristic

Let $G = \mathrm{CSp}_{2n}(q)$, $S = \mathrm{Sp}_{2n}(q)$ with $q = p^f$, $n \geq 2$, $p > 2$. Let $M := \mathbb{F}_q^{2n}$ be the natural module for G, endowed with a symplectic form $(\cdot,\cdot)$ with Gram matrix $\mathrm{diag}\left(\begin{pmatrix} 0 & 1 \\ -1 & 0 \end{pmatrix}, \begin{pmatrix} 0 & I_{n-2} \\ -I_{n-2} & 0 \end{pmatrix}\right)$ with respect to a basis $(e_1, \ldots, e_{2n})$. We can extend $(\cdot,\cdot)$ by linearity to $\tilde{M} := M \otimes_{\mathbb{F}_q} \mathbb{F}_{q^2}$, and define the map

$$\bar{\ } : \sum_{i=1}^{2n} x_i e_i \in \tilde{M} \mapsto \sum_{i=1}^{2n} x_i^q e_i$$

on $\tilde{M}$. Fix $\vartheta \in \mathbb{F}_{q^2}$ with $\vartheta^{q-1} = -1$. Observe that S stabilizes the Hermitian form $u \circ v := \vartheta(u, \bar{v})$ on $\tilde{M}$. This yields an embedding $S \hookrightarrow \mathrm{GU}_{2n}(q)$, which we will use in the proof of Proposition 5.4.

We consider the parabolic subgroup $P = St_S(\langle e_1\rangle_{\mathbb{F}_q})$ of S. Then $Q = O_p(P)$ is a group of order q^{2n-1} of special type. The center $Z(Q)$ has order q and contains $q-1$ transvections t_c, where $c \in \mathbb{F}_q^\bullet$ and t_c has matrix $\left(\begin{pmatrix} 1 & c \\ 0 & 1 \end{pmatrix}, I_{2n-2}\right)$ with respect to the given basis. It is easy to see that Q has a complement

$$\{\mathrm{diag}(a, a^{-1}, X) \mid a \in \mathbb{F}_q^\bullet, X \in \mathrm{Sp}_{2n-2}(q)\}$$

in P. Our assumption on (n, q) guarantees that S is the universal cover of $\mathrm{PSp}_{2n}(q)$.

The transvections t_c are conjugate in G, but they lie in two conjugacy classes in S. We can identify $\mathrm{IBr}_r(Z(Q))$ with $\mathbb{F}_q$ by assigning to each $a \in \mathbb{F}_q$ the character

$$\lambda_a : t_c \mapsto \epsilon^{\mathrm{Tr}_{\mathbb{F}_q/\mathbb{F}_p}(ac)},$$

where $\epsilon = \exp(2\pi i/p)$. Let Ω_+, resp. Ω_-, be the set of characters corresponding to the squares, resp. non-squares, in $\mathbb{F}_q^\bullet$. Let $\mathbb{F}$ be an algebraically closed field of characteristic $r \neq p$ and V be any irreducible $\mathbb{F}S$-module of dimension > 1. Then $\Omega(V)$ is one of Ω_+, Ω_-, or $\Omega := \mathrm{IBr}_r(Z(Q)) \setminus \{1\} = \Omega_+ \cup \Omega_-$. In any case, $d_\lambda(V) \leq 2\dim(V)/q^{n-1}(q-1)$ if $\lambda \in \Omega(V)$.

If V is extendible to G, then clearly $\Omega(V) = \Omega$ and all $d_\lambda(V)$ are the same. Actually we can say more about the case where $\Omega(V) \neq \Omega$. According to [TZ], G has exactly two irreducible complex modules of dimension $(q^n - 1)/2$ and two irreducible complex modules of dimension $(q^n + 1)/2$. All these four modules are called (irreducible) *Weil modules*. One of the main results of [GMST] is

Theorem 5.1 ([GMST]). *Let V be a nontrivial irreducible $\mathbb{F}S$-module with $\Omega(V) \neq \Omega$. Then V is the reduction modulo r of a complex Weil module of S.*

Any V as in Theorem 5.1 will also be called a *Weil module* (in characteristic r).

Proposition 5.2. *Suppose $n \geq 2$, $p > 2$, $p \neq \mathrm{char}(\mathbb{F})$. Let V, W be irreducible $\mathbb{F}S$-modules of dimension > 1 such that $V \otimes W$ is irreducible.*

(i) *If $q \geq 7$, then both of V, W are Weil modules.*

(ii) *If $q = 5$, then at least one of V, W is a Weil module, and both of them are not invariant under* $\mathrm{CSp}_{2n}(q)$.

Proof. (i) Assume the contrary, say W is not a Weil module. Then $\Omega(W) = \Omega$ and so we can choose $\lambda \in \Omega(V)$ such that $\lambda^{-1} \in \Omega(W)$. Consider the complement

$$D := \{\mathrm{diag}(a, a^{-1}, X) \mid a \in \mathbb{F}_q^\bullet, X \in \mathrm{Sp}_{2n-2}(q)\}$$

of Q in P. Consider the following subgroups

$$A := \{\mathrm{diag}(1, 1, X) \mid X \in \mathrm{Sp}_{2n-2}(q)\}, \quad B := \{\mathrm{diag}(a, a^{-1}, I_{n-2}) \mid a \in \mathbb{F}_q^\bullet\}$$

of D. Then A normalizes Q and $C := QA$ centralizes $Z(Q)$. Also, $P = BC$ and B centralizes C/Q. Hence, $\dim(V) \cdot \dim(W) \leq (S : P) d_\lambda(V) d_{\lambda^{-1}}(W)$ by Proposition 2.4. Thus

$$q^{2n-2}(q-1)^2/4 \leq \frac{\dim(V) \cdot \dim(W)}{d_\lambda(V) \cdot d_{\lambda^{-1}}(W)} \leq \frac{q^{2n}-1}{q-1},$$

a contradiction since $q \geq 7$.

(ii) First we consider the case where $\Omega(V) = \Omega(W) = \Omega$. Then we can choose $\lambda \in \Omega(V)$ such that $\dim(W_{\lambda^{-1}})$ is smallest among all $\dim(W_\mu)$, $\mu \in \Omega(W)$. Clearly, $\dim(V)/d_\lambda(V) \geq q^{n-1}(q-1)/2$ and $\dim(W)/d_{\lambda^{-1}}(W) \geq q^{n-1}(q-1)$. Using the same arguments as in (i) we come to the inequality

$$q^{2n-2}(q-1)^2/2 \le \frac{\dim(V)\cdot\dim(W)}{d_\lambda(V)\cdot d_{\lambda^{-1}}(W)} \le \frac{q^{2n}-1}{q-1},$$

a contradiction since $q = 5$.

Next we consider the case where W is invariant under $\mathrm{CSp}_{2n}(q)$. Then $\Omega(W) = \Omega$ and $\dim(W_\mu)$ is the same for all $\mu \in \Omega(W)$; hence

$$\dim(W)/d_{\lambda^{-1}}(W) \ge q^{n-1}(q-1).$$

Now we can use the same arguments as in (i) and again come to the contradiction that

$$q^{2n-2}(q-1)^2/2 \le \frac{\dim(V)\cdot\dim(W)}{d_\lambda(V)\cdot d_{\lambda^{-1}}(W)} \le \frac{q^{2n}-1}{q-1}. \qquad \square$$

The same proof as above shows

Proposition 5.3. *Suppose $n \ge 1$, $p > 2$, and $q = p^f \ge 5$. Then $\mathrm{CSp}_{2n}(q)$ has property $\mathcal{P}(r)$ for any r.*

In view of Proposition 5.2, one has to study the tensor product of Weil modules of $S = \mathrm{Sp}_{2n}(q)$. As mentioned above, S has four complex irreducible Weil modules, whose characters are denoted by ξ, ξ^* (of degree $(q^n+1)/2$), and by η, η^* (of degree $(q^n-1)/2$). If $r \ne 2, p$, then the reductions modulo r of these four characters are absolutely irreducible, and distinct (look at the character values at transvections). If $r = 2$, then $\xi \,(\mathrm{mod}\,2) = 1 + \eta\,(\mathrm{mod}\,2)$, $\xi^*\,(\mathrm{mod}\,2) = 1 + \eta^*\,(\mathrm{mod}\,2)$, and $\eta\,(\mathrm{mod}\,2)$ and $\eta^*\,(\mathrm{mod}\,2)$ are absolutely irreducible and distinct. If V is any irreducible module, then let $S^2(V)$, resp. $\Lambda^2(V)$, be the quotient (of codimension 0 or 1) of $\mathrm{Sym}^2(V)$, resp. of $V \wedge V$, by the fixed point subspace. If V is not self-dual, then let $A(V)$ be the natural submodule of codimension 1 in $V \otimes V^*$. Let

$$\mathcal{M} = \{V \otimes W,\ S^2(V),\ \Lambda^2(V),\ A^2(V) \mid V, W \text{ Weil modules of } \mathrm{Sp}_{2n}(q)\}.$$

Proposition 5.4. (i) *Suppose $q \ge 11$ and $r \ne p$. Then no $M \in \mathcal{M}$ can be irreducible.*

(ii) *Suppose $q = 7, 9$. Then the only modules $M \in \mathcal{M}$ which can be irreducible are $S^2(\eta)$, $S^2(\eta^*)$, $\Lambda^2(\xi)$, and $\Lambda^2(\eta^*)$. Moreover, these modules are irreducible if $r = 0$.*

(iii) *Suppose $q = 5$. Then the only modules $M \in \mathcal{M}$ which can be irreducible are $S^2(\xi) = S^2(\xi^*)$, $\Lambda^2(\xi)$, $\Lambda^2(\xi^*)$, $S^2(\eta)$, $S^2(\eta^*)$, $\Lambda^2(\eta) = \Lambda^2(\eta^*)$, $\xi\eta = \xi^*\eta^*$. Moreover, these modules are irreducible if $r = 0$.*

(iv) *Suppose $q = 3$. Then the only modules $M \in \mathcal{M}$ which can be irreducible are $A(\xi)$, $A(\eta)$, $\xi\eta^*$, $\xi^*\eta$, $S^2(\xi) = S^2(\xi^*)$, $\Lambda^2(\xi)$, $\Lambda^2(\xi^*)$, $S^2(\eta)$, $S^2(\eta^*)$, $\Lambda^2(\eta) = \Lambda^2(\eta^*)$, $\xi\eta = \xi^*\eta^*$. Moreover, these modules are irreducible if $r = 0$.*

Proof. The discussion before this proposition shows that it is enough to prove this proposition for complex Weil modules.

We can embed S naturally in $\mathrm{GL}_{2n}(q)$. The group $\mathrm{GL}_{2n}(q)$ has a so-called (reducible) rational Weil character τ of degree q^{2n}, and $\tau|_S$ is just the permutation character of S acting on the natural module $\mathbb{F}_q^{2n}$; in particular, $(\tau|_S, \tau|_S)_S = 2q+2$. In [T] we will show that τ_S is the sum of twice the trivial character, an irreducible character of degree $(q^n-1)(q^n+q)/2(q-1)$ (of multiplicity 1), an irreducible character of degree $(q^n+1)(q^n-q)/2(q-1)$ (of multiplicity 1), two irreducible characters of degree $(q^{2n}-1)/2(q-1)$ (each of multiplicity 1), and $(q-3)/2$ irreducible characters of degree $(q^{2n}-1)/(q-1)$ (each of multiplicity 2).

As mentioned at the beginning of the section, one can also embed S naturally in $\mathrm{GU}_{2n}(q)$. The group $\mathrm{GU}_{2n}(q)$ has a so-called (reducible) $\mathbb{Q}$-valued Weil character σ of degree q^{2n}, and $(\sigma|_S, \sigma|_S)_S = 2q+2$. It will be shown in [T] that σ_S is the sum of an irreducible character of degree $(q^n+1)(q^n+q)/2(q+1)$ (of multiplicity 1), an irreducible character of degree $(q^n-1)(q^n-q)/2(q+1)$ (of multiplicity 1), two irreducible characters of degree $(q^{2n}-1)/2(q+1)$ (each of multiplicity 1), and $(q-1)/2$ irreducible characters of degree $(q^{2n}-1)/(q+1)$ (each of multiplicity 2).

We distinguish two cases.

1) $q \equiv 1 \pmod 4$. In this case the characters ξ, ξ^* are real, and the characters η, η^* are real-valued but have Schur index 2 over $\mathbb{R}$.

One can show that $\tau|_S = (\xi+\eta)^2 = (\xi^*+\eta^*)^2$ and $\sigma|_S = (\xi+\eta)(\xi^*+\eta^*)$ (see [T]). The largest dimension of irreducible constituents of σ_S is $(q^{2n}-1)/(q+1)$, which is less than $(q^n-1)^2/4$; hence none of the characters $\xi\xi^*$, $\eta\eta^*$, $\xi\eta^*$, and $\xi^*\eta$ is irreducible.

Next we suppose $q \geq 9$. Then the largest dimension of irreducible constituents of τ_S is $(q^{2n}-1)/(q-1)$, which is less than the dimension of $S^2(\xi)$ and $\xi\eta$. Also, the dimension of $\Lambda^2(\eta)$ cannot be equal to the dimension of any irreducible constituents of $\tau|_S$. Finally, the only case when the dimension of $S^2(\eta)$ and $\Lambda^2(\xi)$ is equal to the dimension of an irreducible constituent of $\tau|_S$ is $q=9$. In fact, if $q=9$, then one can work out the following decomposition of the characters in $\mathcal{M}$ in terms of the irreducible constituents of $\tau|_S$:

$$\xi\eta = \overline{\frac{q^{2n}-1}{q-1}} + \overline{\frac{q^{2n}-1}{q-1}}, \quad S^2(\xi) = \overline{\frac{q^{2n}-1}{2(q-1)}} + \overline{\frac{(q^n-1)(q^n+q)}{2(q-1)}},$$

$$\Lambda^2(\xi), S^2(\eta) = \overline{\frac{q^{2n}-1}{q-1}}, \quad \Lambda^2(\eta) = \overline{\frac{q^{2n}-1}{2(q-1)}} + \overline{\frac{(q^n+1)(q^n-q)}{2(q-1)}},$$

where $\overline{d}$ denotes an irreducible character of degree d.

Finally, suppose $q=5$. Then we have the obvious decomposition

$$\tau_S = (\xi+\eta)^2 = S^2(\xi) + \Lambda^2(\xi) + S^2(\eta) + \Lambda^2(\eta) + 2\xi\eta + 2\cdot 1_S.$$

Since $(\tau_S, \tau_S)_S = 12$, each summand in this decomposition is irreducible. Here are the degrees of these summands:

$$(\xi\eta)(1) = \frac{q^{2n}-1}{q-1}, \quad S^2(\xi)(1) = \frac{(q^n-1)(q^n+q)}{2(q-1)},$$

$$\Lambda^2(\xi)(1) = S^2(\eta)(1) = \frac{q^{2n}-1}{2(q-1)}, \quad \Lambda^2(\eta)(1) = \frac{(q^n+1)(q^n-q)}{2(q-1)}.$$

2) $q \equiv 3\,(\mathrm{mod}\,4)$. Then $\xi^* = \overline{\xi}$, $\eta^* = \overline{\eta}$, and both ξ, η are non-real.

One can show that $\tau|_S = (\xi+\eta)(\xi^*+\eta^*)$ and $\sigma|_S = (\xi+\eta)^2 = (\xi^*+\eta^*)^2$ (see [T]). First we look at $\tau|_S$. Suppose $q \geq 7$. In this case, the largest dimension of any irreducible constituent of τ_S is equal to $(q^{2n}-1)/(q-1)$, which is less than $(q^n-1)(q^n-3)/4$; hence none of the characters $A(\xi)$, $A(\eta)$, $\xi\eta^*$, and $\xi^*\eta$ is irreducible. Next suppose $q = 3$. Then we have the obvious decomposition

$$\tau_S = A(\xi) + A(\eta) + \xi\eta^* + \xi^*\eta + 2 \cdot 1_S.$$

Since $(\tau_S, \tau_S)_S = 8$, each summand in this decomposition is irreducible. Here are the degrees of these summands:

$$A(\xi)(1) = \frac{(q^n-1)(q^n+q)}{2(q-1)}, \quad A(\eta)(1) = \frac{(q^n+1)(q^n-q)}{2(q-1)},$$

$$(\xi\eta^*)(1) = (\xi^*\eta)(1) = \frac{q^{2n}-1}{2(q-1)}.$$

Next we consider $\sigma|_S$ and suppose $q \geq 7$. Then the largest dimension of irreducible constituents of σ_S is $(q^{2n}-1)/(q+1)$, which is less than the dimension of $S^2(\xi)$ and $\xi\eta$. Also, the dimension of $\Lambda^2(\eta)$ cannot be equal to the dimension of any irreducible constituents of $\sigma|_S$. Finally, the only case when the dimension of $S^2(\eta)$ and $\Lambda^2(\xi)$ is equal to the dimension of an irreducible constituent of $\sigma|_S$ is $q = 7$. In fact, if $q = 7$, then one can work out the following decomposition of the characters in $\mathcal{M}$ in terms of the irreducible constituents of $\sigma|_S$:

$$\xi\eta = \overline{\frac{q^{2n}-1}{q+1}} + \overline{\frac{q^{2n}-1}{q+1}}, \quad S^2(\xi) = \overline{\frac{q^{2n}-1}{2(q+1)}} + \overline{\frac{(q^n+1)(q^n+q)}{2(q+1)}},$$

$$\Lambda^2(\xi), S^2(\eta) = \overline{\frac{q^{2n}-1}{q+1}}, \quad \Lambda^2(\eta) = \overline{\frac{q^{2n}-1}{2(q+1)}} + \overline{\frac{(q^n-1)(q^n-q)}{2(q+1)}}.$$

Finally, suppose $q = 3$. Then we have the obvious decomposition

$$\sigma_S = (\xi+\eta)^2 = S^2(\xi) + \Lambda^2(\xi) + S^2(\eta) + \Lambda^2(\eta) + 2\xi\eta.$$

Since $(\sigma_S, \sigma_S)_S = 8$, each summand in this decomposition is irreducible. Here are the degrees of these summands:

$$(\xi\eta)(1) = \frac{q^{2n}-1}{q+1}, \quad S^2(\xi)(1) = \frac{(q^n+1)(q^n+q)}{2(q+1)},$$

$$\Lambda^2(\xi)(1) = S^2(\eta)(1) = \frac{q^{2n}-1}{2(q+1)}, \quad \Lambda^2(\eta)(1) = \frac{(q^n-1)(q^n-q)}{q+1}. \qquad \square$$

Propositions 5.2 and 5.4 together imply

Corollary 5.5. *Suppose $n \geq 2$, $p > 2$, and $q \geq 7$. Then the group $\mathrm{Sp}_{2n}(q)$ has property $\mathcal{P}(r)$ for any $r \neq p$.*

6. Orthogonal groups in characteristic 2

Let $S = \Omega^\epsilon_{2n}(q)$ with $q = 2^f > 2$, $n \geq 4$, $\epsilon = \pm$. Let $\mathbb{F}_q^{2n}$ be the natural module for S, endowed with a non-degenerate quadratic form whose bilinear form has Gram matrix $\mathrm{diag}\left(\begin{pmatrix} 0 & I_2 \\ I_2 & 0 \end{pmatrix}, \begin{pmatrix} 0 & I_{n-2} \\ I_{n-2} & 0 \end{pmatrix}\right)$ with respect to a basis $(e_1, \dots, e_{2n})$. We may assume that $\langle e_1, e_2 \rangle_{\mathbb{F}_q}$ is totally singular. Consider the parabolic subgroup $P = St_S(\langle e_1, e_2 \rangle_{\mathbb{F}_q})$ of S. Then $Q = O_p(P)$ is a group of order q^{4n-7} of special type. The center $Z(Q)$ has order q and contains $q-1$ elements t_c, where $c \in \mathbb{F}_q^\bullet$. It is easy to see that Q has a complement

$$D := \{\mathrm{diag}(A, {}^tA^{-1}, X) \mid A \in \mathrm{GL}_2(q), X \in \Omega^\epsilon_{2n-4}(q)\}$$

in P, and all the elements t_c are conjugate in P. Our assumption on (n, q) guarantees that S is the universal cover of $\Omega^\epsilon_{2n}(q)$.

Let V be any irreducible $\mathbb{F}S$-module of dimension > 1. Since the subgroup

$$\{\mathrm{diag}(a, 1, a^{-1}, 1, I_{2n-4}) \mid a \in \mathbb{F}_q^\bullet\}$$

acts transitively on $\mathrm{IBr}_r(Z(Q)) \setminus \{1\}$, $\Omega(V) = \mathrm{IBr}_r(Z(Q)) \setminus \{1\}$, and all $d_\lambda(V)$ arc the same. In particular, $d_\lambda(V) \leq \dim(V)/q^{2n-4}(q-1)$.

Proposition 6.1. *Suppose $q = 2^f \geq 4$ and $n \geq 4$. Then the group $\Omega^\epsilon_{2n}(q)$ has property $\mathcal{P}(r)$ for any $r \neq 2$.*

Proof. Assume that $S = \Omega^\epsilon_{2n}(q)$ has irreducible $\mathbb{F}S$-modules V, W of dimension > 1 such that $V \otimes W$ is irreducible. Fix $\lambda \in \Omega(V)$ and consider the subgroup

$$B := \{\mathrm{diag}(aI_2, a^{-1}I_2, I_{2n-4}) \mid a \in \mathbb{F}_q^\bullet\}$$

in P. Also, let C be the (unique) normal subgroup of type $Q \cdot (\mathrm{SL}_2(q) \times \Omega^\epsilon_{2n-4}(q))$ in P. Then C normalizes Q and centralizes $Z(Q)$. Furthermore, $P = BC$, and B

centralizes C/Q. Hence $\dim(V)\cdot\dim(W) \leq (S:P)d_\lambda(V)d_{\lambda^{-1}}(W)$ by Proposition 2.4. Thus

$$q^{4n-8}(q-1)^2 \leq \frac{\dim(V)\cdot\dim(W)}{d_\lambda(V)\cdot d_{\lambda^{-1}}(W)} \leq \frac{(q^n-\epsilon)(q^{2n-2}-1)(q^{n-2}+\epsilon)}{(q-1)(q^2-1)},$$

a contradiction since $q \geq 4$ and $n \geq 4$. □

7. Orthogonal groups in odd characteristic

Let $S = \mathrm{Spin}_n^\epsilon(q)$ with $q = p^f \geq 5$, $p > 2$ a prime, $n \geq 7$, $\epsilon = \pm$. Let $\mathbb{F}_q^n$ be the natural module for $\bar{S} := \Omega_n^\epsilon(q)$, endowed with a non-degenerate quadratic form, whose bilinear form has Gram matrix $\mathrm{diag}\left(\begin{pmatrix} 0 & I_2 \\ I_2 & 0 \end{pmatrix}, J\right)$ with respect to a basis $(e_1,\dots,e_n)$ for a non-degenerate diagonal matrix J. We may assume that $\langle e_1, e_2\rangle_{\mathbb{F}_q}$ is totally singular. Consider the parabolic subgroup $\bar{P} = St_{\bar{S}}(\langle e_1, e_2\rangle_{\mathbb{F}_q})$ of $\bar{S}$. Then $Q = O_p(\bar{P})$ is a group of order q^{2n-7} of special type. Namely, Q is the subgroup

$$\left\{ [A,B] := \begin{pmatrix} I_2 & B & -{}^tA\cdot J \\ 0 & I_2 & 0 \\ 0 & A & I_{n-4} \end{pmatrix} \;\middle|\; \begin{array}{l} A \in M_{n-4,2}(\mathbb{F}_q),\ B \in M_{2,2}(\mathbb{F}_q), \\ B + {}^tB + {}^tAJA = 0 \end{array} \right\}.$$

Each matrix $[A,B] \in Q$ can be identified with the triple $[X,Y,z]$, where $A = (X, -2J^{-1}Y)$ and $B = z\begin{pmatrix} 0 & 1 \\ -1 & 0 \end{pmatrix} - \frac{1}{2}{}^tAJA$, $X, Y \in \mathbb{F}_q^{n-4}$ and $z \in \mathbb{F}_q$. This identification shows that Q is isomorphic to the group described before Lemma 2.3 with $m = n-4$. The center $Z(Q)$ has order q and contains $q-1$ elements t_c, where $c \in \mathbb{F}_q^\bullet$. Let $\mathrm{GL}_2^{\bullet 2}(q)$ be the subgroup of all $A \in \mathrm{GL}_2(q)$ with $\det(A)$ being a square. By [KL, 2.7.2], the element $\mathrm{diag}(A, {}^tA^{-1}, I_{n-4})$, $A \in \mathrm{GL}_2(q)$, belongs to $\Omega_n^\epsilon(q)$ if and only if $\det(A)$ is a square. Hence Q has a complement D in P which is isomorphic to $\left(\mathrm{GL}_2^{\bullet 2}(q) \times \Omega_{n-4}^\epsilon(q)\right)\cdot 2$, and all the elements t_c are conjugate in P. Let C be the (unique) normal subgroup of type $\mathrm{SL}_2(q) \times \Omega_{n-4}^\epsilon(q)$ in D and let $\bar{P}' = Q\cdot C$.

Let P and P' be complete inverse images of $\bar{P}$ and $\bar{P}'$ in S. Then P is the parabolic subgroup corresponding to the second node of the Dynkin diagram of S. S and P certainly contain a subgroup $N \simeq \mathrm{Spin}_{n-4}^\epsilon(q)$ (which corresponds to the subdiagram containing all but the first and the second nodes). If a root α corresponds to the first node, then S and P also contain the subgroup M (generated by the root subgroups) corresponding to $\pm\alpha$. It is known that $M \simeq \mathrm{SL}_2(q)$, since the rank of S is bigger than 1. Clearly $[M,N] = 1$ and $M \cap N = 1$ (as S is of simply connected type). Since M and N are perfect, they are contained in P'. Also, we can lift Q to $O_p(P)$, which we also denote by Q. Now it is clear that $P' = Q\cdot(M \times N)$. Observe that $M \times N$ acts trivially on $Z(Q)$.

Let V be any irreducible $\mathbb{F}S$-module of dimension > 1. Since the complete inverse image of the subgroup $\{\mathrm{diag}(a, 1, a^{-1}, 1, I_{n-4}) \mid a \in \mathbb{F}_q^\bullet\}$ in S acts transitively on $\mathrm{IBr}_r(Z(Q))\setminus\{1\}$, $\Omega(V) = \mathrm{IBr}_r(Z(Q))\setminus\{1\}$ and all $d_\lambda(V)$ are the same. In particular, $d_\lambda(V) \leq \dim(V)/q^{n-4}(q-1)$.

Proposition 7.1. *Suppose $q = p^f \geq 5$ and $n \geq 7$. Then the group $\mathrm{Spin}_n^\epsilon(q)$ has property $\mathcal{P}(r)$ for any $r \neq p$.*

Proof. Assume that $S = \mathrm{Spin}_n^\epsilon(q)$ has irreducible $\mathbb{F}S$-modules V, W of dimension > 1 such that $V \otimes W$ is irreducible. Fix $\lambda \in \Omega(V)$ and consider the subgroup L of S generated by the root subgroups corresponding to the first and second nodes of the Dynkin diagram of S. Then $L \simeq \mathrm{SL}_3(q)$, and L contains M. Inside L we can choose a torus $B \simeq \mathbb{Z}_{q-1}$ which centralizes M and N. By Proposition 2.4, $\dim(V) \cdot \dim(W) \leq (S : P'B) d_\lambda(V) d_{\lambda^{-1}}(W)$. Also, $P/(P'B) \simeq \mathbb{Z}_2$. Consequently,

$$q^{4k-8}(q-1)^2 \leq \frac{\dim(V) \cdot \dim(W)}{d_\lambda(V) \cdot d_{\lambda^{-1}}(W)} \leq \frac{2(q^k - \epsilon)(q^{2k-2} - 1)(q^{k-2} + \epsilon)}{(q-1)(q^2-1)}$$

if $n = 2k$, and

$$q^{4k-6}(q-1)^2 \leq \frac{\dim(V) \cdot \dim(W)}{d_\lambda(V) \cdot d_{\lambda^{-1}}(W)} \leq \frac{2(q^{2k} - 1)(q^{2k-2} - 1)}{(q-1)(q^2-1)}$$

if $n = 2k + 1$. Both of these inequalities cannot hold since $q \geq 5$ and $n \geq 7$. □

8. Exceptional groups of Lie type

In this section we consider exceptional groups G of Lie type, i.e. of type ${}^2B_2$, ${}^3D_4$, G_2, ${}^2G_2$, F_4, ${}^2F_4$, E_6, ${}^2E_6$, E_7, or E_8, defined over a finite field $\mathbb{F}_q$. Let $\ell(G)$ denote the smallest integer $t > 1$ such that G has a projective irreducible representation of degree t over a field of characteristic not dividing q. Lower bounds for $\ell(G)$ were found by Landazuri and Seitz in [LS] and improved by Seitz and Zalesskii in [SZ]. Recently, the Landazuri–Seitz–Zalesskii bounds have been improved in a number of cases, cf. [GT], [Hof1], [Hof2]. Here we would like to exhibit the exact value of $\ell(G)$ in the case $G = G_2(q)$ and q is divisible by 3.

Proposition 8.1. *Suppose that $G = G_2(q)$, where $3|q$. Then $\ell(G) = q^2(q^2+1)$ if $q > 3$, and 14 if $q = 3$.* *

Proof. The statement is well-known if $q = 3$, cf. [Atlas] and [JLPW]. Suppose $q = 3^f > 3$ and r is any prime other than 2 and 3. The Brauer trees for cyclic r-blocks of G are described in [Sh]. All the non-cyclic r-blocks have maximal defect,

*After the paper had been submitted, we learned from G. Hiss that Proposition 8.1 was already proved in his Habilitationsschrift. We are thankful to him for this remark.

and the decomposition numbers for these blocks are computed in [H]. These two results readily imply that $\varphi(1) \geq q^4 + q^2 + 1$ for all nontrivial $\varphi \in \mathrm{IBr}_r(G)$. Finally, the 2-modular characters of G are determined in [HS], from which it follows that the smallest degree of a nontrivial 2-modular character is $q^2(q^2+1)$. □

Observe that $\ell(G_2(q)) = q(q^2-1)$ if 3 is coprime to q (and $q \neq 2, 4$).

The Landazuri–Seitz–Zalesskii bounds for ${}^3D_4(q)$, ${}^2E_6(q)$, and for $F_4(q)$ with q odd, have recently been improved in [MMT]. This leaves the cases of ${}^2F_4(q)$ and $F_4(q)$ with q even as the only cases where the gap between the Landazuri–Seitz–Zalesskii bounds and the complex bound (as found in [Lu]) is still considerably large.

Proposition 8.2. *Let G be the universal cover of a finite simple exceptional group of Lie type, defined over a field $\mathbb{F}_q$. Then G has property $\mathcal{P}(r)$ for any r not dividing q, except possibly for the following cases:*

(i) $q \leq 3$;

(ii) *G is of type F_4 or ${}^2F_4$, and $2|q$.*

Proof. 1) First we consider the case where G is of type G_2 (and $3|q$), ${}^2G_2$, or ${}^2B_2$. The subcase of ${}^2B_2(8)$ can be checked by hand using [Atlas] and [JLPW]. In all the remaining subcases G is simple, and $\ell(G) = q^2(q^2+1)$ (by Proposition 8.1), $q(q-1)$, and $(q-1)\sqrt{q/2}$, respectively. On the other hand, the largest degree of any irreducible complex character (and therefore of any irreducible Brauer character in cross characteristic) of G is $(q+1)(q^2+q+1)(q^3+1)$, $(q^2-1)(q+\sqrt{3q}+1)$, and $(q-1)(q+\sqrt{2q}+1)$, respectively. Since this degree is clearly smaller than $\ell(G)^2$, G has property $\mathcal{P}(r)$.

2) Now we suppose that G is the universal cover of an exceptional group of Lie type $E_8(q)$, $E_7(q)$, $E_6(q)$, ${}^2E_6(q)$, $F_4(q)$ (and q is odd), ${}^3D_4(q)$, or $G_2(q)$ (and 3 is coprime to q), respectively. Then G has a subgroup Q of special type, of order q^{1+56}, q^{1+32}, q^{1+20}, q^{1+20}, q^{1+14}, q^{1+8}, and q^{1+4}, respectively (cf. for instance Lemma 4.3 of [LS]). In all cases $Z(Q)$ is an elementary abelian group of order q. If $P = N_G(Q)$ and $P' = C_P(Z(Q))$, then P/Q contains a subgroup $(P'/Q) * A$ of index $\gcd(2, q-1)$, where $A \simeq \mathbb{Z}_{q-1}$. Correspondingly, P'/Q is the (simply connected) group of type $E_7(q)$, $\mathrm{Spin}^+_{12}(q)$, $\mathrm{SL}_6(q)$, $\mathrm{SU}_6(q)$, $\mathrm{Sp}_6(q)$, $\mathrm{SL}_2(q^3)$, and $\mathrm{SL}_2(q)$. Also, P acts transitively on $Z(Q) \setminus \{1\}$ and on $\mathrm{IBr}_r(Z(Q)) \setminus \{1\}$ (to see this observe that every long root group is contained in an $\mathrm{SL}_3(q)$-subgroup). Let V and W be any two $\mathbb{F}G$-modules of dimension > 1 such that $V \otimes W$ is irreducible. Then the same arguments as the ones we used for $\mathrm{SL}_n(q)$ in the proof of Proposition 3.2 show that

$$\dim(V) \cdot \dim(W) \leq (G : P) d_\lambda(V) d_{\lambda^{-1}}(W) \cdot \gcd(2, q-1).$$

But $\dim(V)/d_\lambda(V)$ and $\dim(W)/d_{\lambda^{-1}}(W)$ are at least $(q-1)\sqrt{|Q|/q}$. Hence we get the following inequality

$$(q-1)^2 \cdot |Q| \leq q \cdot (G : P) \cdot \gcd(2, q-1),$$

which cannot happen in any of the above cases when $q \geq 4$. □

Proposition 8.3. *Let $q = 2^f > 2$ and $G = F_4(q)$ or ${}^2F_4(q)$. Then G has property $\mathcal{P}(0)$.*

Proof. Suppose $\alpha, \beta \in \mathrm{Irr}(G)$ are of degree > 1 and such that $\gamma := \alpha \otimes \beta$ is irreducible. Let ϕ_m be the mth cyclotomic polynomial in q.

1) Suppose that G is of type F_4. By [S2, (2.1)],

$$\gamma(1) \leq (q^{12}-1)(q^8-1)(q^6-1)(q^2-1)/(q-1)^4.$$

On the other hand, $\alpha(1) \geq \ell(G) \geq q^7(q^3-1)(q-1)/2$. Hence

$$\beta(1) \leq \frac{2(q^{12}-1)(q^8-1)(q^6-1)(q^2-1)}{q^7(q^3-1)(q-1)^5} < \phi_1\phi_3\phi_4\phi_6\phi_8\phi_{12},$$

since $q \geq 4$. The five smallest nontrivial degrees d_i, $1 \leq i \leq 5$, of complex irreducible characters of G are computed in [Lu], and the fifth degree d_5 is exactly $\phi_1\phi_3\phi_4\phi_6\phi_8\phi_{12}$. Hence $\beta(1)$ is one of the four smallest degrees. The same thing is true for $\alpha(1)$.

Suppose that $\alpha(1) = d_4 = q^2\phi_3^2\phi_6^2\phi_{12}$. Since $|G|$ is not divisible by ϕ_{12}^2 and d_2 is divisible by ϕ_{12}, $\beta(1)$ is neither d_2 nor d_4. Since $|G|$ is not divisible by ϕ_6^3 and d_3 is divisible by ϕ_6, $\beta(1)$ cannot be d_3. Finally, since $|G|$ is not divisible by ϕ_3^3 and d_1 is divisible by ϕ_3, $\beta(1)$ cannot be d_1, a contradiction.

Thus $\alpha(1)$ is one of the three smallest degrees. In particular, $\alpha(1)$ is divisible by ϕ_8. By symmetry, the same thing is true for $\beta(1)$. Hence $\gamma(1)$ and $|G|$ are divisible by ϕ_8^2, again a contradiction.

2) Next suppose that $G = {}^2F_4(q)$. The three smallest nontrivial degrees d_i, $1 \leq i \leq 3$, of complex irreducible characters of G are computed in [Lu], where $d_3 = \phi_1\phi_4^2\phi_{12}$ if $q \geq 32$, and $d_3 = 13778800$ if $q = 8$. By [S2, (2.1)],

$$\gamma(1) \leq (q^6+1)(q^4-1)(q^3+1)(q-1)/(q-\sqrt{2q}+1)^2.$$

On the other hand, $\alpha(1) \geq \ell(G) \geq q^4(q-1)\sqrt{q/2}$. Hence

$$\beta(1) \leq \frac{(q^6+1)(q^4-1)(q^3+1)(q-1)}{(q-\sqrt{2q}+1)^2 q^4(q-1)\sqrt{q/2}} < d_3,$$

since $q \geq 8$. This implies that $\beta(1)$ is d_1 or d_2. The same thing is true for $\alpha(1)$. But d_1 and d_2 are divisible by ϕ_6; hence $\gamma(1)$ and $|G|$ are divisible by ϕ_6^2, a contradiction. □

One can show that if G is as in Proposition 8.3 and $\alpha, \beta \in \mathrm{IBr}_r(G)$ are nontrivial characters such that $\alpha \otimes \beta$ is irreducible, then both α and β belong to unipotent blocks.

References

[A] M. Aschbacher, On the maximal subgroups of the finite classical groups, Invent. Math. 76 (1984), 469–514.

[Ar] Z. Arad, On products of conjugacy classes and irreducible characters in finite groups, Proc. Symp. Pure Math. 47 (1987), Pt. II, 3–10.

[Atlas] J. H. Conway, R. T. Curtis, S. P. Norton, R. A. Parker, and R. A. Wilson, An ATLAS of Finite Groups, Clarendon Press, Oxford 1985.

[BK1] C. Bessenrodt and A. S. Kleshchev, On Kronecker products of complex representations of the symmetric and alternating groups, Pacific J. Math. 190 (1999), 201–223.

[BK2] C. Bessenrodt and A. S. Kleshchev, On tensor products of modular representations of symmetric groups, Bull. London Math. Soc. 32 (2000), 292–296.

[BK3] C. Bessenrodt and A. S. Kleshchev, Irreducible tensor products over alternating groups, J. Algebra 228 (2000), 536–550.

[DD] R. Dipper and J. Du, Harish-Chandra vertices, J. Reine Angew. Math. 437 (1993), 101–130.

[F] W. Feit, The Representation Theory of Finite Groups, North-Holland Publishing Company, Amsterdam 1982.

[FT] P. Ferguson and A. Turull, Prime characters and factorizations of quasi-primitive characters, Math. Z. 190 (1985), 583–604.

[GM] D. Gluck and K. Magaard, Character and fixed space ratios of Brauer characters of finite classical groups, J. Algebra 204 (1998), 188–201.

[GK] R. Gow and A. S. Kleshchev, Connections between the representations of the symmetric group and the symplectic group in characteristic 2, J. Algebra 221 (1999), 60–89.

[GMST] R. M. Guralnick, K. Magaard, J. Saxl, and Pham Huu Tiep, Cross characteristic representations of odd characteristic symplectic groups and unitary groups, submitted.

[GT] R. M. Guralnick and Pham Huu Tiep, Low-dimensional representations of special linear groups in cross characteristics, Proc. London Math. Soc. (3) 78 (1999), 116–138.

[H] G. Hiss, On the decomposition numbers of $G_2(q)$, J. Algebra 120 (1989), 339–360.

[HS] G. Hiss and J. Shamash, 2-blocks and 2-modular characters of the Chevalley groups $G_2(q)$, Math. Comp. 59 (1992), 645–672.

[Hof1] C. Hoffman, Projective representations in cross characteristic for some classical groups, J. Algebra 229 (2000), 666–677.

[Hof2] C. Hoffman, Projective representations for some exceptional finite groups of Lie type, in: Modular Representation Theory of Finite Groups (Michael J. Collins, Brian J. Parshall, Leonard L. Scott, eds.) Walter de Gruyter, Berlin 2001, 223–230.

[HL] R. Howlett and G. Lehrer, On Harish-Chandra induction and restriction for modules of Levi subgroups, J. Algebra 165 (1994), 172–183.

[Hup] B. Huppert, Endliche Gruppen I, Springer-Verlag, Berlin 1967.

[Is] I. M. Isaacs, Irreducible products of characters, J. Algebra 223 (2000), 630–646.

[JLPW] C. Jansen, K. Lux, R. A. Parker, and R. A. Wilson, An ATLAS of Brauer Characters, Oxford University Press, Oxford 1995.

[JJ] J. Graham and G. James, On a conjecture of Gow and Kleshchev concerning tensor products, J. Algebra 227 (2000), 767–782.

[KL] P. Kleidman and M. Liebeck, The Subgroup Structure of the Finite Classical Groups, London Math. Soc. Lecture Note Ser. 129, Cambridge University Press, 1990.

[LST] W. Lempken, B. Schröder, and Pham Huu Tiep, Symmetric squares, spherical designs, and lattice minima, J. Algebra 240 (2001), 185–208.

[LS] V. Landazuri and G. Seitz, On the minimal degrees of the projective representations of the finite Chevalley groups, J. Algebra 32 (1974), 418–443.

[Lu] F. Lübeck, Smallest degrees of complex characters of exceptional groups of Lie type, Comm. Algebra, to appear.

[MMT] K. Magaard, G. Malle, and Pham Huu Tiep, Irreducibility of tensor squares, symmetric squares, and alternating squares, Pacific J. Math., to appear.

[S1] G. M. Seitz, The maximal subgroups of classical algebraic groups, Mem. Amer. Math. Soc. 67 (1987), 365.

[S2] G. M. Seitz, Cross-characteristic embeddings of finite groups of Lie type, Proc. London Math. Soc. (3) 60 (1990), 166–200.

[SZ] G. M. Seitz and A. E. Zalesskii, On the minimal degrees of projective representations of the finite Chevalley groups. II, J. Algebra 158 (1993), 233–243.

[Sh] J. Shamash, Blocks and Brauer trees for the groups $G_2(2^k)$, $G_2(3^k)$, Comm. Algebra 20 (1992), 1375–1387.

[T] Pham Huu Tiep, Dual pairs and low-dimensional representations of finite classical groups, in preparation.

[TZ] Pham Huu Tiep and A. E. Zalesskii, Some characterizations of the Weil representations of the symplectic and unitary groups, J. Algebra 192 (1997), 130–165.

[Z] A. V. Zelevinsky, Representations of finite classical groups: A Hopf algebra approach, Lecture Notes in Math. 869, Springer-Verlag, Berlin 1981.

[Zs] K. Zsigmondy, Zur Theorie der Potenzreste, Monatsh. Math. Phys. 3 (1892), 265–284.

Department of Mathematics, Wayne State University, Detroit, MI 48202, U.S.A.
E-mail: kaym@math.wayne.edu

Department of Mathematics, University of Florida, Gainesville, FL 32611, U.S.A.
E-mail: tiep@math.ufl.edu